Marine Ecology: Current and Future Developments

(Volume 2)

Monitoring Artificial Materials and Microbes in Marine Ecosystems: Interactions and Assessment Methods

Edited by

Toshiyuki Takahashi

Department of Chemical Science and Engineering, National Institute of Technology (KOSEN), Miyakonojo College, Miyakonojo City, Japan

Marine Ecology: Current and Future Developments

Volume # 2

Monitoring Artificial Materials and Microbes in Marine Ecosystems: Interactions and Assessment Methods

Editor: Toshiyuki Takahashi

ISSN (Online): 2661-4685

ISSN (Print): 2661-4677

ISBN (Online): 978-981-14-3725-0

ISBN (Paperback): 978-981-14-3724-3

need for a court order if at any point you breach any terms of this License Agreement. In no event will any delay or failure by Bentham Science Publishers in enforcing your compliance with this License Agreement constitute a waiver of any of its rights.

3. You acknowledge that you have read this License Agreement, and agree to be bound by its terms and conditions. To the extent that any other terms and conditions presented on any website of Bentham Science Publishers conflict with, or are inconsistent with, the terms and conditions set out in this License Agreement, you acknowledge that the terms and conditions set out in this License Agreement shall prevail.

Bentham Science Publishers Pte. Ltd.
80 Robinson Road #02-00
Singapore 068898
Singapore
Email: subscriptions@benthamscience.net

BENTHAM SCIENCE

CONTENTS

PREFACE

Marine ecosystems provide human communities with organic and inorganic sources of food and food additives. Moreover, marine ecosystems provide recreational space used for sports, relaxation and culture. To enjoy the important benefits provided by marine ecosystems for many years to come, we must understand and conserve marine ecology, as well as make an effort to implement and follow sustainable natural resource management.

To conserve and take control marine ecosystems, it is important to evaluate the various components to support marine ecosystems. Unfortunately, about 20% of coral reefs have already disappeared globally. In addition, seaweed beds depletion has become common in coastal areas over the last several decades. Putting flesh on the above information, marine ecosystems are also supported by plankton, which might be underrepresented because of their invisibility to the naked eye. This loss of marine diversity, and hence the diversity within them, have considerably reduced marine-ecosystem services.

To provide systematic and comprehensive coverage of these topics, from marine ecosystems to environmental monitoring, with a focus on evaluating the role of microorganisms particularly, we have assimilated the expertise of an excellent group of international experts, including marine ecologists, marine biologists, environmental scientists, microbiologists, material scientists, and industrial engineers. This book summarizes fundamental knowledge about marine ecology, as well as exploring the applied aspects of marine ecosystems. In addition, this book presents practical approaches to detect and evaluate microbes, even in marine environments. A significant feature of this book is the description of pilot and challenging approaches focused on improving marine ecosystems, such as using steelmaking byproducts as materials for rehabilitation.

This book will benefit both academics and other professionals, including marine ecologists, microbiologists, environmental engineers, and engineers associated with industrial applications, such as bioremediation and algal biorefineries. As shown in the Table of Contents, this book is separated into three parts. "Part 1" is of utility to beginners, because it includes an informative guide of each presented field. Consequently, this book could be used as an informative book for undergraduate courses in marine biology, ecology, and microbiology. After beginning with fundamental topics, including seaweed and coral reefs for entry-level beginners, this book expands to incorporate the importance of microbes and their associations in marine ecosystems, including interactions of marine organisms with artificial materials.

Although the Great East Japan Earthquake, which struck Japan on March 11, 2011, was an unfortunate disaster, the subsequent changes, including the availability of nutrient salts, to the coastal environment were analyzed in detail. This analysis includes information on how the disaster impacted the coastal ecosystem. Therefore, "Part 1" presents this information within the context of an actual example of sustainable environmental assessment, ahead of evaluating microbial parameters in the latter parts of the book.

"Part 2" presents high-quality and constructive information with technical components for practical application. To conserve marine environments and achieve a green design for artificial materials used in the marine environment, it is important to detect and evaluate marine organisms precisely, particularly easily missed microorganisms. Instead of simply presenting the basic concepts about these microorganisms in the field of marine ecology, this book introduces readers to methodologies used to collect, detect, and evaluate microorganisms as constituents of marine ecosystems, because it is easy to overlook elements that are invisible to the naked eye. Our book presents highly sensitive *in situ* analytical methods using optical and scanning probe microscopy-based imaging, single-cell detection, and DNA-sequence-based techniques. Furthermore, several of the highly sensitive and specific techniques presented in the book could be applied to evaluating microbes in a variety of situations, including diagnostics in clinical situations, as well as industrial applications, such as water treatment. Thus, "Part 2" is also a useful book text for university students at both the undergraduate and graduate level.

Finally, "Part 3" presents the results of several demonstration experiments that could be used in practice by experts and professionals. To conserve marine ecosystems, artificial materials or structures could be used to facilitate the recovery of depleted seaweed beds in coastal areas. This book also describes existing knowledge of how aquatic organisms interact with artificial materials. Using the biofouling of cooling industrial pipes by biofilms as an example, this book presents the mechanisms of biofouling on surface materials, as well as providing clues towards controlling this problem.

After explaining the chemical properties of steelmaking slag in the aqueous environment, "Part 3" of this book presents a pilot approach towards improving marine ecosystems with steelmaking byproducts. This topic is also an important case study towards understanding how microbes, visible-scale organisms, and metallic materials interact. Consequently, this book provides useful insights towards resolving the loss of marine-ecosystem services caused by decreased cover of seaweed beds and coral reefs.

Overall, this book helps readers to evaluate microbial communities in marine environments, providing many insights on methods that could be used to help recover marine environments and marine-ecosystem services.

Finally, I thank many co-authors of this book for their dedication in sharing their expertise in both a concise and accessible way.

Toshiyuki Takahashi
Department of Chemical Science and Engineering,
National Institute of Technology (KOSEN),
Miyakonojo College,
Miyakonojo City,
Japan

List of Contributors

Akiko Ogawa　　Department of Chemistry and Biochemistry, National Institute of Technology (KOSEN), Suzuka College, Shirako-cho, Suzuka, Mie 510-0294, Japan

Bogdan I. Gerashchenko　　R.E. Kavetsky Institute of Experimental Pathology, Oncology and Radiobiology, National Academy of Sciences of Ukraine, 45 Vasylkivska Street, Kyiv 03022, Ukraine

Chika Kosugi　　Advanced Technology Research Laboratories, Nippon Steel Corporation, 20-1 Shintomi, Futtsu, Chiba 293-8511, Japan

Dana M. Barry　　Department of Electrical & Computer Engineering at Clarkson University, Potsdam, N.Y., USA
The State University of New York, Canton, in Canton, New York 13617, USA

Futoshi Iwata　　Graduate School of Integrated Science and Technology, Shizuoka University, Johoku, Naka-ku, Hamamatsu, 432-8561, Japan

Hideyuki Kanematsu　　Department of Materials Science and Engineering, National Institute of Technology (KOSEN), Suzuka College, Shirako-cho, Suzuka, Mie 510-0294, Japan

Hiroyuki Matsuura　　Department of Materials Engineering, The University of Tokyo, 7-3-1 Hongo, Bunkyo, Tokyo 113-8656, Japan

Hotaka Kai　　Department of Chemistry and Biochemistry, National Institute of Technology (KOSEN), Suzuka College, Shirako-cho, Suzuka, Mie 510-0294, Japan

Katsuhiko Sano　　D & D Corporation, 7870-21 Sakura-cho, Yokkaichi, Mie 512-1211, Japan

Minato Wakisaka　　Graduate School of Life Science and Systems Engineering, Kyushu Institute of Technology, 2-4 Hibikino, Wakamatsu-ku, Kitakyushu, Fukuoka 808-0196, Japan

Mitsuo Yamamoto　　Graduate School of Agricultural and Life Sciences, The University of Tokyo, 1-1-1 Yayoi, Bunkyo-ku, Tokyo 113-8657, Japan

Naoko Isomura　　Department of Bioresources Engineering, National Institute of Technology (KOSEN), Okinawa College, 905 Henoko, Nago, Okinawa 905-2192, Japan

Nobumitsu Hirai　　Department of Chemistry and Biochemistry, National Institute of Technology (KOSEN), Suzuka College, Shirako-cho, Suzuka, Mie 510-0294, Japan

Romaidi　　Biology Department, Science and Technology Faculty, State Islamic University of Malang, Jalan Gajayana 50, Malang 65144, Indonesia

Ryo Inoue　　Graduate School of International Resource Sciences, Akita University, 1-1 Tegatagakuen-machi, Akita, Akita 010-8502, Japan

Seiji Yokoyama　　Department of Mechanical Engineering, Toyohashi University of Technology, 1-1 Hibarigaoka, Tempaku-cho, Toyohashi, Aichi 441-8580, Japan

Shuji Kawakami　　Department of Construction Systems Engineering, National Institute of Technology (KOSEN), Anan College, 265 Aoki, Minobayashi-cho, Anan, Tokushima 774-0017, Japan

Sumihiro Koyama　　R & D section, Able Co. Ltd., 7-9 Nishi-Gokencho, Shinjyuku-ku, Tokyo 162-0812, Japan

Takeshi Kougo Department of Materials Science and Engineering, National Institute of Technology (KOSEN), Suzuka College, Shirako-cho, Suzuka, Mie 510-0294, Japan

Tatsuya Ueki Marine Biological Laboratory, Graduate School of Integrated Sciences for Life, Hiroshima University, 2445 Mukaishima, Onomichi, Hiroshima 722-0073, Japan

Toshiaki Kato Advanced Technology Research Laboratories, Nippon Steel Corporation, 20-1 Shintomi, Futtsu, Chiba 293-8511, Japan, Chuo-ku, Tokyo 104-0031, Japan Nippon Steel Eco-Tech Corporation, 1-18-1 Kyobashi

Toshiyuki Takahashi Department of Chemical Science and Engineering, National Institute of Technology (KOSEN), Miyakonojo College, 473-1 Yoshio-cho, Miyakonojo, Miyazaki 885-8567, Japan

Tri Kustono Adi Chemistry Department, Science and Technology Faculty, State Islamic University of Malang, Jalan Gajayana 50, Malang 65144, Indonesia

Tsuyoshi Yamaguchi Department of Civil and Environmental Engineering, National Institute of Technology (KOSEN), Matsue College, 14-4 Nishiikuma-cho, Matsue, Shimane 690-8518, Japan

Yoshihiro Suzuki Department of Civil and Environmental Engineering, University of Miyazaki, 1-1 Gakuen Kibabadai Nishi, Miyazaki, Miyazaki 889-2192, Japan

Part 1: Marine Ecosystem and Environmental Monitoring

CHAPTER 1

Coral Reef Ecosystems in Marine Environments

Naoko Isomura[*]

Department of Bioresources Engineering, National Institute of Technology (KOSEN), Okinawa College, Nago, Japan

Abstract: Japan has different climatic zones running from the south to the north of the country, which include subtropical to cool continental zones. Within these zones, 78 genera and 415 species of corals are found. Coral reefs are continuing to decline worldwide owing to large-scale bleaching events that have accompanied climate change. In Japan, the decline of reef-building corals occurred from 1998 to 2007, with many corals also dying during a large-scale bleaching event in 2016. In this chapter, I introduce our studies regarding coral reproduction, "Hybridization and speciation in *Acropora*" and "Elucidation of synchronous spawning mechanism in *Acropora*." The importance and contribution of coral reproduction to the restoration of coral reefs are discussed because the influence of bleaching on reproduction is not limited only to the existing coral colonies, but it might also affect next generation colonies and the maintenance of coral populations.

Keywords: Bleaching, Corals, Multi-specific synchronous spawning, Reproduction.

INTRODUCTION

Hermatypic corals that form coral reefs are mostly species belonging to Scleractinia, Hexacorallia, Anthozoa, and Cnidaria, although some belong to Octocorallia and Hydrozoa [1]. Many corals inhabit the shallow sea, and they show intracellular symbiosis with dinoflagellate algae.

Japan has different climatic zones running from the south to the north of the country, which include subtropical to cool continental zones. Within these zones, 78 genera and 415 species have been identified to date [1 - 4]. In our previous study [5], we divided the area inhabited by corals into three parts. The "coral reef region" (24–30° N) has high coral species diversity and includes the Yaeyama Archipelago, Okinawa Islands, Amami Archipelago, Tokara Archipelago, and

---

[*] **Corresponding author Naoko Isomura:** Department of Bioresources Engineering, National Institute of Technology (KOSEN), Okinawa College, Nago, Japan; Tel: +81-980-55-4135; Fax: +81-980-55-4012; E-mail: iso@okinawa-ct.ac.jp

Toshiyuki Takahashi (Ed.)

Ogasawara Islands. The "non-coral reef region" (30–33° N) shows moderate coral species diversity and includes Kyushu, Shikoku, and the Kii Peninsula. In this region, corals assemble without making a reef structure. The "peripheral region" (33–35°N) shows undeveloped coral assemblages with low coral species diversity, and includes the Izu Peninsula, Boso Peninsula, and Izu Islands (Fig. **1**). The number of species decreases with increasing latitude owing to lower temperatures, with the number of species in each region estimated at 415 in the coral reef region, 200 in the non-coral reef region, and 55 in the peripheral region.

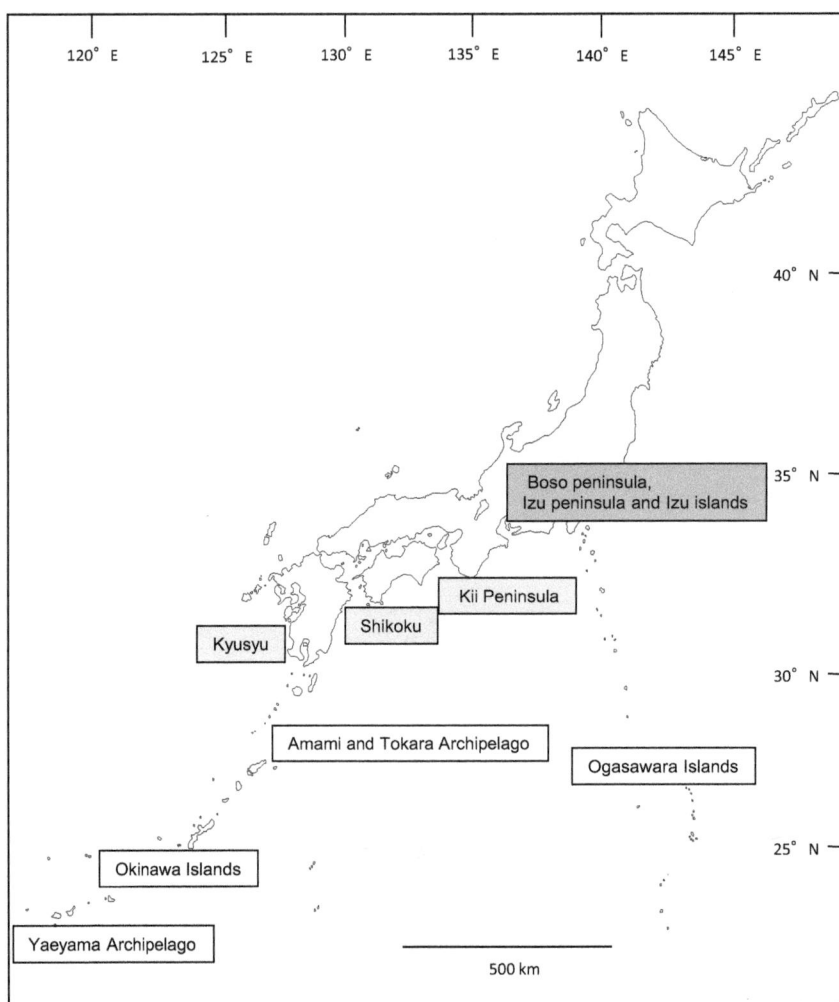

Fig. (1). Map of Japan showing the three coral regions (coral reef region, non-coral reef region, and peripheral region). Squares surrounded by a black line with white color show the coral reef region; squares with pale gray color show the non-coral reef region and the square with dark gray color shows the peripheral region (modified from Isomura and Fukami 2018 [5]).

In recent years, coral reefs have continued to decline worldwide owing to large-scale bleaching events that have accompanied climate change or by becoming prey for the crown-of-thorns starfish (*Acanthaster* spp.). In Japan, the decline of reef-building corals has continued after the large-scale bleaching in 1998, and Kawagoe [6] reported that approximately 70% of corals in Sekisei Lagoon (a barrier reef lying between the coasts of Ishigaki, Taketomi, and Iriomote Island) died owing to a large-scale bleaching event in 2016. Corals that previously dominated the subtropical zone have extended their habitat northwards in recent years, and the seaweed beds have been increasingly replaced by corals in fishing grounds [7]. Meanwhile, the area that the coral has joined by moving north has become a shelter of coral species that are reduced in the temperate areas [8]. The northern part of the peripheral region is a "marginal zone," which never existed previously or had an extremely small quantity of coral populations, and is now where coral communities are expected to behave differently from those in subtropical areas.

Currently, we are mainly conducting research on "Hybridization and speciation in *Acropora*" and "Elucidation of synchronous spawning mechanism in *Acropora*." The genus *Acropora* is the largest group of reef corals and consists of approximately 150 species worldwide. This group of corals consists of very important organisms that are the foundation for the construction of coral reef ecosystems. The response and/or behavior of the genus *Acropora* under climate change conditions is expected to affect not only corals but also the organisms that use coral reefs as their habitats as well as the environment surrounding them. In this chapter, I introduce our studies of *Acropora* and discuss the future of corals including *Acropora*.

Species Diversity of *Acropora via* Hybridization

Among scleractinian corals (hereinafter referred to as "corals"), the genus *Acropora* is an important component in tropical, subtropical, and temperate areas, and contains approximately 150 species. Due to its large number of species, "hybrid speciation" is believed to have occurred within this genus [9]. Evidence of hybrid speciation in *Acropora* has been shown as follows: (1) many colonies of intermediate forms have been observed and (2) many species showed inter-specific synchronized spawning, which causes multi-specfic crossing. Examples of. (1) are the three coral species in the Caribbean Sea, *A. cervicornis, A. palmata,* and *A. prolifera,* with *A. prolifera* shown to be a hybrid of *A. cervicornis* and *A. palmata* [10]. Although genetic information confirmed these hybrids, it was predicted that this species was a hybrid because it contained two other intermediate forms. Regarding point (2), multi-specfic synchronous spawning has been reported on the Great Barrier Reef [11] and inter-specific hybridization experiments have been conducted at various locations, which have revealed that they can cross between genetically similar species [12, 13]. Although these studies reveal the ability of crossbreeding and hybrid formation in *Acropora*, actual "hybrid speciation" has not been confirmed to date. In addition to the above conditions (1) and (2), two more conditions may be required as follows: (3) hybrids can survive at any stage of life history as well as the parent species, and (4) hybrids are fertile, which is required for the success of the next generation. Moreover, two sub conditions of point (4) exist as follows: (4a) mating between hybrids is established more frequently than crossing with other species or parent species, and (4b) under inter-specific synchronous spawning conditions, inter-specific breeding is equivalent to or more effective than intra-specific crossing.

We focused on two species, *A. florida* (Fig. **2A**) and *A. intermedia* (Fig. **2B**) in Akajima Island, Okinawa, to study the hybrid speciation of *Acropora*. These two species can hybridize artificially and are distinguished morphologically because *A. florida* has a bottle-brushed morphology (Fig. **2A**), in which, a number of short branches protrude from the main thick branches, whereas *A. intermedia* has a long branching morphology (Fig. **2B**). Previous studies in the Caribbean suggest that hybrids are likely to exhibit an intermediate form of the parent species; however, it is expected that hybrids can be easily discriminated, such as in these two species with distinctly different colony shapes. In fact, putative hybrids showing intermediate forms of both species were confirmed from Akajima Island (Fig. **2C**). In 2007, two types of cross breeding were performed. The obtained larvae (confirmed hybrids) were cultivated for 7 years and it was confirmed for the first time globally that these confirmed hybrids spawned gametes and were fertile (Fig. **2D**). Furthermore, from the results of cross breeding experiments over some years, it was possible to verify the above four conditions that are required for hybrid speciation to occur in *A. florida* and *A. intermedia* [14 - 16].

From 2007, the authors bred hybrids of artificially produced *A. florida* and *A. intermedia* outdoors, and measured their morphological traits in 2011 and 2014. The morphological traits of the parent species colonies were also measured during the same period, and morphological analysis using the multinomial logit model was performed. The hybrids showed an intermediate morph between the two parents and the traits were similar to the mother species [17]. In addition, the mother species was estimated from the morphology of the putative hybrids that were found in the field. Therefore, the parent species of the hybrids could be estimated more accurately together with the results from the gene analyses.

Furthermore, it was confirmed that there was an intermediate form of colonies that did not have a clear branching structure similar to *A. intermedia*, although it had a bottle-brushed form that is a feature of *A. florida* (Fig. **2E**). This colony had a larger axial corallite and possessed the characteristics of *A. gemmifera* (Fig. **2F**), which inhabited the same area as the other two species. We are now examining the ability of hybrid formation in *A. gemmifera* using *A. florida* and *A. intermedia* to verify how many species of *Acropora* can form hybrids under multi-specific synchronous spawning in Okinawa.

Fig. (2). Photographs of the *Acropora* species that were used in our study. **A**: *A. florida*, **B**: *A. intermedia*, **C**: Putative hybrid considered to originate from *A. florida* and *A. intermedia*, **D**: Spawning of real hybrid by *A. florida* and *A. intermedia*, **E**: Putative hybrid considered to be related to *A. gemmifera*, **F**: *A. gemmifera*. Scale bars = 10 cm. This figure is modified from Isomura 2018 (Umiushi Tsushin, 98, 8-10, written in Japanese).

Although this is only a preliminary analysis, we examined the multiple exon regions of *A. florida*, *A. intermedia*, *A. gemmifera,* and putative hybrids, and found the following: (1) the haplotype was shown to be shared among *A. florida*, *A. intermedia*, *A. gemmifera,* and the putative hybrids, (2) another haplotype was shared between *A. intermedia* and the putative hybrids, and (3) another haplotype was shared only among the *A. florida* colonies. These results indicate that *A. gemmifera* may be related to hybrid production and introgression. Further detailed genetic analysis will clarify how many hybrids exist and how much introgression is occurring among these species. Moreover, hybrid speciation will be elucidated, which has been a longstanding mystery of coral research, by integrating crossing experiments, distribution surveys, and genetic analyses.

Search for Endocrine System Regulating Multi-specific Synchronous Spawning in *Acropora*

The genus *Acropora* develops their gametes throughout the year, and they spawn eggs and sperm synchronously during the summer season. Synchronization of these reproductive activities was thought to be due to changes in the external environment such as the lunar age, water temperature, and the tide. Corals receive changes in the external environment as stimuli and regulate gametogenesis and spawning by hormones and/or neurotransmitters. However, the physiological mechanism for synchronous spawning is almost unknown. Sex steroid hormones are produced from cholesterol, and sex hormones containing male and female sex hormones are formed as metabolites from cholesterol. Among these hormones, estrogen is known as the female hormone and plays an important role in reproductive activities such as the formation of ovaries and uteri. Estrogens act *via* the estrogen receptor (hereinafter referred to as "ER"). After estrogen is secreted, it enters the nucleus, binds to the ER, and regulates the transcription of reproductive genes. Invertebrates already have the same ER gene as that of humans [18] and the association between sexual maturation and the ER has been reported in oysters [19]. In scleractinian corals, estrogen has been detected in *Euphyllia ancora* and the estrogen level increases during the spawning months [20].

Catecholamine acts as a neurotransmitter and regulates important functions such as locomotion, biological rhythm, blood pressure, and the reproductive and endocrine systems. Noradrenaline, a type of catecholamine, functions via a noradrenaline transporter (hereinafter referred to as NAT). Even in invertebrates such as bivalves, catecholamines have been reported to be related to reproduction and spawning [21]. It has been reported that dopamine, a catecholamine, inhibited the spawning of *Acropora* [22]. Due to next generating sequencing, the whole genome decoding of *A. digitifera* was performed [23], and this genomic information can be used in studies of reef corals. We found both the ER and NAT genes using the genome database of *A. digitifera*.

Recently, we began to examine the expression kinetics of the ER and NAT genes in gametogenesis and spawning in *A. intermedia* to elucidate a part of the endocrine mechanism concerning the reproductive activity of *Acropora*. Although only a preliminary result, it was suggested that *A. intermedia* secreted much estrogen during the initial formation of the oocyte and regulated the amount of secretion during the development of the oocyte. In contrast, the expression level of the NAT gene decreased toward spawning and recovered after spawning. Noradrenaline and dopamine, a precursor of noradrenaline, are known to inhibit spawning and reproductive activity in corals and invertebrates [22, 24]. It has also been suggested that noradrenaline suppresses spawning and that a reduction of noradrenaline secretion triggered spawning in *A. intermedia*. In previous studies, catecholamines containing noradrenaline have been detected as substances in *Acropora* spp. before and after spawning [25]. Moreover, it is necessary to detect the proteins and substances related to reproduction in coral tissues, and to the clarify changes that take place before and after spawning and between seasons.

Contribution of Coral Reproduction to the Restoration of Coral Reefs

The NOAA (2015) predicted that a global bleaching event, which was the third largest in recorded history, would occur from 2015 to 2016 [26], with this predicted global bleaching event beginning at the Mariana Islands in 2014 and expanding to the South Pacific and Indian Ocean in 2015, and the northern Great Barrier Reef in 2016 [27]. Subsequent to this, a mass coral-bleaching event occurred in Ryukyu Islands, Japan, during the summer of 2016. Severe damage was observed, especially around Miyako Island and in Sekisei Lagoon [6, 28].

The amount of energy resources available to the coral hosts may decrease by the loss or reduction of zooxanthellae during bleaching. Energy resources are allocated among the processes of growth, survival, and reproduction to maximize evolutionary fitness [29], and thus the effects of bleaching on reproduction may be expected. The effects of bleaching on reproduction have been reported in Japan [30, 31] as have the effects of climate change and/or ocean acidification on reproduction [32, 33].

The influence of bleaching on reproduction is not limited to the present coral colonies; it may also affect next generation colonies and population maintenance. The recovery of *Acropora* depends on the density of nearby adult colonies [34]; therefore, adequate fecundity of adult colonies that are sexually mature will be important for both population fitness and recruitment to self and/or other populations. However, we have only limited information regarding coral reproduction, even for the well-studied genus *Acropora*, although there are more than 400 corals species around Japan. Further studies on coral ecology and reproduction are required to understand how corals are affected by ongoing climate change and how to minimize the influence of the declining coral cover in the future.

CONSENT FOR PUBLICATION

Not applicable.

CONFLICT OF INTEREST

The author confirms that this chapter contents have no conflict of interest.

ACKNOWLEDGEMENTS

Declare none.

REFFERENCES

[1]   Nishihira M, Veron JEN. Hermatypic corals of Japan. Tokyo: Kaiyusha 1995; p. 440. (in Japanese)

[2]   Veron JEN. Corals of the world, Aust. Inst. Mar. Sci. Townsville. 2000a; 1: p. 463.

[3]   Veron JEN. Corals of the world, Aust. Inst. Mar. Sci. Townsville. 2000b; 2: p. 429.

[4]   Veron JEN. Corals of the world, Aust. Inst. Mar. Sci. Townsville. 2000c; 3: p. 490.

[5]   Isomura N, Fukami H. Coral Reproduction in Japan.Coral Reef Studies of Japan. Springer Nature Singapore Pte Ltd. 2018; pp. 95-110.
[http://dx.doi.org/10.1007/978-981-10-6473-9_7]

[6]   Kawagoe H. Mass coral bleaching in 2016 reported by the monitoring sites 1000 project. J Jpn Coral Reef Soc 2017; 19: 21-8.
[http://dx.doi.org/10.3755/jcrs.19.21]

[7]   Kumagai NH, García Molinos J, Yamano H, Takao S, Fujii M, Yamanaka Y. Ocean currents and herbivory drive macroalgae-to-coral community shift under climate warming. Proc Natl Acad Sci USA 2018; 115(36): 8990-5.
[http://dx.doi.org/10.1073/pnas.1716826115] [PMID: 30126981]

[8]   Nakabayashi A, Yamakita T, Nakamura T, *et al.* The potential role of temperate Japanese regions as refugia for the coral *Acropora hyacinthus* in the face of climate change. Sci Rep 2019; 9(1): 1892.
[http://dx.doi.org/10.1038/s41598-018-38333-5] [PMID: 30760801]

[9]   Veron JEN. Corals in space and time: the biogeography and evolution of the Scleractinia. Ithaca: Cornell University Press 1995; p. 321.

[10]   Vollmer SV, Palumbi SR. Hybridization and the evolution of reef coral diversity. Science 2002; 296(5575): 2023-5.
[http://dx.doi.org/10.1126/science.1069524] [PMID: 12065836]

[11]   Babcock RC, Bull GD, Harrison PL, *et al.* Synchronous spawning of 105 scleractinian coral species on the Great Barrier Reef. Mar Biol 1986; 90(3): 379-94.
[http://dx.doi.org/10.1007/BF00428562]

[12]   Willis B L, Babcock R C, Harrison P L, Wallace C C. Experimental hybridization and breeding incompatibilities within the mating systems of mass spawning reef corals. Coral Reefs 1997; 16(Suppl.): S53-65.
[http://dx.doi.org/10.1007/s003380050242]

[13]   Hatta M, Fukami H, Wang W, *et al.* Reproductive and genetic evidence for a reticulate evolutionary history of mass-spawning corals. Mol Biol Evol 1999; 16(11): 1607-13.
[http://dx.doi.org/10.1093/oxfordjournals.molbev.a026073] [PMID: 10555292]

[14]   Isomura N, Iwao K, Fukami H. Possible natural hybridization of two morphologically distinct species of *Acropora* (Cnidaria, Scleractinia) in the Pacific: fertilization and larval survival rates. PLoS One 2013; 8(2)e56701
[http://dx.doi.org/10.1371/journal.pone.0056701] [PMID: 23457605]

[15]   Isomura N, Iwao K, Morita M, Fukami H. Spawning and fertility of F1 hybrids of the coral genus *Acropora* in the Indo-Pacific. Coral Reefs 2016; 35(3): 851-5.
[http://dx.doi.org/10.1007/s00338-016-1461-9]

[16]   Kitanobo S, Isomura N, Fukami H, Iwao K, Morita M. The reef-building coral *Acropora* conditionally hybridize under sperm limitation. Biol Lett 2016; 12(8)20160511
[http://dx.doi.org/10.1098/rsbl.2016.0511] [PMID: 27555653]

[17]   Fukami H, Iwao K, Kumagai NH, Morita M, Isomura N. Maternal inheritance of F1 hybrid morphology and colony shape in the coral genus *Acropora.* PeerJ 2019; 7e6429
[http://dx.doi.org/10.7717/peerj.6429] [PMID: 30809440]

[18]   Katsu Y, Kubokawa K, Urushitani H, Iguchi T. Estrogen-dependent transactivation of amphioxus steroid hormone receptor *via* both estrogen and androgen response elements. Endocrinology 2010; 151(2): 639-48.
[http://dx.doi.org/10.1210/en.2009-0766] [PMID: 19966182]

[19]   Ni J, Zeng Z, Ke C. Sex steroid levels and expression patterns of estrogen receptor gene in the oyster *Crassostrea angulata* during reproductive cycle. Aquaculture 2013; 376: 105-16.
[http://dx.doi.org/10.1016/j.aquaculture.2012.11.023]

[20]   Twan WH, Hwang JS, Chang CF. Sex steroids in scleractinian coral, *Euphyllia ancora*: implication in mass spawning. Biol Reprod 2003; 68(6): 2255-60.
[http://dx.doi.org/10.1095/biolreprod.102.012450] [PMID: 12606339]

[21]   Martinez G, Rivera A. Role of monoamines in the reproductive process of *Argopecten pupuratus.* Inver Rep Dev 1994; 25(2): 167-74.
[http://dx.doi.org/10.1080/07924259.1994.9672381]

[22]   Isomura N, Yamauchi C, Takeuchi Y, Takemura A. Does dopamine block the spawning of the acroporid coral Acropora tenuis? Sci Rep 2013; 3: 2649.
[http://dx.doi.org/10.1038/srep02649] [PMID: 24026104]

[23]   Shinzato C, Shoguchi E, Kawashima T, *et al.* Using the *Acropora digitifera* genome to understand coral responses to environmental change. Nature 2011; 476(7360): 320-3.
[http://dx.doi.org/10.1038/nature10249] [PMID: 21785439]

[24]   Khotimchenko YS. Effect of noradrenaline, dopamine and adrenolytics on growth and maturation of the sea urchin, *Strongylocentrotus nudus* Agassiz. Int J Inver Rep Dev 1982; 4(6): 369-73.
[http://dx.doi.org/10.1080/01651269.1982.10553445]

[25]   Taira J, Higa I, Tsuchida E, Isomura N, Iguchi A. Neurotransmitters in hermatypic coral, *Acropora* spp., and its contribution to synchronous spawning during reproductive event. Biochem Biophys Res Commun 2018; 501(1): 80-4.
[http://dx.doi.org/10.1016/j.bbrc.2018.04.170] [PMID: 29689267]

[26]   NOAA. NOAA declares third ever global coral bleaching event [Internet Article] 2015. Available from: https://www.noaa.gov/media-release/noaa-declares-third-ever-global-coral-bleaching-event.

[27]   Hughes TP, Kerry JT, Álvarez-Noriega M, *et al.* Global warming and recurrent mass bleaching of corals. Nature 2017; 543(7645): 373-7.
[http://dx.doi.org/10.1038/nature21707] [PMID: 28300113]

[28] Nakamura T. Mass coral bleaching event in Sekisei lagoon observed in the summer of 2016. J Jpn Coral Reef Soc 2017; 19: 29-40. [in Japanese with English abstract].
[http://dx.doi.org/10.3755/jcrs.19.29]

[29] Ramirez Llodra E. Fecundity and life-history strategies in marine invertebrates. Adv Mar Biol 2002; 43: 87-170.
[http://dx.doi.org/10.1016/S0065-2881(02)43004-0] [PMID: 12154615]

[30] Hirose M, Hidaka M. Reduced reproductive success in scleractinian corals that survived the 1998 bleaching in Okinawa. Galaxea 2000; 2: 17-21.
[http://dx.doi.org/10.3755/jcrs.2000.17]

[31] Omori M, Fukami H, Kobinata H, Hatta M. Significant drop of fertilization of *Acropora* corals in 1999: An after-effect of heavy coral bleaching? Limnol Oceanogr 2001; 46: 704-6.
[http://dx.doi.org/10.4319/lo.2001.46.3.0704]

[32] Morita M, Suwa R, Iguchi A, *et al.* Ocean acidification reduces sperm flagellar motility in broadcast spawning reef invertebrates. Zygote 2010; 18(2): 103-7.
[http://dx.doi.org/10.1017/S0967199409990177] [PMID: 20370935]

[33] Iguchi A, Suzuki A, Sakai K, Nojiri Y. Comparison of the effects of thermal stress and CO_2-driven acidified seawater on fertilization in coral Acropora digitifera. Zygote 2015; 23(4): 631-4.
[http://dx.doi.org/10.1017/S0967199414000185] [PMID: 24847859]

[34] Hughes TP, Baird AH, Dinsdale EA, *et al.* Supply-side ecology works both ways: The link between benthic adults, fecundity, and larval recruits. Ecology 2000; 81: 2241-9.
[http://dx.doi.org/10.1890/0012-9658(2000)081[2241:SSEWBW]2.0.CO;2]

CHAPTER 2

Marine Polymers as Ecofriendly Alternatives to Petroleum-Based Plastics

Minato Wakisaka

Graduate School of Life Science and Systems Engineering, Kyushu Institute of Technology, Kitakyushu, Japan

Abstract: Nanofibers with diameters between 100 nm to 200 nm were easily prepared from various water-soluble marine polysaccharides by combining ultrasonic atomization with freeze casting. Scanning electron microscopy demonstrated considerable differences in fiber diameter and morphology between the types as well as the concentrations of polymers. Nanofibers with uniform orientation were obtained by rapid freezing. The anti-bacterial activity of chitosan nanofibers is suitable for their use as food packaging material. Biodegradable composites of chitosan nanofibers and cellulose paper could be a solution to the problem of ocean pollution from single-use petroleum-based plastics.

Keywords: Freeze casting, Nanofiber, Polysaccharides.

INTRODUCTION

Nanofiber production from polysaccharides has attracted a great deal of research attention because of properties of the nanofibers, including biodegradability and biocompatibility [1].

Marine polysaccharides, which are present in many marine organisms, are very important biological macromolecules. Chitin is a long-chain polymer of N-acetylglucosamine (Fig. **1A**). It is one of the most abundant polysaccharides isolated from the shells of crabs and shrimps. Chitosan is a natural cationic polysaccharide (Fig. **1B**) produced commercially by deacetylation of chitin. Other types of polysaccharides are rich in seaweed, such as carrageenan from red sea-weeds and alginates from brown seaweeds. Carrageenan consists of linear sulfated polysaccharides. There are three main varieties that differ based on their degree of sulfation. κ-carrageenan has one sulfate group per disaccharide (Fig. **1C**). Alginate is a natural anionic polysaccharide with a linear structure (Fig. **1D**)

Corresponding author Minato Wakisaka: Graduate School of Life Science and Systems Engineering, Kyushu Institute of Technology, Kitakyushu, Japan; Tel/Fax: +81-093-695-6066; E-mail: wakisaka@life.kyutech.ac.jp

Toshiyuki Takahashi (Ed.)

composed of β-(1–4) linked D-mannuronic acid and its epimer comprised of α-guluronic acid units. Along its polymer chain, alginate has regions rich in sequential mannuronic acid units, guluronic acid units, and regions in which both monomers are equally prevalent.

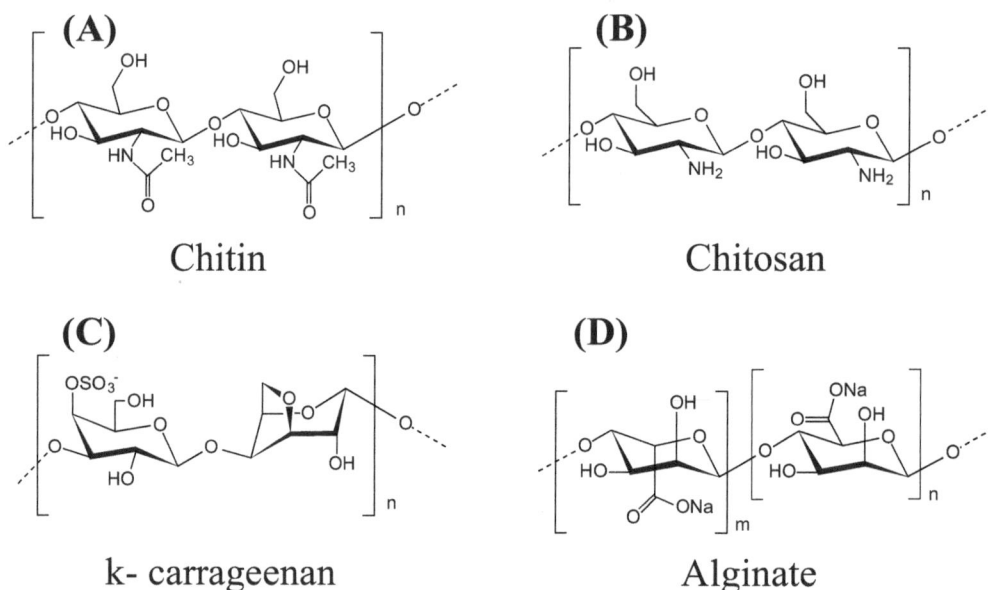

Fig. (1). Chemical structure of marine polysaccharides.

Spinning methods, such as electrospinning [2] or wet spinning, are commonly used for nanofiber fabrication. Wet spinning requires complex processes and is expensive. In addition to the risk of electric shock, electrospinning using high-voltage electrical fields is limited by the difficulty in controlling fiber orientation and is sensitive to the conductivity of the spinning dope solution.

A novel and unique technique of fabricating nanofibers with a uniform orientation combines ultrasonic atomization and freeze casting. In this study, the applicability of this technique to various marine polysaccharides was investigated. Furthermore, biodegradable composites of chitosan nanofiber with cellulose paper were developed. The potential of these composites as food packaging material was evaluated. Such a biodegradable packaging material could provide a solution to the problem of ocean pollution with single-use petroleum-based plastics.

MATERIALS AND METHODS

Chitosan 10 powder (deacetylation degree: 80%, MW = approximately 22 kDa), sodium alginate, and κ-carrageenan were all purchased from Wako Pure Chemical Industries, Ltd. (Osaka, Japan). All the chemicals were of analytical grade, for laboratory research and investigational use only, and were used as received.

Chitosan hydrogels (4wt%) were obtained by dissolving chitosan powder (4.0 g) and l-lactic acid (2.0 ml) in distilled water (94.0 ml). The hydrogels were diluted to the desired low concentrations with distilled water. Alginate was dissolved in distilled water at 25°C to obtain aqueous solutions of concentration 0.1wt%, 0.2wt%, 0.4wt%, and 0.8wt%. κ-carrageenan was dissolved in hot water at 85°C.

The polymer solution was poured into the reservoir of the model HX8111/32 Sonicare AirFloss (PSAF) device (Philips, Amsterdam, Netherlands) and atomized several times until the nozzle produced a full solution mist. A stainless-steel dish was filled with liquid nitrogen, ice/NaCl, or dry ice/ethanol and covered with a silicon wafer until the wafer surface just began to freeze. The PSAF nozzle tip was 5 cm from the wafer center and flush with its edge. It was kept parallel to the wafer surface while one single layer was atomized along this edge. Next, the wafer was immediately placed into an FDU-1200 freeze dryer (EYELA, Tokyo, Japan) and lyophilized under vacuum for approximately 5 hr at −52°C (Fig. **2**).

Fig. (2). Schematic illustration of ultrasonic atomization-freeze casting process for nanofiber fabrication.

For morphological observation, nanofibers were removed from the wafer using conductive tapes after freeze–drying. Fiber samples were immediately coated with a conductive platinum layer by sputtering using a model E-1030 device (HITACHI, Tokyo, Japan). Their morphologies were examined by wet scanning electron microscopy (HITACHI). Scanning electron micrographs were evaluated using ImageJ software (NIH, Bethesda. MD, USA). At least 100 isolated nanofibers were randomly selected and their diameters were measured.

RESULTS AND DISCUSSION

Uniformly oriented, small diameter chitosan nanofibers were successfully produced by combining ultrasonic atomization and freeze casting [3], as shown in

Fig. (3). In our preliminary studies, chitosan concentrations exceeding 4wt% produced a lamellar or porous structure upon freeze–drying. Here, lower concentrations ranging between 0.1 and 2.5wt% were chosen with the aim of obtaining fibers. Fiber diameters showed no obvious change with increasing concentration before sharply increasing at 0.8wt%. The 0.1 to 0.2wt% chitosan solutions yielded thinner fibers with an average diameter of 63.56 ± 24.05 nm. However, these fibers presented short, sparse, and disordered structures mixed with a small number of particles (Fig. **3A**). Above 0.4wt% chitosan, most continuous fibers adopted the same orientation (Fig. **3A**). Similarly oriented fibers with an average diameter of 102.98 ± 23.98 nm were obtained at 0.8wt% (Fig. **3B**), which was slightly larger than that obtained at 0.4wt%. When the chitosan concentration increased, the number of fibers decreased, but their diameters were greater. A lamellar structure appeared at 1.4wt% and completely replaced the fibers at 1.6wt% (Fig. **3C**). However, ultrasonic atomization largely maintained the fiber structure, delaying the emergence of this lamellar structure. At 2.5wt% chitosan, honeycomb-like compartments appeared (Fig. **3D**).

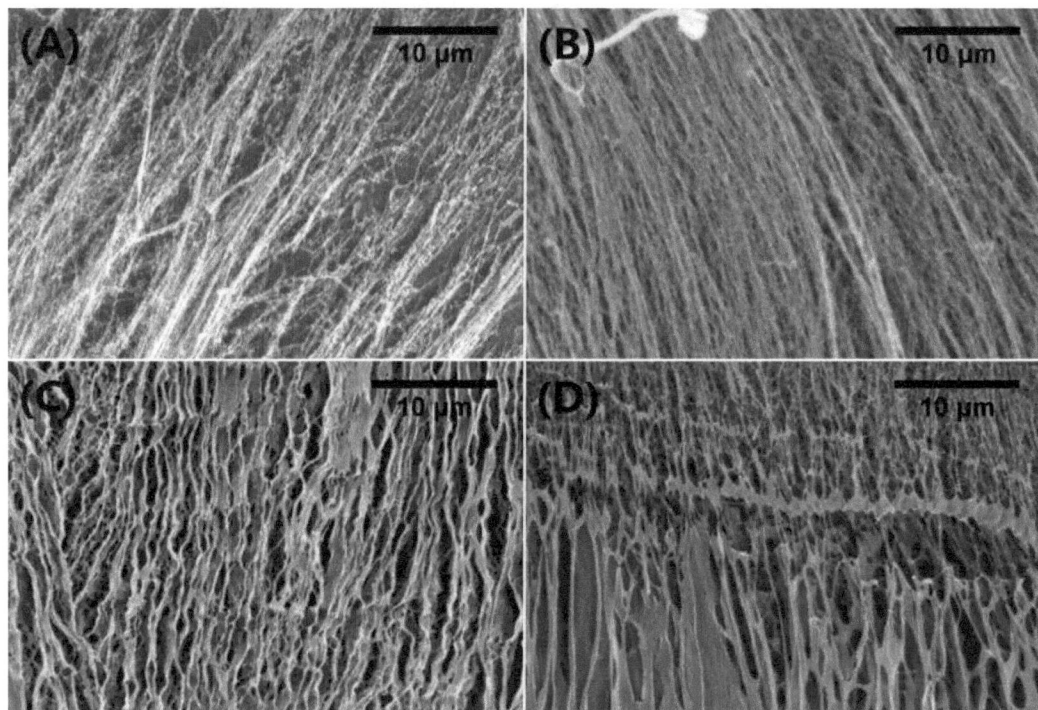

Fig. (3). SEM images of nanofibers at different chitosan concentrations. (**A**) 0.1wt%, (**B**) 0.8wt%, (**C**) 1.6wt%, and (**D**) 2.5wt%.

The effects of parameters other than chitosan concentration, such as the atomization tool and temperature, were also investigated. Uniformly oriented, small diameter nanofibers with an average diameter of 91.58 ± 17.36 nm were obtained using PSAF for ultrasonic atomization. A network structure of randomly oriented fibers with an average diameter of 216.76 ± 76.86 nm was obtained using a common household sprayer. When the atomization tool was changed to an airbrush, air pressure affected the fiber structure. Randomly oriented fibers with an average diameter of 161.36 ± 40.67 nm were observed at 0.4 MPa. A comparison of different atomization tools revealed that the uniform fiber orientation relied on the ability of the tool to provide a solution mist at a high initial speed. Temperature impacts fiber structures by directly affecting the freezing rate and indirectly by changing the solution viscosity. Changes in freezing rates were investigated using different solutions and freezing temperatures. The average fiber diameter was 91.58 ± 17.36 nm at 0°C and slightly increased to 111.16 ± 23.10 nm at 25°C. When the solution temperature increased to 37°C, the average fiber diameter also increased to 184.88 ± 61.31 nm. This might have been due to the decreased freezing rate with increasing solution temperature, because slower freezing rates generate larger ice crystals, which act as a template for the formation of larger chitosan fiber morphologies. However, the influence of solution temperature on the fiber structure was not as pronounced as that of the freezing temperature. Randomly oriented fibers with an average fiber diameter of 287.88 ± 114.30 nm were obtained when an ice/NaCl mixture (approximately -15°C) was used instead of liquid nitrogen. The average fiber diameter decreased to 209.64 ± 47.88 nm when a dry ice/ethanol mixture (approximately -50 °C) was used as the coolant. Liquid nitrogen was the most suitable coolant for rapid freezing necessary to produce uniformly oriented, small diameter fibers.

The applicability of this nanofiber fabrication technique combining ultrasonic-atomization and freeze casting was also assessed for other marine polymers, including alginate and κ-carrageenan [4].

Alginate demonstrated an excellent fiber formation property similar to that of chitosan. At 0.1wt%, a small network structure of randomly oriented fibers with an average diameter of 118.81 ± 27.72 nm formed (Fig. **4A**). When the concentration increased to 0.2wt%, nanofibers with uniform orientation at average diameter of 137.93 ± 45.22 nm were obtained (Fig. **4B**). Average diameter increased along with its concentration, 180.22 ± 51.04 nm at 0.4wt% (Fig. **4C**) and 244.77 ± 55.74 nm at 0.8wt% (Fig. **4D**), which resulted in formation of a lamellar-like structure.

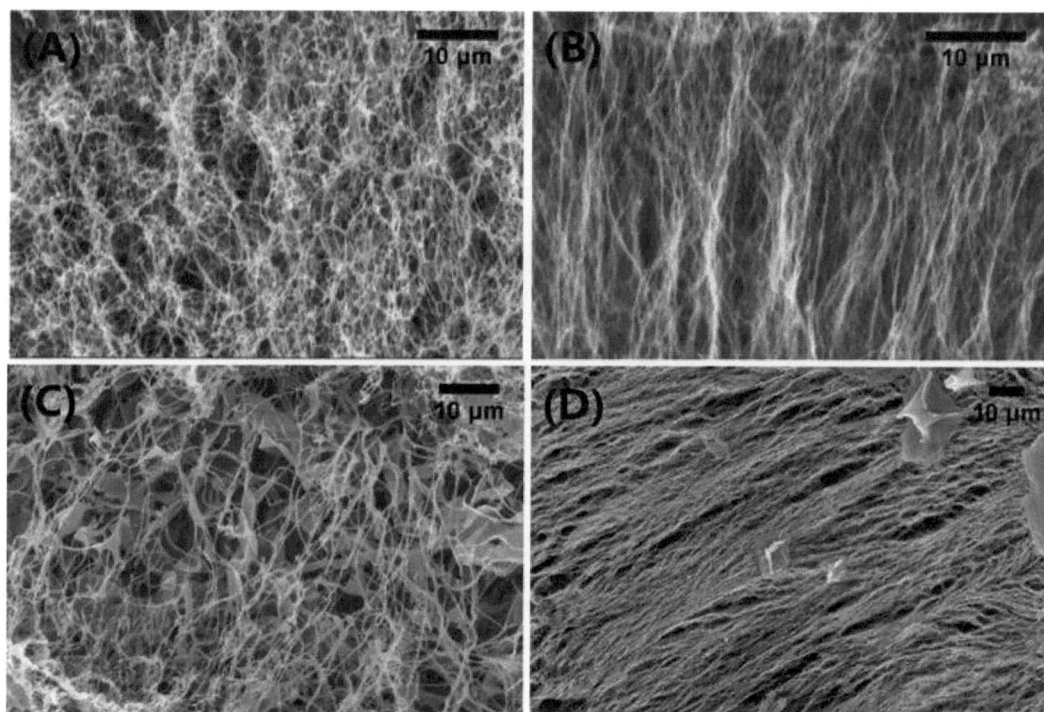

Fig. (4). SEM images of nanofiber at different alginate concentrations. **(A)** 0.1wt%, **(B)** 0.2wt%, **(C)** 0.4wt%, and **(D)** 0.8wt%.

This lamellar-like structure more obviously observed with κ-carrageenan. Only a small amount of fiber structure with an average diameter of 418.98 ± 99.01 nm was observed (Fig. **5A**). When increasing concentrations of κ-carrageenan were used, the thickness of the lamellar-like structure increased and a porous morphology resulted (Fig. **5B-D**).

These morphological changes were considered to be due to the viscosity of the polymer solution. A highly viscous solution might lower the speed of ice crystal growth, resulting in a more random structure. Fig. (**6**) displays a schematic illustration of the ice segregation-induced self-assembly [5]. The combined ultrasonic atomization-freeze casting approach benefits from synergistic effects. When a solution with lower viscosity is applied, the ultrasonic atomization provides a high initial speed, thus producing a thin polymer solution layer on the cooled surface for unidirectional rapid freezing. Next, acting as templates, small and ordered ice crystals segregate the polymers and freeze–drying subsequently removes these ice crystals without damaging the resulting fibrous structured polymer (Fig. **6A**). Both small diameter and uniform orientation resulted from the high freezing rate. Normally, higher freezing rates generate smaller ice crystals and smaller polymer template morphologies, unlike lower freezing rates [6]. In the presence of liquid nitrogen, the entire freezing process lasted less than 0.01 s, promoting the alignment into nanofiber structures under the template effects of the growing ice crystals. The high initial velocity ultrasonic atomization spread the solution into a thin liquid layer on the cold surface, enabling rapid and unidirectional freezing. High-viscosity solutions produce a lower initial atomization velocity than do their low-viscosity counterparts through the nozzle tip. As a result, the high-viscosity solutions form a thicker and uneven liquid film on the cooled surface, increasing the freezing time. Furthermore, strong intermolecular forces hinder the normal growth of ice crystals during freeze casting, reducing the driving force of the fiber formation. These forces alter the crystal growth direction, making it difficult to achieve a uniform template or fibrous morphologies.

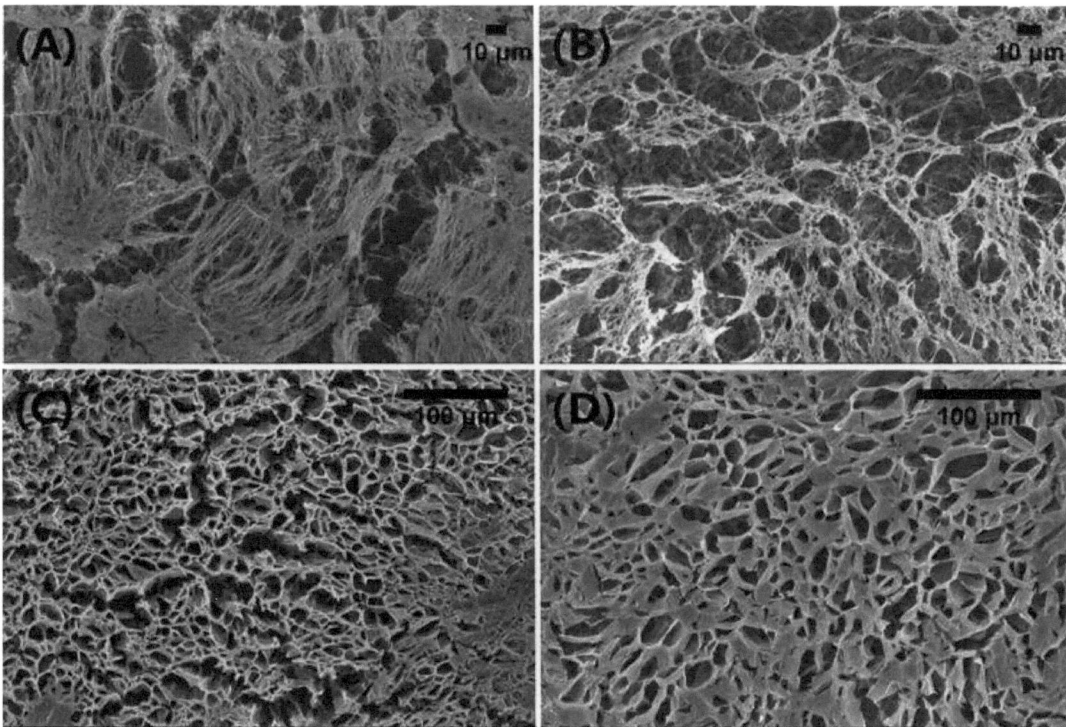

Fig. (5). SEM images of nanofiber at different κ-carrageenan concentrations. (**A**) 0.1wt%, (**B**) 0.2wt%, (**C**) 0.4wt%, and (**D**) 0.8wt%.

The anti-bacterial property of chitosan nanofibers was compared using different types of gram-negative and positive bacteria (Fig. **7**). A halo (zone of growth inhibition) was clearly observed for gram-negative and positive bacteria that cause food poisoning. The areas of the halos decreased using the different chitosan nanofibers (Sample 1 and 2). Cellulose paper combined with chitosan nanofibers (Sample 4 to 6) also inhibited bacterial growth in cases. Cellulose paper alone did not inhibit bacterial growth (Sample 3). Considering the difference of molecular structure between cellulose and chitosan, the amino moiety of chitosan could contribute to the anti-bacterial property [7].

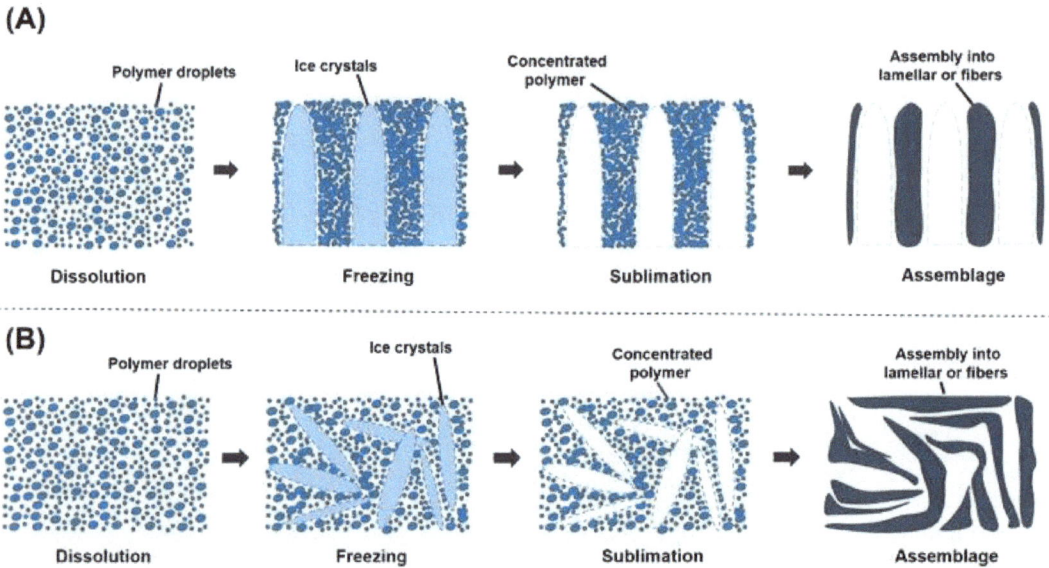

Fig. (6). Schematic illustration of **(A)** fiber formation by rapid freezing, **(B)** lamellar or porous structure formation by slow freezing.

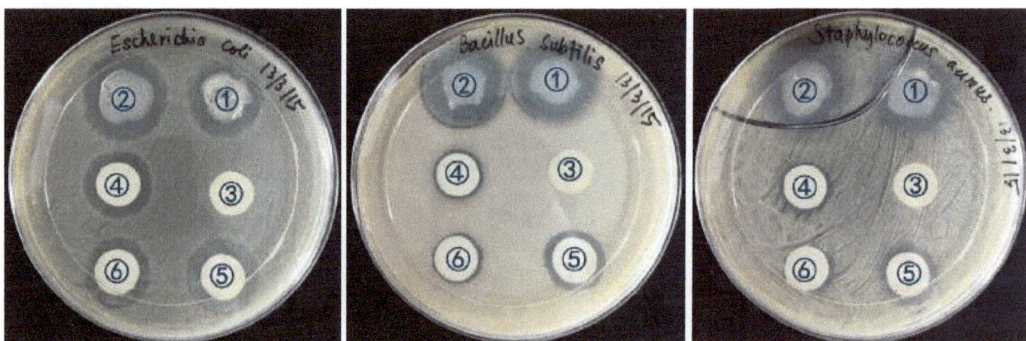

Fig. (7). Anti-bacterial test with various packaging material. (Left: *Escherichia coli*, Middle: *Bacillus subtilis*, Right: *Staphylococcus aureus*).

The applicability of chitosan nanofibers as a food packaging material was explored for strawberry preservation (Fig. **8**). After 5 days of preservation, strawberries without any packaging had changed color and were severely damaged, with some rotten spots observed in berries wrapped with plastic or filter paper. On the contrary, strawberries packaged with filter paper coated with chitosan nanofibers maintained an excellent appearance and good quality, with no loss in nutritional value, such as vitamin C content.

	No package	Plastic wrap	Filter paper	Filter paper with chitosan nanofiber
Samples (after 5 day)				
Sensory evaluation	X	O	X	O
Rotten degree	X	X	X	O
Weight loss rate	X	O	X	O
Vitamin C content	X	O	X	O

Fig. (8). Comparison of the effectiveness of various packaging materials used in strawberry preservation.

Other advantages of this packaging material combining cellulose filter paper with chitosan nanofiber included its superior mechanical property during usage and biodegradability after disposal. This biodegradable functional packaging material would help alleviate the pollution of marine ecosystems caused by single-use petroleum-based plastics.

CONCLUSION

Nanofibers with diameters between 100 nm and 200 nm were easily prepared from various marine polysaccharides by combining ultrasonic atomization with freeze casting. Scanning electron microscopy demonstrated that fiber diameter and morphology varied considerably between the types as well with the polymer concentrations. Nanofibers of uniform orientation were obtained from alginate and chitosan with lower concentrations below 0.4wt%. The simple combination of ultrasonic atomization and freeze casting can be applied to nanofiber fabrication for water-soluble polysaccharides and for other natural and synthetic polymers.

As for industrial applications, continuous production or scale-up of this nanofiber fabrication method could be achieved using a rotating drum shaped reactor, as shown in Fig. (**9**). In this process, a solution containing polysaccharides is atomized onto the drum shaped reactor that is filled with liquid nitrogen for rapid freezing. The thin layer of frozen sample is scraped from the surface and freeze dried to obtain the nanofibers. This concept has already been applied for the production of carbon nanofibers from lignin [8].

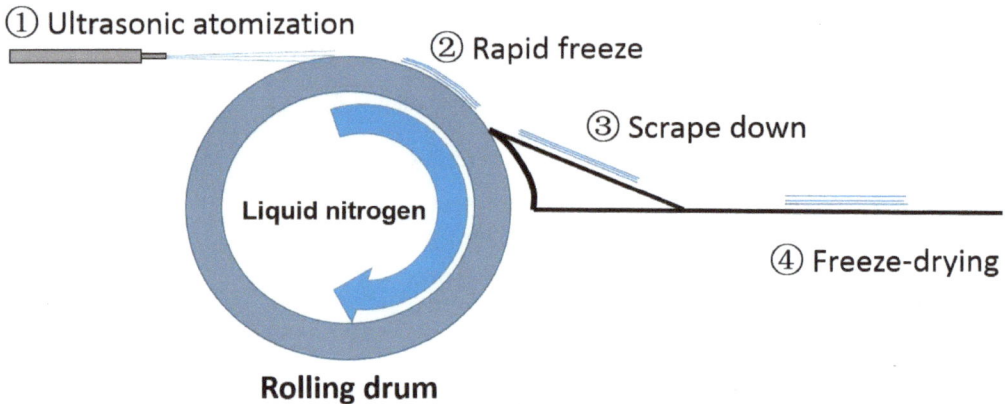

Fig. (9). Schematic diagram of continuous nanofiber fabrication by the freeze casting method.

The functional application of chitosan nanofibers as an anti-bacterial food packaging material was successfully demonstrated. This potential of nanofibers from marine polymers needs to be further explored to fully utilize the unique properties of functional moieties.

Finally, the biodegradable composites made from biodegradable plastics reinforced with nanofibers produced from renewable resources could provide solution to ocean pollution caused by conventional single-use plastics derived from petroleum.

CONSENT FOR PUBLICATION

Not applicable.

CONFLICT OF INTEREST

The author confirms that this chapter contents have no conflict of interest.

ACKNOWLEDGEMENTS

Declare none.

REFERENCES

[1]   Boddohi S, Kipper MJ. Engineering nanoassemblies of polysaccharides. Adv Mater 2010; 22(28): 2998-3016.
        [http://dx.doi.org/10.1002/adma.200903790] [PMID: 20593437]

[2]   Geng X, Kwon OH, Jang J. Electrospinning of chitosan dissolved in concentrated acetic acid solution. Biomaterials 2005; 26(27): 5427-32.
        [http://dx.doi.org/10.1016/j.biomaterials.2005.01.066] [PMID: 15860199]

[3]   Wang Y, Wakisaka M. Chitosan nanofibers fabricated by combined ultrasonic atomization and freeze casting. Carbohydr Polym 2015; 122: 18-25.
        [http://dx.doi.org/10.1016/j.carbpol.2014.12.080] [PMID: 25817638]

[4]   Wakisaka M, Wang Y. Polysaccharide nanofibers fabricated by combined ultrasonic atomization and freeze casting. Kobunshi Ronbunshu 2016; 73: 233-7.
        [http://dx.doi.org/10.1295/koron.2015-0082]

[5]   Gutiérrez MC, Ferrer ML, del Monte F. Ice-templated materials: Sophisticated structures exhibiting enhanced functionalities obtained after unidirectional freezing and ice-segregation-induced self-assembly. Chem Mater 2008; 20(3): 634-48.
        [http://dx.doi.org/10.1021/cm702028z]

[6]   Mukai SR, Onodera K, Yamada I. Studies on the growth of ice crystal templates during the synthesis of a monolithic silica microhoneycomb using the ice templating method. Adsorption 2011; 17(1): 49-54.
        [http://dx.doi.org/10.1007/s10450-010-9286-2]

[7]   Kim KW, Thomas RL, Lee C, Park HJ. Antimicrobial activity of native chitosan, degraded chitosan, and O-carboxymethylated chitosan. J Food Prot 2003; 66(8): 1495-8.
        [http://dx.doi.org/10.4315/0362-028X-66.8.1495] [PMID: 12929845]

[8]   Spender J, Demers AL, Xie X, *et al.* Method for production of polymer and carbon nanofibers from water-soluble polymers. Nano Lett 2012; 12(7): 3857-60.
        [http://dx.doi.org/10.1021/nl301983d] [PMID: 22716198]

CHAPTER 3

Application of Marine Polymers for Seagrass Bed Restoration and Marine Ecosystem Services Preservation

Minato Wakisaka[*]

Graduate School of Life Science and Systems Engineering, Kyushu Institute of Technology, Kitakyushu, Japan

Abstract: An effective and simple technique for seagrass bed restoration is required since the area of seagrass bed, one of the most productive marine ecosystems in Japan, is decreasing. The physiology of the plant *Zostera marina*, in relation to seed germination, was investigated in detail. Optimal seed storage condition and pretreatment for breaking seed dormancy increased the seed germination rate. Planting of biodegradable gel-coated seeds for *Zostera marina* seagrass bed restoration was proposed.

Keywords: Seagrass bed restoration, Seed Germination, *Zostera marina*.

INTRODUCTION

Seagrass/seaweed bed is one of the most productive marine ecosystems. Reproduction of seagrass and seaweed is different from each other, which grows from the seed and spore, respectively. The habitat of *Zostera marina* seagrass bed is within the ambient coastal sea sand (Fig **1**). Ecosystem services provided by seagrass/seaweed bed include multiple functions, such as providing a cradle for aquatic life, purifying water, acting as carbon dioxide sink, and so on (Fig **2**). Carbon dioxide captured by marine ecosystems is called "Blue carbon," which is comparable to "Green carbon" captured by terrestrial ecosystems.

Areas of seagrass/seaweed bed are drastically decreasing owing to anthropogenic activities, like creation of landfills and water pollution caused by the inflow of wastewater.

For example, 70% of the seagrass bed area in Setouchi inland sea, Japan has diminished from 1960s to 1990s [1]. The severe impact of the diminishing ecosystem

---

[*] **Corresponding author Minato Wakisaka:** Graduate School of Life Science and Systems Engineering, Kyushu Institute of Technology, Kitakyushu, Japan; Tel/Fax: +81-093-695-6066; E-mail: wakisaka@life.kyutech.ac.jp

Toshiyuki Takahashi (Ed.)

services provided by the seagrass/seaweed bed is now being widely recognized not only by the fishermen but also by the people of local communities.

Fig. (1). Seagrass bed of *Zostera marina* (left) and its seed (right).

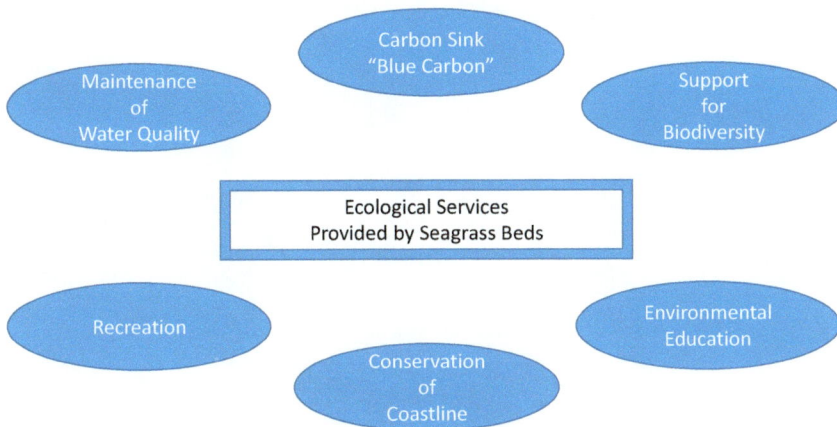

Fig. (2). Ecological services provided by seagrass beds.

Currently, efforts for seagrass bed restoration are being made not only in Japan but all over the world [2]. Even though versatile methodologies of seedling or nursery planting were tried, the results were far below expectation in most of the cases owing to the low germination rate of the seed or fixation of *Zostera marina* seeds to sea sand. Furthermore, substantial manpower is required for the collection of seeds, maintenance of nursery, planting, monitoring, and so on. Thus, there is a requirement for an effective and simple methodology for the restoration of the seagrass bed.

Our approach for the promotion of seed fixation to sand, without loss due to ocean current, has been to coat the seed with biodegradable gel, as shown in Fig (**3**). The understanding of the physiology of *Zostera marina*, especially seed germination, is lacking relative to that of terrestrial plants. The *Zostera marina* seed undergoes a period of dormancy, and reportedly, many studies have focused on the dormancy break of the seeds of *Zostera* species [2 - 4]. In general, the uptake of water is the trigger for seed dormancy break and transition towards germination. Various pre-treatments, which are known to be effective in inducing seed germination, such as plant hormones, osmotic stress, and physical stimuli, or the pulse power, were applied to *Zostera marina* in this study. Suitable biodegradable material for the coating of *Zostera marina* seed was selected, and its impact on germination was also investigated.

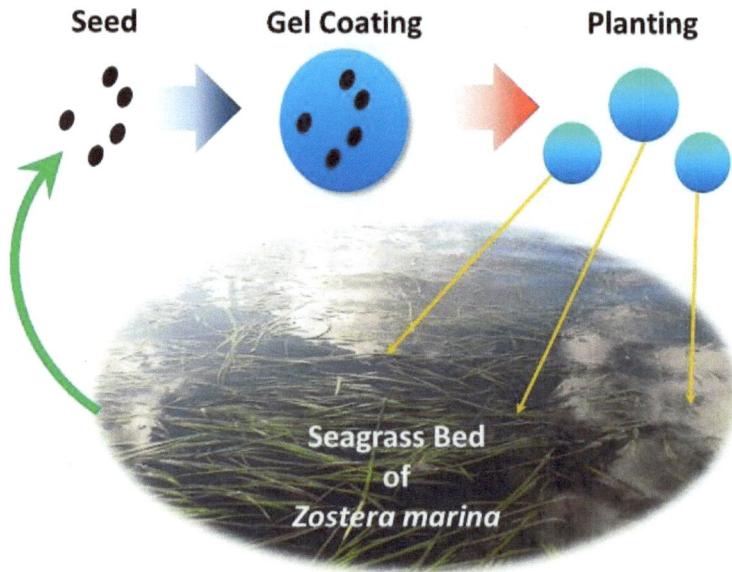

Fig. (3). Planting of gel-coated seeds for seagrass bed restoration of *Zostera marina*.

MATERIALS AND METHODS

Zostera marina grown in the Tsuyazaki coast in Fukutsu, Fukuoka was monitored from May 2011 to Dec 2012. The seeds collected from that area were selected using saturated saline water and kept in sterilized artificial seawater at 5°C, under dark condition until usage. For the determination of the effect of osmotic pressure, the seeds were immersed in freshwater, at 5, 15, 30, and 50°C for 24 h, and compared with the seeds immersed in polyvinyl alcohol (PVA), D-mannitol, and glycerin, which have an osmotic pressure of 0.7-1.3 MPa. The effect of gibberellin, a plant hormone known to be involved in dormancy break and seed germination in terrestrial plants, was investigated. Pulse power, known to stimulate plant growth, was also applied within the range of 0-50 kV/cm. Germination of seeds embedded in sea sand, sea mud, and carboxymethyl cellulose (CMC) gel was compared. Twenty seeds were used for each group and their germination at 5°C and 15°C was compared under dark condition.

RESULTS AND DISCUSSION

As shown in Fig (**4**), 60% of the seeds were successfully germinated when immersed in freshwater at 30°C, while seed germination was delayed at the other temperatures and the germination ratio was also decreased. On the contrary, seed germination ratio decreased to < 20% when immersed in sugar alcohol, as shown in Fig (**5**). Higher rate of germination was achieved with freshwater since seeds were exposed to hypotonic condition, which triggered water uptake necessary for seed dormancy break and germination.

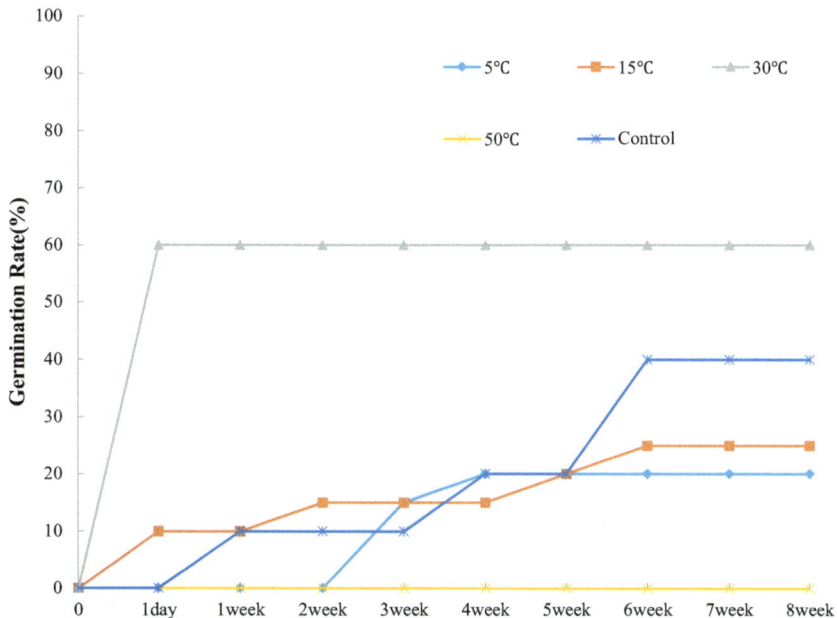

Fig. (4). Effect of freshwater immersion and temperature on seed germination.

The plant hormone gibberellin was not effective in inducing dormancy break and seed germination at 0.001-1 mM in *Zostera marina* seeds.

Pulse power, a physical stimulus, was applied. Repetitively operated compact pulsed power generators with a moderate peak power are developed for agricultural and the food processing applications. These applications are mainly based on biological effects and are used for germination control of plants like *Basidiomycota* and *Arabidopsis*, inactivation of bacteria in soil and liquid medium (hydroponics), extraction of juice from fruits and vegetables, and decontamination of air and liquid [5]. As shown in Fig. (**6**), the optimum intensity for seed germination and growth of *Zostera marina* was approximately 10 kV/cm.

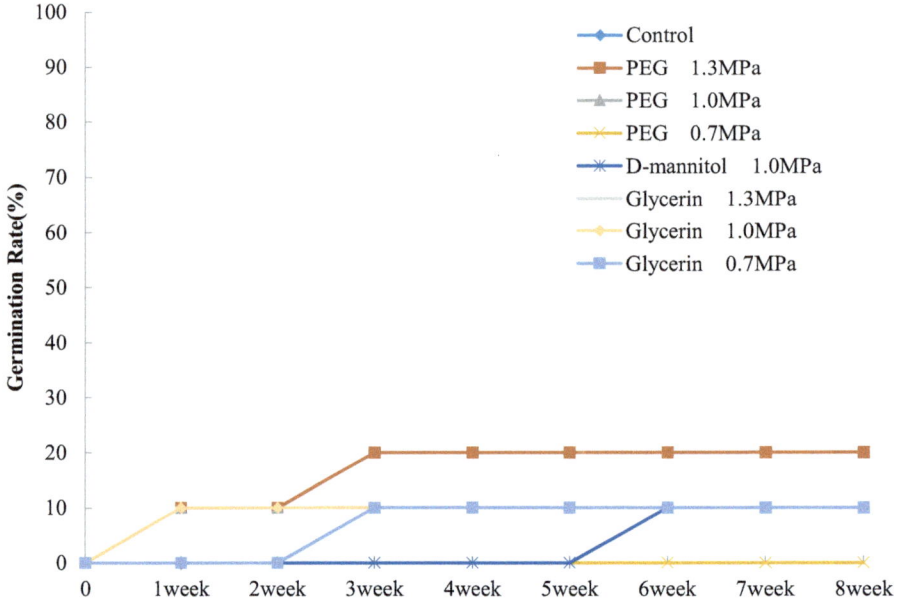

Fig. (5). Effect of osmotic stress on seed germination.

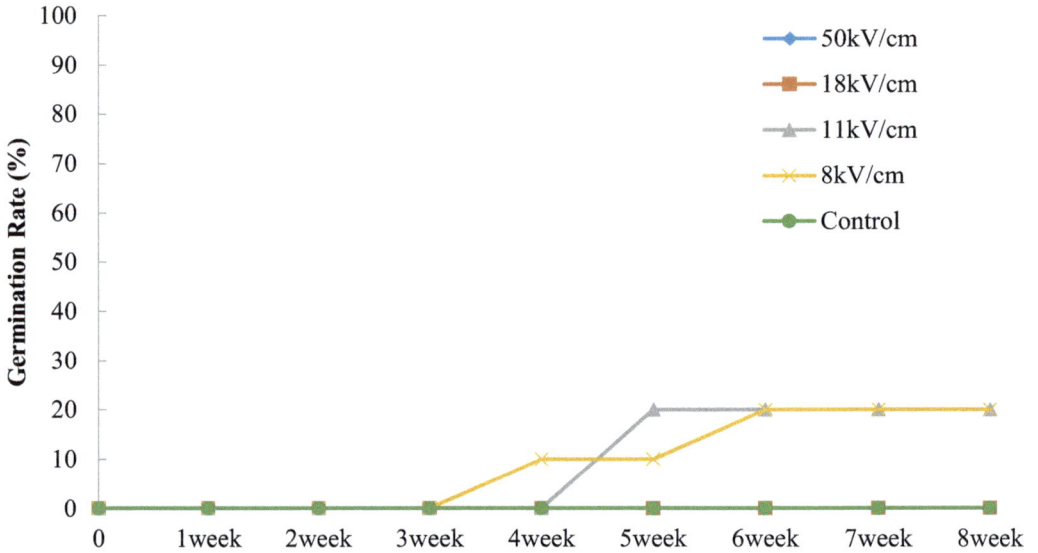

Fig. (6). Effect of pulse power application on seed germination.

The CMC gel, as the substrate for the seedling bed, was most suitable for seed germination relative to other gelling agents, such as agar, alginic acid, chitosan, and PVA. Agar gel was hard and tough, whereas alginic acid turned into a very hard gel upon reaction with divalent metal ions in seawater, making them unsuitable for seed coating. Chitosan gel could not maintain its shape in seawater and PVA gel floated on seawater and could not be used as an anchor.

As shown in Fig (**7**), although germination was observed in seeds embedded into sea sand or sea mud, no seed germination was observed when the seeds were immersed into seawater without the embedding material. Therefore, mechanical stress induced by embedding possibly affects seed germination. The *Zostera marina* seeds are well-adapted to anoxic conditions for germination [6].

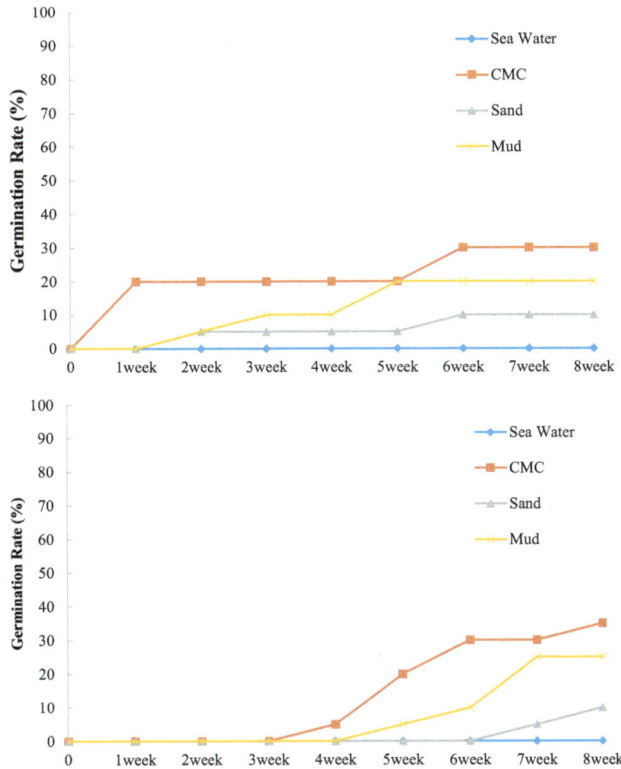

Fig. (7). Effect of embedding material on seed germination. (Upper) seeds stored at 15°C, (Lower) seeds stored at 5°C.

Jarvis and Moore investigated the effects of seed source, sediment type, and burial depth on mixed-annual and perennial *Zostera marina* seed germination and reported a reduced germination of *Zostera marina* seeds buried ≥5 cm in coarse sediments, which may represent a possible bottleneck in successful sexual reproduction, likely affecting the resilience to and recovery from disturbance for both perennial and mixed-annual *Zostera marina* beds [7].

The storage temperature also affects seed germination. Comparison of seed germination at 5°C and 15°C revealed that germination started earlier at the higher temperature, although the final germination rates were almost the same, around 40%, after 8 weeks.

Furthermore, longer storage time at lower temperature resulted in a lower germination ratio, as shown in Fig. (**8**). Dooley also reported that seed viability is reduced with storage: 78% in 1 year to 32% in 4 years [8].

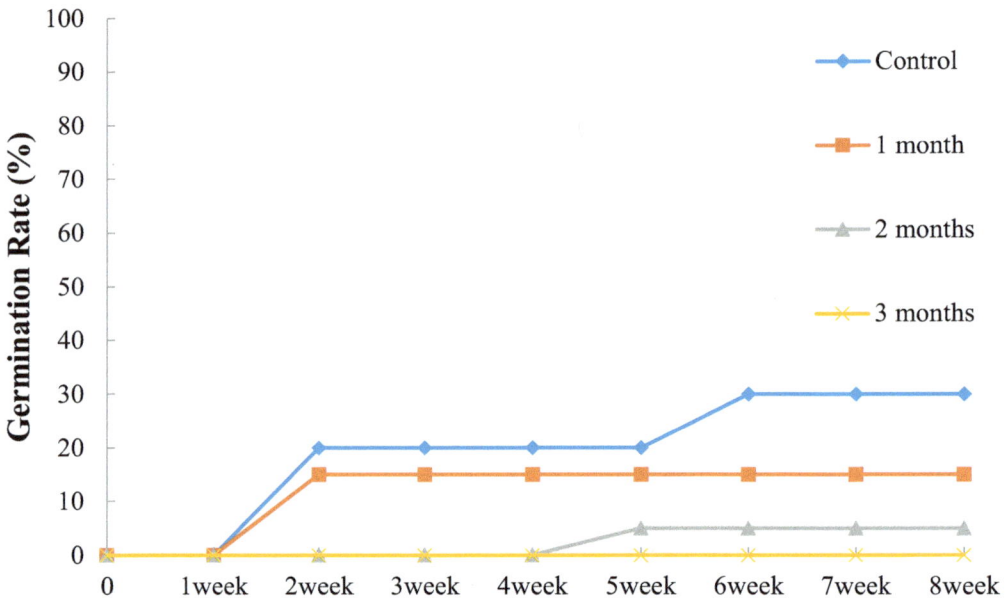

Fig. (8). Effect of storage period on seed germination.

CONCLUSION

In our study, we established that proper storage conditions and pre-treatment were effective in enhancing the germination of *Zostera marina* seeds. Coating with CMC gel provides an alternative for effective seagrass bed restoration.

The physiological mechanism of dormancy break and germination of *Zostera marina* seeds need to be explored further, as an understanding of the mechanism will be helpful for successful seagrass bed restoration.

CONSENT FOR PUBLICATION

Not applicable.

CONFLICT OF INTEREST

The author confirms that this chapter contents have no conflict of interest.

ACKNOWLEDGEMENTS

Declare none.

REFFERENCES

[1]   Terawaki T, Yoshikawa K, Yoshida G, Uchimura M, Iseki K. Ecology and restoration techniques for Sargassum beds in the Seto Inland Sea, Japan. Mar Pollut Bull 2003; 47(1-6): 198-201.
[http://dx.doi.org/10.1016/S0025-326X(03)00054-7] [PMID: 12787620]

[2]   Marion SR, Orth RJ. Innovative techniques for large scale seagrass restoration using *Zostera marina* (eelgrass) seeds. Restor Ecol 2010; 18(4): 514-26.
[http://dx.doi.org/10.1111/j.1526-100X.2010.00692.x]

[3]   Harrison PG. Mechanisms of seed dormancy in an annual population of *Zostera marina* (eelgrass) from the Netherlands. Can J Bot 1991; 69(9): 1972-6.
[http://dx.doi.org/10.1139/b91-247]

[4]   Morita T, Miyamatsu A, Fujii M, *et al.* Germination in *Zostera japonica* is determined by cold stratification, tidal elevation and sediment type. Aquat Bot 2011; 95(3): 234-41.
[http://dx.doi.org/10.1016/j.aquabot.2011.07.003]

[5]   Conacher C A, Poiner I R, Butler J, Pun S, Tree D J. Germination, storage and viability testing of seeds of *Zostera capricorni* Aschers. from a tropical bay in Australia. Aquat Bot 1994; 49(1): 47-58.
[http://dx.doi.org/10.1016/0304-3770(94)90005-1]

[6]   Takaki K, Ihara S. Agricultural and food processing applications of pulsed power technology. IEEJ Trans Fund Mater 2009; 129(7): 439-45.

[7]   Jarvis JC, Moore KA. Effects of seed source, sediment type, and burial depth on mixed-annual and perennial Zostera marina L. seed germination and seedling establishment. Estuaries Coasts 2015; 38(3): 964-78.
[http://dx.doi.org/10.1007/s12237-014-9869-3]

[8]   Dooley F D, Wyllie-Echeverria S, Van Volkenburgh E. Long-term seed storage and viability of Zostera marina. Aquat Bot 2013; 111: 130-4.
[http://dx.doi.org/10.1016/j.aquabot.2013.06.006]

CHAPTER 4

Relationship between Algal Blooms and Marine-Ecosystem Services

Toshiyuki Takahashi[*]

Department of Chemical Science and Engineering, National Institute of Technology (KOSEN), Miyakonojo College, Miyakonojo, Japan

Abstract: Although the status of microalgae as a primary producer in the aquatic environment is well-known, some algae are also an important biomass supporting our varied fishery resources as prey in fisheries industries. Algal blooms occur due to environmental changes in factors, such as temperature, salinity, nutrient concentrations, and light availability. The blooms occasionally result in serious damage to human activities including fisheries. Starting with information regarding several algal toxins from harmful algae, this chapter discusses the common characteristics of an algal bloom and several case studies of algal blooms causing fish-kill in coastal waters and aquaculture farms in Japan. In addition to low dissolved oxygen concentration induced by excessive proliferation of algae, some harmful algae produce distinctive toxins which contaminate marine foods like shellfish. Therefore, this chapter also outlines general features of algal toxins. The properties of bloom-forming algae described in this chapter require the fisheries industry and their related sectors to constantly monitor harmful algal blooms and verify whether the shellfish accumulates harmful algae.

Keywords: *Chattonella*, *Fibrocapsa*, Microalgae, Microcystin, Nodularin, Harmful algae, *Heterosigma*.

INTRODUCTION

Not only seaweed but also phytoplankton and microbes such as cyanobacteria, as primary producers, support marine ecosystems [1 - 3]. Microalgae are an important biomass that support our rich and varied fishery resources because they have been used as prey for clam and crustacea in the fisheries industries, and as prey for zooplankton such as rotifers in farming fisheries [4, 5]. Microalgae have attracted attention for use in various applications, such as agriculture, food, pharmaceuticals, environmental cleanup, and biofuels [6 - 10] (Fig. **1**).

[*] **Corresponding author Toshiyuki Takahashi:** Department of Chemical Science and Engineering, National Institute of Technology, Miyakonojo College, Miyakonojo, Japan; Tel; +81-986-47-1219; Fax; +81-986-47-1231; E-mail: mttaka@cc.miyakonojo-nct.ac.jp

Algal blooms occur naturally triggered by certain environmental changes in factors, such as water temperature, salinity, nutrient levels, and duration of sunshine. Algal blooms might be antithetical to the industrial usefulness presented above. An algal bloom is a phenomenon where phytoplankton or microalgae proliferate abnormally, resulting in the collapse of the equilibrium among species in marine ecosystems. Some algal blooms described as red tide are harmful. In addition to their changing the color of the sea, they often damage aquatic ecosystems, human health, and the economy of farming fisheries.

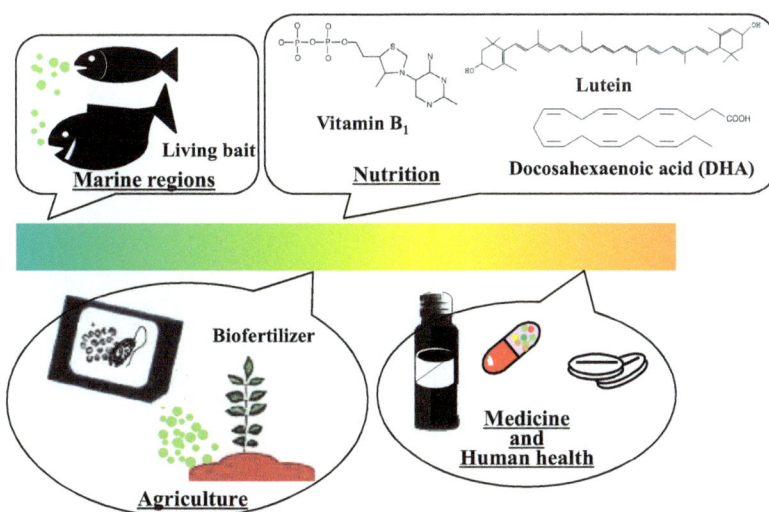

Fig. (1). Potential of microalgae for industrial applications. Source [10].

This chapter presents characteristic features of algal blooms. Furthermore, this chapter summarizes the influences of algal blooms on human activities using fisheries as an example.

CHARACTERISTICS OF ALGAL BLOOMS

Common characteristics of an algal bloom include an extremely high density of algae, which is dominated by only a single or a few algal species, visible accumulation of algae, and degradation of water quality. Algal blooms, however, are natural phenomena in marine and freshwater ecosystems. They are induced as responses to environmental factors such as nutrient, light availability, pH, wind intensity, water currents, and water temperatures [11, 12]. The proliferation of both toxic and non-toxic algal species causing algal blooms (Fig. **2**) can have deleterious effects on the following ecological, public health, and economic situations.

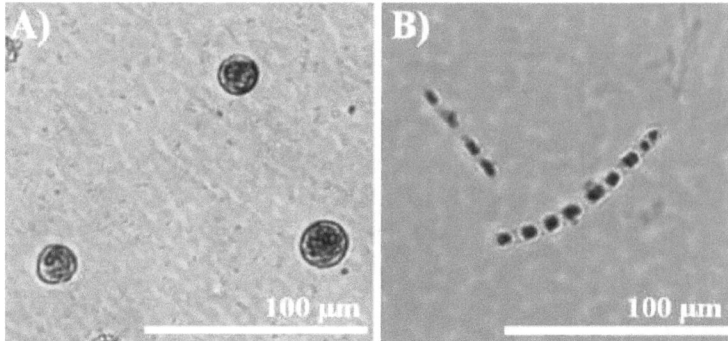

Fig. (2). Examples of algal species causing algal blooms.
Bloom-forming raphidophytes *Heterosigma akashiwo* (**A**) and bloom-forming diatom *Skeletonema* sp. (**B**).

Low dissolved oxygen concentration induced by excessive proliferation of algae triggers the subsequent death of aquatic animals and food-web disruption. In addition to the low oxygen levels, both the clogging of fish gills and production of certain toxins by harmful algae are direct causes of death of aquatic animals. Even if proliferation is of non-toxic algae, their blooms result in oxygen depletion and loss of biodiversity in the phytoplankton community structure, and reduce the amount of light reaching the benthos [13, 14].

In addition, some toxins produced by harmful algae cause allergic reactions and several illnesses in humans and other organisms. These toxins accumulate, travel up the food-web, and finally cause illness or death in humans and other animals [13]. Representative toxins from harmful algae are cyanotoxins called microcystins, amnestic shellfish poison from some diatoms, paralytic shellfish poison from some dinoflagellates and cyanobacteria, neurotoxic shellfish poison, and diarrheic shellfish poison. Cyanobacteria are included as causative organisms in this chapter, although they are taxonomically photosynthetic bacteria.

The issues mentioned above also cause economic concerns. Mass mortality of aquatic animals including fish and the change in color of the sea impairs scenery and spoils the view of the sea. The change in the seascape induces loss of recreational revenue at tourist destinations along their coasts. To reduce toxic damage from contaminated water including algal toxins, the relevant local government or administrator must bear the expenses for an additional drinking water treatment cost. Contamination of fishery and aquaculture and the following bans on export of aquatic food understandably result in immense economic loss.

CASE EXAMPLE OF ALGAL BLOOMS [15]

As harmful algal blooms have increased in frequency, intensity, and duration in many parts of the world, their blooms often cause concern in Japan. Several algal species such as *Chattonella* sp., *Fibrocapsa* sp., and *Heterosigma* sp. have been well known as bloom-forming algae. *C. antiqua, C. marina,* and *C. verruculosa* belonging to *Chattonella*, *F. japonica* to *Fibrocapsa,* and *H. akashiwo* to *Heterosigma* have particularly caused algal blooms including fish-kill in coastal waters and aquaculture farms in Japan. Eutrophication control of the water environment directly affects the occurrence frequency and intensity of these blooms. Therefore, in addition to control total volume of water for reduction in pollution load, both chemical oxygen demand (COD) and total phosphorus (T-P) concentration have been monitored in major ocean areas.

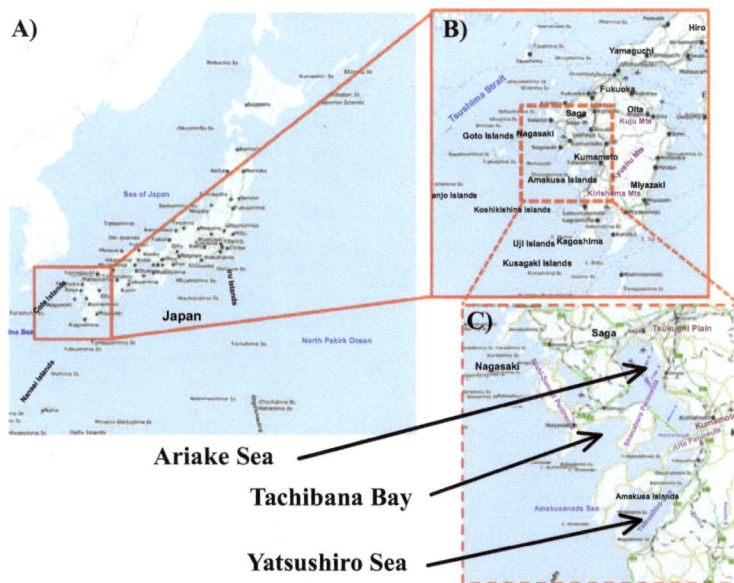

Fig. (3). Geographical location of case studies regarding algal blooms using Google Maps. **(A)** Map of Japan, **(B)** magnification of the Kyushu region in panel A, and **(C)** magnification of the study areas regarding algal blooms in the Kyushu region.

To gain an understanding of the recent situation regarding algal blooms, this chapter shows case examples of algal blooms in the Kyushu region of Japan (Fig. **3**). Fig. (**4**) especially shows an algal bloom occurrence in the Ariake Sea, Yatsushiro Sea, and Tachibana Bay where aquafarming of fish and marine culture of seaweed has been actively performed as a local industry. While the frequency of occurrence of an algal bloom in Tachibana Bay has been almost constant, those in both the Ariake Sea and Yatsushiro Sea have increased since about 1998.

Although the mean values of the number of algal bloom occurrences during 1970-1980 were 15.0/year in the Ariake Sea and 8.0/year in the Yatsushiro Sea, those during 2000-2015 have increased to 35.6/year in the Ariake Sea and 16.5/year in the Yatsushiro Sea. Thus, the recent mean values of the number of algal bloom occurrences during 2000-2015 have increased to twice as many as those during 1970-1980.

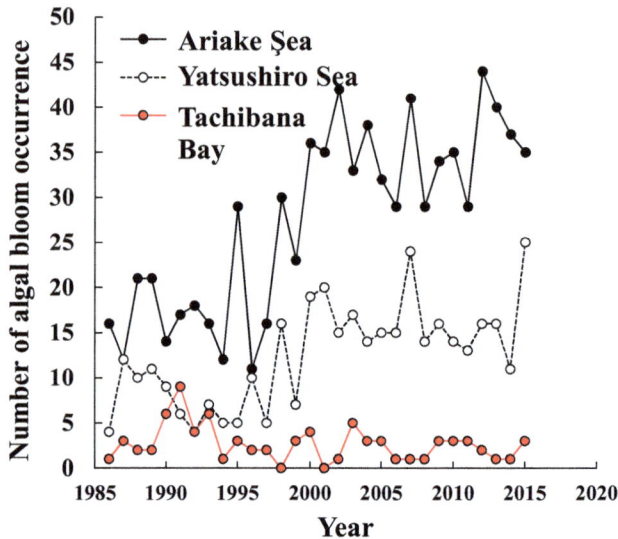

Fig. (4). Number of algal bloom occurrences in the Ariake Sea, Yatsushiro Sea, and Tachibana Bay in Japan modified from [15].

Variation in causative species constituting each algal bloom expresses as differences in both physiological and ecological features, as does divergence between each effect on other aquatic organisms. Fig. (**5**) shows the historical change in causative species constituting each algal bloom in the Ariake Sea in which the most frequent algal bloom occurrence has been during the survey period presented in Fig. (**4**). An algal bloom does not always comprise homogeneous species. Diatom blooms have occurred with the most increased frequency in the Ariake Sea throughout the period of the survey; raphidophytes and dinoflagellates have been the second- or the third-highest algae.

Although algal blooms comprising raphidophytes were few until about 1988, the frequency of their blooms has increased considerably since 1998 with a frequency comparable to that of the blooms of dinoflagellates. The representative species in raphidophytes in the Ariake Sea are *Heterosigma akashiwo, Chattonella antiqua, C. marina,* and *C. ovata.* Both diatoms and dinoflagellates in comparison with raphidophytes have also increased during the same period (Fig. **6**). Except for the increase in the proportion of raphidophytes in recent years, component species of algal blooms in recent times have been rarely different from those during the 1970s-1980s.

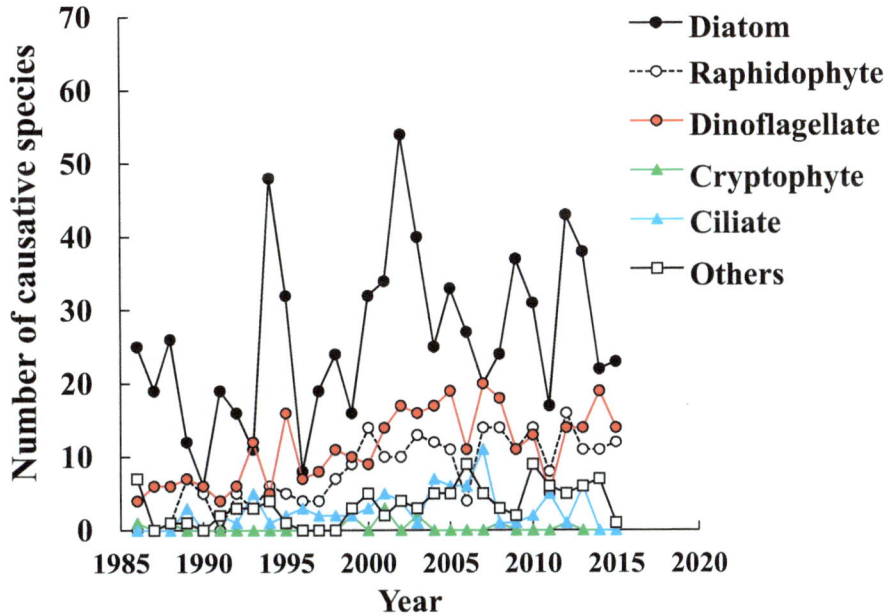

Fig. (5). Historical change in the number of causative species in each algal bloom in the Ariake Sea modified from [15].
Some algal blooms generally comprise heterogeneous species. Note that the number of algal species constituting each algal bloom is not necessarily consistent with the frequency of algal blooms shown in Fig. (4).

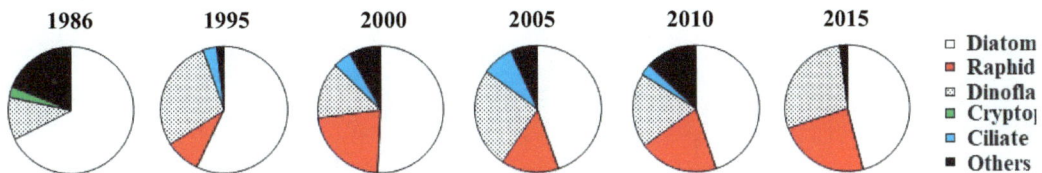

Fig. (6). Rate analyses of causative species in each algal bloom in the Ariake Sea using time-series. Each datum from Fig. (5) is reconstructed as each pie chart.

FEATURES OF EACH ALGAL SPECIES CONSTITUTING AN ALGAL BLOOM [15]

Diatoms in both coastal and inner bays have been an important algal species as an essential producer in the marine ecological chain. Therefore, a certain amount of proliferation of diatoms might seem to be unavoidable. Not only small diatoms but also several large ones such as *Rhizosolenia imbricata* and *Eucampia zodiacus* have mainly caused damage to laver culture in the Ariake Sea.

Heterosigma akashiwo and *Chattonella* sp. such as *C. antiqua, C. marina,* and *C. ovata* cause algal blooms constituting raphidophytes in the Ariake Sea. Their proliferation tends to cause damage to fish.

Ceratium sp. such as *C. furca* and *C. fusus* and *Akashiwo sanguinea* cause algal blooms constituting dinoflagellates in the Ariake Sea. They do not tend to trigger serious red tide problems. *Karenia mikimotoi* and *Cochlodinium* sp. such as *C. polykrikoides* as bloom-forming dinoflagellates, however, have tended to cause serious damage to the fisheries industry.

TOXINS OF SEVERAL ALGAL BLOOMS

Microcystins are the most common of the toxins and often result in the poisoning of animals and humans. The name "Microcystin" is derived from the species *Microcystis aeruginosa*. These cyclic heptapeptide hepatotoxins produced by cyanobacteria are stable in the water environment such as in both warm and cold water. Many different microcystins, including microcystin-LR ($C_{49}H_{74}N_{10}O_{12}$) [16, 17] (Fig. 7) and microcystin-LA ($C_{46}H_{67}N_7O_{12}$) [18] have been found. These microcystins can inhibit protein phosphatase inside hepatocytes. By disrupting the normal balance of phosphate groups in cytoskeletal proteins, microcystins control the regulation of cytoskeleton at molecular levels.

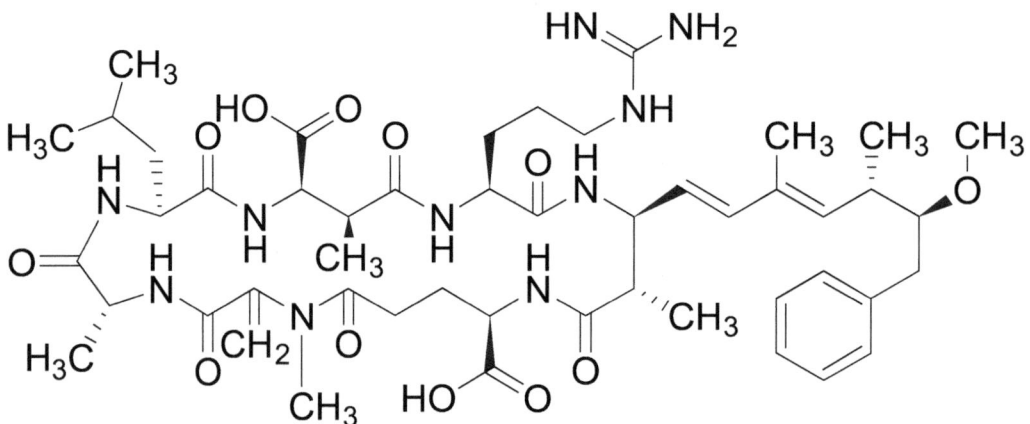

Fig. (7). Chemical structure of a microcystin (Microcystin-LR). Source [17].

There is another type of cyanobacteria toxin called "Nodularin" (Fig. **8**) [19 - 21] produced by *Nodularia spumigena*. Although the chemical structures of nodularins are closely related to that of microcystins consisting seven amino acids, nodularins comprise only five amino acids in the peptide ring [19, 21]. In addition to the similarity of chemical structure among these toxins, nodularins have the hepatotoxic effect induced by the potent inhibition of protein phosphatase [19, 20].

Fig. (8). Chemical structure of a nodularin. Source [21].

Protein phosphatase is generally critical for normal regulation of cell division. Therefore, there is a concern that an exposure of these toxins may play a role in the initiation of cancer in the liver and other chronic disorders of the gastrointestinal tract [19].

Many other poisons including amnestic shellfish poisoning (ASP), paralytic shellfish poisoning (PSP), diarrheic shellfish poisoning, ciguatera fish poisoning, and neurotoxic shellfish poisoning are serious illnesses caused by eating aquatic food like shellfish contaminated by toxic marine algae.

To cite a few examples, a biotoxin called domoic acid (Fig. **9**) [22], that is produced by the diatom *Pseudo-nitzschia* sp., and is a causative substance of ASP [23, 24]. Several shellfish eat these microalgae and consequently retain the toxin. After consuming these shellfish contaminated by domoic acid, gastroenteritis occurs and neurological symptoms including memory loss appear in severe cases [25]. Different from ASP, PSP results from the consumption of mussels, clams, and oysters contaminated by potent neurotoxins called saxitoxin. These toxins are produced by toxic dinoflagellates such as *Alexandrium, Gymnodinium,* and

Pyrodinium in the marine environment and some cyanobacteria in the freshwater environment [26 - 30]. The toxins also pass through the food web in water environments and accumulate in the vector organisms. The PSP toxins, which can block sodium channels in neurons [30, 31], cause paralysis in humans. Therefore, the fisheries industry and their related sectors must constantly monitor the density of harmful algal blooms and control the accumulation of these algae in shellfish.

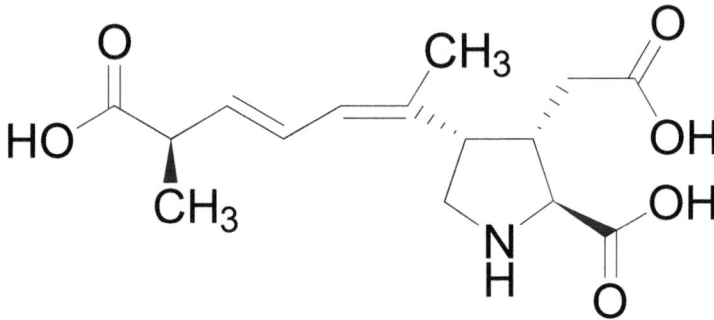

Fig. (9). Chemical structure of domoic acid. [22].

CONCLUSIONS

Some microalgae are necessary biomass to support our varied fishery resources because of their application as prey in fisheries industries. Algal blooms induced by environmental changes occasionally results in serious damage to human activities including fisheries and human health. In fact, harmful algal blooms have increased in many parts of the world including Japan. Several algal species such as *Chattonella* sp., *Fibrocapsa* sp. and *Heterosigma* sp. have particularly caused algal blooms including fish-kill in coastal waters and aquaculture farms in Japan.

To prevent algal blooms from serious losses on ecological, public health and economic situations, the fishery industries and their related sectors must monitor the density of algal blooms constantly. In addition, they have to control the accumulation of harmful algae in shellfish. To precisely evaluate the behaviors of bloom-forming algae, a method to selectively and efficiently detect only these organisms from the marine environment is necessary.

CONSENT FOR PUBLICATION

Not applicable.

CONFLICT OF INTEREST

The author confirms that this chapter contents have no conflict of interest.

ACKNOWLEDGEMENTS

Declare none.

REFERENCES

[1]   Glöckner FO, Stal LJ, Sandaa R-A, *et al.* Marine Microbial Diversity and its role in Ecosystem Functioning and Environmental Change. Marine Bord Position Paper 17. Ostend, Belgium: Marine Board-ESF 2012; pp. 1-80.

[2]   Herbert RA. Nitrogen cycling in coastal marine ecosystems. FEMS Microbiol Rev 1999; 23(5): 563-90.
[http://dx.doi.org/10.1111/j.1574-6976.1999.tb00414.x] [PMID: 10525167]

[3]   Azam F, Malfatti F. Microbial structuring of marine ecosystems. Nat Rev Microbiol 2007; 5(10): 782-91.
[http://dx.doi.org/10.1038/nrmicro1747] [PMID: 17853906]

[4]   Okauchi M. Selection of algal strains with high-growth rate as food organisms and development of their effective production techniques. Nippon Suisan Gakkaishi 2014; 80(3): 323-6.
[http://dx.doi.org/10.2331/suisan.80.323]

[5]   Spolaore P, Joannis-Cassan C, Duran E, Isambert A. Commercial applications of microalgae. J Biosci Bioeng 2006; 101(2): 87-96.
[http://dx.doi.org/10.1263/jbb.101.87] [PMID: 16569602]

[6]   Arashida R. Characteristics of the microalgae Euglena and its applications in foods and ecological fields. Jpn Soc Photosyn Res 2012; 22: 33-8.

[7]   Faheed FA, Fattah Z A-E. Effect of *chlorella vulgaris* as bio-fertilizer on growth parameters and metabolic aspects of lettuce plant. J Agricul Soc Sci 2008; 4: 165-9.

[8]   Chaturvedi V, Nikhil K. Effect of algal bio-fertilizer on the *vigna radiate*: A critical review. Int J Eng Res Appl 2016; 6: 85-94.

[9]   Chisti Y. Biodiesel from microalgae. Biotechnol Adv 2007; 25(3): 294-306.
[http://dx.doi.org/10.1016/j.biotechadv.2007.02.001] [PMID: 17350212]

[10]   Takahashi T. Applicability of automated cell counter with a chlorophyll detector in routine management of microalgae. Sci Rep 2018; 8(1): 4967.
[http://dx.doi.org/10.1038/s41598-018-23311-8] [PMID: 29563559]

[11]   der Merwe D V, Price K P. Harmful algal bloom characterization at ultra-high spatial and temporal resolution using small unmanned aircraft systems. Toxins 2015; 7: 1065-78.
[http://dx.doi.org/10.3390/toxins7041065]

[12]   Kislik C, Dronova I, Kelly M. UAVs in support of algal bloom research: A review of current applications and future opportunities. Drones 2018; 2: 35.
[http://dx.doi.org/10.3390/drones2040035]

[13]   Erdner DL, Dyble J, Parsons ML, *et al.* Centers for oceans and human health: A unified approach to the challenge of harmful algal blooms. Environ Health 2008; 7 (Suppl. 2): S2.
[http://dx.doi.org/10.1186/1476-069X-7-S2-S2] [PMID: 19025673]

[14]   Moore SK, Trainer VL, Mantua NJ, *et al.* Impacts of climate variability and future climate change on harmful algal blooms and human health. Environ Health 2008; 7 (Suppl. 2): S4.
[http://dx.doi.org/10.1186/1476-069X-7-S2-S4] [PMID: 19025675]

[15]   Ministry of the Environment. Council report about Ariake sea and Yatsushiro sea 2017. https://www.env.go.jp/council/20ari-yatsu/report20170331/report20170331_all.pdf

[16]   Wickstrom M, Haschek W, Henningsen G, Miller LA, Wyman J, Beasley V. Sequential ultrastructural

and biochemical changes induced by microcystin-LR in isolated perfused rat livers. Nat Toxins 1996; 4(5): 195-205.
[http://dx.doi.org/10.1002/(SICI)(1996)4:5<195::AID-NT1>3.0.CO;2-B] [PMID: 8946394]

[17]   Merck KGaA. Product catalog.   https://www.sigmaaldrich.com/catalog/substance/microcystinlr solution9951710104337211?lang=ja&region=JP

[18]   Oliveira N B, Schwartz C A, Bloch C Jr, Paulino L, Pires O R Jr. Bioacumulation of Cyanotoxins in Hypophthalmichthys molitrix (Silver Carp) in Paranoá Lake, Brasilia-DF, Brazil. Bull Environ Contam Toxicol 2013; 90(3): 308-13.
[http://dx.doi.org/10.1007/s00128-012-0873-7]

[19]   Pearson L, Mihali T, Moffitt M, Kellmann R, Neilan B. On the chemistry, toxicology and genetics of the cyanobacterial toxins, microcystin, nodularin, saxitoxin and cylindrospermopsin. Mar Drugs 2010; 8(5): 1650-80.
[http://dx.doi.org/10.3390/md8051650] [PMID: 20559491]

[20]   Meili N, Christen V, Fent K. Nodularin induces tumor necrosis factor-alpha and mitogen-activated protein kinases (MAPK) and leads to induction of endoplasmic reticulum stress. Toxicol Appl Pharmacol 2016; 300: 25-33.
[http://dx.doi.org/10.1016/j.taap.2016.03.014] [PMID: 27061667]

[21]   Merck KGaA. Product catalog. https://www.sigmaaldrich.com/catalog/product/sigma/n5148?lang= ja&region=JP

[22]   Merck KGaA. Product catalog. https://www.sigmaaldrich.com/catalog/product/sigma/d6152?lang= ja&region=JP

[23]   Jakobsen B, Tasker A, Zimmer J. Domoic acid neurotoxicity in hippocampal slice cultures. Amino Acids 2002; 23(1-3): 37-44.
[http://dx.doi.org/10.1007/s00726-001-0107-5] [PMID: 12373516]

[24]   Bates SS, Hubbard KA, Lundholm N, Montresor M, Leaw CP. *Pseudo-nitzschia, Nitzschia*, and domoic acid: New research since 2011. Harmful Algae 2018; 79: 3-43.
[http://dx.doi.org/10.1016/j.hal.2018.06.001] [PMID: 30420013]

[25]   Todd ECD. Domoic Acid and Amnesic Shellfish Poisoning - A Review. J Food Prot 1993; 56(1): 69-83.
[http://dx.doi.org/10.4315/0362-028X-56.1.69] [PMID: 31084045]

[26]   Morse EV. Paralytic shellfish poisoning: a review. J Am Vet Med Assoc 1977; 171(11): 1178-80.
[PMID: 924835]

[27]   Lefebvre KA, Bill BD, Erickson A, *et al.* Characterization of intracellular and extracellular saxitoxin levels in both field and cultured *Alexandrium* spp. samples from Sequim Bay, Washington. Mar Drugs 2008; 6(2): 103-16.
[http://dx.doi.org/10.3390/md6020103] [PMID: 18728762]

[28]   Campbell K, Rawn DFK, Niedzwiadek B, Elliott CT. Paralytic shellfish poisoning (PSP) toxin binders for optical biosensor technology: problems and possibilities for the future: a review. Food Addit Contam Part A Chem Anal Control Expo Risk Assess 2011; 28(6): 711-25.
[http://dx.doi.org/10.1080/19440049.2010.531198] [PMID: 21623494]

[29]   Usup G, Kulis DM, Anderson DM. Growth and toxin production of the toxic dinoflagellate Pyrodinium bahamense var. compressum in laboratory cultures. Nat Toxins 1994; 2(5): 254-62.
[http://dx.doi.org/10.1002/nt.2620020503] [PMID: 7866660]

[30]   Cusick K D, Sayler G S. An overview on the marine neurotoxin, saxitoxin: Genetics, molecular targets, methods of detection and ecological functions. Drugs 2013; 11: 991-1018.

[31]   O'Neill K, Musgrave IF, Humpage A. Low dose extended exposure to saxitoxin and its potential neurodevelopmental effects: A review. Environ Toxicol Pharmacol 2016; 48: 7-16.
[http://dx.doi.org/10.1016/j.etap.2016.09.020] [PMID: 27716534]

CHAPTER 5

Some Background on Conventional Environmental Evaluations

Hotaka Kai[*]

Department of Chemistry and Biochemistry, National Institute of Technology, Suzuka College, Suzuka, Japan

Abstract: Many studies have reported the changes in environmental conditions by using various water analysis (and environmental analysis) techniques. In these reports, "the necessity for water analysis," "the basic knowledge for water analysis," "the requirements, conditions, and methods for water analysis," a "plan for water analysis," and whether the entities to be measured satisfy the "Environmental Quality Standards" and the "Effluent Standards" are summarized.

Keywords: Effluent Standards, Environmental analysis, Environmental condition, Environmental Quality Standards.

INTRODUCTION

This chapter introduces the necessity to establish environmental analysis for water systems, including a standard (indicator or parameter), basic knowledge, selection of suitable methods for water analysis, and some outlines for actual environmental analyses. Moreover, we will review the most recent articles about environmental analysis for water systems and explain the related important factors.

EFFECTIVE WATER UTILIZATION AND THE NECESSITY OF ANALYSIS

Water resources, such as tap water and well water, are required for our day-to-day activities, such as cooling, washing, and agriculture. Furthermore, water resources can be linked directly to the existing state of natural resources, such as river water and seawater. Thus, there are many types of water sources available. Water is one of the most important and essential resources. Water does not stay in the same place but circulates throughout the Earth while changing its form, such as river water, lake water, underground water, tap water, wastewater, and seawater.

---

[*] **Corresponding author Hotaka Kai:** Department of Chemistry and Biochemistry, National Institute of Technology, Suzuka College, Suzuka, Japan; Tel: +81-59-368-1821; Fax: +81-59-368-1820; E-mail: kai@chem.suzuka-ct.ac.jp

Furthermore, only 2.5% of water on the Earth's surface is fresh water, which can be utilized in our daily life and in most industrial activities.

For effective water utilization, it is necessary to understand the quality, characteristics, and behavior of water that circulates in various forms on the Earth's surface. The chemical characteristics and quality of water after use are often significantly different from the ones before use. Water is difficult to reuse because it might be contaminated or polluted. Thus, used and polluted water requires a suitable treatment process to change it to clean and fresh water, such that after the treatment, the water returns to its fresh status in the natural cycle. With respect to various methods for water analysis, we can comprehend the condition of water before and after use in terms of the water in a natural environment, in public supply, and in the process of treatment. Through water analysis, we can classify and reserve water resources accordingly; thus, we can develop an effective water utilization method, which could benefit our daily lives and various industrial activities.

ENVIRONMENTAL QUALITY STANDARD AND EFFLUENT STANDARDS

"How do we retain the quality of public waters?" "How do we utilize our water resources in an effective manner currently and in the future?" The answers to these questions may differ depending on how water quality and value are assessed. In Japan, several laws have been passed for water pollution control (with the Environmental Basic Law having priority). "Environmental Quality Standards" (Table **1**) and "Effluent Standards" (Table **2**) are set.

Table 1. Environmental Quality Standards [2].

Chemical	Permissible Limit
Cadmium and its compounds	0.01 mg Cd/l
Cyanide compounds	Not detectable
Lead and its compounds	0.01 mg Pb/l
Arsenic and its compounds	0.01 mg As/l
Mercury and its compounds	0.0005 mg Hg/l
Alkyl mercury compounds	Not detectable
PCBs	Not detectable
Trichloroethylene	0.03 mg/l
Tetrachloroethylene	0.01 mg/l
Dichloromethane	0.02 mg/l

(Table 1) cont.....

Chemical	Permissible Limit
Carbon Tetrachloride	0.002 mg/l
1, 2-Dichloro ethane	0.004 mg/l
1, 1-Dichloro ethylene	0.02 mg/l
cis-1, 2-Dichloro ethylene	0.04 mg/l
1, 1, 1-Trichloro ethane	1 mg/l
1, 1, 2-Trichloro ethane	0.006 mg/l
1, 3-Dichloropropene	0.002 mg/l
Thiram	0.006 mg/l
Simazine	0.003 mg/l
Thiobencarb	0.02 mg/l
Benzene	0.01 mg/l
Selenium and its compounds	0.01 mg Se/l
Boron and its compounds	1 mg B/l
Fluorine and its compounds	0.8 mg F/l
Nitrate nitrogen and Nitrite nitrogen	10 mg/l

"Environmental Quality Standards" are set to protect human health and conserve the living environment. They are guidelines that list the desirable quality standards. In addition, these standards should be maintained and contain administrative policy objectives. Furthermore, they promise to apply utmost efforts to meet the demand of the government (article 16-4, Environmental Basic Law [1]). At the same time, these standards do not show maximum permissible limits or tolerable limits for protecting human health and conserving the living environment. Furthermore, people do not need to permit the pollution up to the limit of these environmental quality standards.

Table 2. Effluent Standards [3].

Chemical	Permissible Limit
Cadmium and its compounds	0.03 mg Cd/l
Cyanide compounds	1 mg CN/l
Organic phosphorus compounds (Parathion, Methyl Parathion, Methyl Demeton and EPN only)	1 mg/l
Arsenic and its compounds	0.1 mg As/l
Mercury and its compounds	0.005 mg Hg/l
Alkyl mercury compounds	Not detectable
PCBs	0.003 mg/l

(Table 1) cont.....

Lead and its compounds	0.1 mg Pb/l
Hexavalent Chromium	0.5 mg Cr/l
Trichloroethylene	0.1 mg/l
Tetrachloroethylene	0.1 mg/l
Dichloromethane	0.2 mg/l
Carbon Tetrachloride	0.02 mg/l
1, 2-Dichloro ethane	0.04 mg/l
1, 1-Dichloro ethylene	1mg/l
cis-1, 2-Dichloro ethylene	0.4 mg/l
1, 1, 1-Trichloro ethane	3 mg/l
1, 1, 2-Trichloro ethane	0.06 mg/l
1, 3-Dichloropropene	0.02 mg/l
Thiram	0.06 mg/l
Simazine	0.03 mg/l
Thiobencarb	0.2 mg/l
Benzene	0.1 mg/l
Selenium and its compounds	0.1 mg Se/l
Boron and its compounds (Non-coastal areas)	10 mg B/l
Boron and its compounds (Coastal areas)	230 mg B/l
Fluorine and its compounds (Non-coastal areas)	8 mg F/l
Fluorine and its compounds (Coastal areas)	15 mg F/l
Ammonia, Ammonium compounds (Total of NH_3-N multiplied by)	100 mg/l
Nitrate and Nitrite compounds (0.4, NO_2-N and NO_3-N)	100 mg/l
1,4-Dioxane	0.5 mg/l

For the area that is not undergoing environmental pollution, it is desirable to retain the environmental quality at least in terms of the current condition. Each of these "Environmental Quality Standards" is basically determined by scientific knowledge; hence, we must strive to acquire new scientific knowledge, which should help with the effective implementation of similar standards. Moreover, we must reset "Environmental Quality Standards" by a relevant scientific decision, *i.e.,* "Environmental Quality Standards" should be revised (for example, changing a standard value and adding new standards). In addition, "Environmental Quality Standards for Groundwater" have been set from the perspective of protecting human health. Therefore, these standards are different from the "Environmental Quality Standards" for public waters. The standards are also administrative policy objectives for measuring and water quality and control. "Effluent Standards" are set under the Water Pollution Control Law for the purpose of protection of national health and for conservation of the living environment.

The purpose of this law will be to (1) regulate water discharge from factories and workplaces into public waters, (2) regulate water permeation underground, and (3) execute measures and proper monitoring with respect to living drainage. In short, "Environmental Quality Standards" are set points (policy objectives), but "Effluent Standards" are regulation values. In other words, in case a person or organization does not live up to the requirements defined by the regulations, penalties will be imposed. Because an effluent is diluted by natural water in the environment and the concentration of the effluent will decrease, each point in the "Effluent Standards" is ten times more concentrated than the corresponding points in the "Environmental Quality Standards." Fig. (**1**) shows the relationship between the two: the "Environmental Quality Standards" and the "Effluent Standards." We can assess whether the results of water analysis meet the "Environmental Quality Standards" and/or the "Effluent Standards". If they do not meet the points under either of the standards, then the concept of "Protection of Human Health" and "Conservation of the Living Environment" might be compromised, and pollution may already affect other unexpected sites. Therefore, regarding the points above, the importance and necessity of water analysis should be obvious.

BASIC KNOWLEDGE FOR WATER ANALYSIS

There are several methods for water analysis. These methods are not universal but using them depends on the quality of the water sample is. We must first understand the qualifying and the measuring range of each water analysis method. We previously acquired basic knowledge in terms of analytical science and mastered basic manipulation in terms of chemical analysis before exploring water analysis. Generally, water samples do not show the same result of water quality analysis at two random time points; moreover, the characteristics of water and its quality vacillate continuously. For instance, chemical quality of the water sample used might change if the water sampling point is shifted slightly in horizontal and vertical directions. In addition, it is necessary to analyze the quality of water in its natural state along with the natural and artificial factors that have control over the quality of water in the sampling area.

There are various microorganisms (*e.g.,* bacteria and plankton) and/or aquatic organisms in water. We can assess the degree of water pollution considering the extent of the presence of these organisms (assessment by an indicator organism). A bacteria test (for example, assessing the number of general bacterial cells and the number of colitis germ legions) is an indispensable analytical technique for drinking water. Components of water (especially organic constituents) undergo some kind of biochemical change (metabolism) caused by microorganisms, bacteria, and/or plankton. If the analysis of these components is necessary for water safety purposes, then the analysis must be carried out forthwith from the point when the water samples are collected. To deal with these matters, the knowledge of biological water treatment systems is necessary.

REQUIREMENTS, CONDITIONS, AND METHODS FOR WATER ANALYSIS

There are several methods for water analysis. However, these methods are not universal for all kinds of water samples. It is desirable to choose practical methods, as long as the methods satisfy the requirements and conditions. These requirements and conditions are stated as follows: (1) "Can be widely applied." The methods that have wide determination ranges and a wide range of concentrations of an interfering substance are the best; (2) "High level of sensitivity." The assays with high sensitivity that can adequately determine a very small quantity of a component are the best; (3) "Convenience and reliability of manipulation." Manipulation that can be mastered in a short period after repeated practice without requiring advanced skills is desirable. In addition, less contingent errors in measurements would be better; (4) "The operation must be rapid." Because many samples need time to be analyzed accurately, it is desirable to have an appropriate assay that requires a short period of time; (5) "It must be practical." Cheap analytical equipment and its easy setup for analysis are better. It is better to choose the assay(s) that satisfies as many of these conditions as possible. However, it is more desirable to be able to choose comprehensively from books, reference literature, analysts' own experience, and advice or/and opinions of those who conduct the joint measurements and those experienced in water quality analysis.

During water quality analysis, quantitative analysis (showing what kind of substance is contained and how much of each substance is contained in water) is required rather than qualitative analysis (showing only what kind of substance is contained in the water) in most cases. Unless quantitative analysis is carried out, the results cannot be checked against the "Environmental Quality Standards" and "Effluent Standards." There are various methods for quantitative analysis. The most common assay is absorption spectrophotometry, which is known to be the official method adopted by the "JIS" (Japanese Industrial Standards). The standards and measurement methods concerning industrial products in Japan are defined in the "Japanese national standards." These standards involve absorption spectrophotometry for many items to be analyzed. In absorption spectrophotometry, the analyte in a solution sample is converted to a light-absorbing substance with an appropriate chemical reagent (color reagent); the amount of light absorbed at a specific wavelength is measured to determine the concentration (the amount) of the target component. Generally, absorption spectrophotometry is a highly sensitive analytical method, with a determination range of approximately 0.1–10 mg/L. Its repeat accuracy can be as good as 0.2% under suitable determination conditions and analysis manipulations. Absorption spectrophotometry is suitable for analyzing large amounts of samples at once. The interference caused by the coexisting components can be dealt with a masking process and most of the analysis manipulations are simple. However, concentration and preservation of the initial samples are often necessary for repeated work for a long period. In addition, there are many points to pay attention to in the process of water analysis by absorption spectrophotometry, which include pH, temperature, the amount and order of added reagents, volume and mixture of the reagent added, exposure time during application and coloring, and a proper procedure against the interference caused by other components.

The atomic absorption photometric method is often applied to water quality analysis for metal determination in contrast to absorption spectrophotometry. Although the quantitative range varies depending on the element type, atomic absorption photometry is a highly sensitive analytical method that can quantify substances in the range approximately 1 µg/L to 1 mg/L. There are 70 elements that can be quantified by atomic absorption analysis. In atomic absorption spectroscopy, atomic spectral lines with narrow widths of elements are used for measurement, thus, absorption in spectra under the influence of proximity lines and coexisting elements is small. However, it is necessary to pay attention to some kinds of interference, such as "chemical interference," "physical interference," and "optical interference." Chemical interference refers to the measurement error that arises due to the interruption of atomization of the target element by the salt, which is difficult to dissolve and is caused by ions other than the target element coexisting in the solution. Physical interference means the measurement error caused by the spraying amount of the solution in the frame due to the physical properties of the solution, such as viscosity, spraying efficiency, and particle size: these properties influence atomization. Optical interference is the measurement disturbance caused by the band spectrum or nearby line(s) of combustion products, such as fuels, coexisting substances, and solvents.

Other analytical methods include capacitance analysis, gravimetric analysis, and equipment analysis by chromatography. Although capacitance analysis is simple in operation and has high accuracy, it is not suitable for the analysis of a very low concentration of the component or very small sample volume. Gravimetric analysis is also excellent in terms of accuracy; however, it is effective only for the components at relatively high concentration and is not suitable for very dilute components. Besides, gravimetric analysis needs long periods of measuring time; hence, it is not suitable for multi-sample analysis. Equipment analysis on an analyzer by chromatography is expensive, and in some cases requires complicated preprocessing. It is necessary to determine the optimal analytical method with a sufficient understanding of the advantages and disadvantages of the adaptive analysis.

PLAN FOR WATER ANALYSIS

During water quality analysis, it is impossible to obtain the necessary water quality data if researchers merely collect water samples and run the analysis on it. It is desirable to perform a series of tasks as shown below. First, researchers need to clearly define the purpose of conducting the water quality analysis. For instance, whether the purpose is for academic research from environmental or medical perspectives or whether it is intended for a survey from a practical standpoint. These surveys are mainly used for such purposes as protection of drinking water and preservation of the natural environment. Therefore, conducting water quality analysis with an unclear purpose is completely unproductive. Next, based on previous studies and survey reports, investigators need to clarify the findings obtained and status of the efforts to analyze water quality from this time forward. Having organized previous studies and having drafted previous plans can help to improve the effectiveness of the water quality analysis conducted. Then, the team members need to examine the entity to be analyzed to achieve the purpose. After that, the participants need to decide on the target area, the schedule (date and time), how to sample the water by a suitable method, perform the necessary pretreatment at the site, and bring the samples back to the laboratory for analysis. Once the analysis results are obtained, the researchers participating in the project need to organize the results and summarize the conclusions for achieving the purpose. Lastly, the researchers need to decide on the final goal of the water quality analysis, for instance, to make an academic publication according to the purpose or to create a report as an example for future surveys. In addition, during planning of a water quality analysis, it is desirable to clarify necessary expenses incurred during the analysis.

EXPLANATION *VIA* THE REVIEW OF THE LATEST ENVIRONMENTAL ANALYSIS RELATED TO THE GREAT EAST JAPAN EARTHQUAKE

The Great Tsunami caused by the earthquake in the Tohoku Region Pacific Offshore (also called the Great East Japan Earthquake) that occurred on March 11, 2011, caused severe damage to coastal areas, such as those in Fukushima, Miyagi, Iwate, and Aomori Prefectures. In the coastal areas of these prefectures, land subsidence caused by the major earthquake, physical damage due to the massive tsunami, and the inflow of the waste due to the earthquake as well as the chemical substances from the land area caused changes in the material flow in water from the land to the coast due to the suspension of infrastructure in water and sewage systems. Furthermore, all these changes led to the destruction of the coastal ecosystem and suspension of the fishing activities, causing an enormous impact on the aquaculture industry. Herewith, in coastal areas damaged by the tsunami, the pattern of material circulation and material flow, which was nurtured for many years before the earthquake, was drastically altered. The water environment centered on the related ecosystem has also changed dramatically. It seems that various influences occurred and several studies on these influences have been conducted.

The maximum seismic activity of 6.0 was observed (on the Japanese seven-stage seismic scale) around a brackish water lake, "Hinuma Lake" in the Ibaraki prefecture, during the Great East Japan Earthquake. Due to the ground subsidence of 20 cm, a 4 m tsunami was seen at the observation point near the estuary of a river connecting Hinuma Lake to the Pacific Ocean; thus, the salinity concentration in Hinuma Lake changed. Alterations in salt concentration disrupted the homeostasis of animals and plants residing there, thus leading to death in worst-case scenarios. Ohuchi *et al.* [3] have reported the factors and causes of influence, which led to a change in the chloride ion concentration of Hinuma Lake before and after the Great East Japan Earthquake in terms of studying the possibility that the ecosystem of Hinuma Lake (registered in the Ramsar site) is also affected.

Ohuchi *et al.* [4] have thoroughly studied the characteristics of Hinuma Lake's natural environment, which enabled them to conduct the research properly based on the preceding studies. The topography of the Hinuma River (flowing from Hinuma Lake to the Pacific Ocean) reveals that invasion of a high-salinity water mass from the Pacific Ocean via extreme tides (~10 times a year) is extremely small; therefore, the chloride ion concentration in Hinuma Lake is stable at an average level of 2–3 g/L. However, when the chloride ion concentration increases, it is argued that the red tide occurs due to the proliferation of blue-green algae: a sea area dinoflagellate (*Prorocentrum* genus), which is not normally seen. It is important to include the above "basic knowledge of water quality analysis" to effectively grasp the features of the topography in advance and to scientifically conclude what the possible cause of the observed phenomena is.

Fig. (1). Equipment for water analysis (left: ion chromatography analyzer [5], middle: ion meter [6], right: burette for titration [7]).

Water sampling for the research was performed from April 2006 to March 2016 on a monthly basis. Based on the research report, Ouchi *et al*. [4] set the sampling dates that do not fall on the invasion of the high-salinity water mass as described above. On the dates when high-salinity water mass invades, the lake is classified as having an unsteady state. Therefore, it is not appropriate to evaluate the quality of Hinuma Lake water in the period of invasion of a high-salinity water mass. Sampling in Hinuma Lake is carried out via a peristaltic pump to fetch the water from the upper layer (0.5 m below the water surface) and the lower layer (0.5 m above the lake's bottom). A peristaltic pump is easy to operate as well as to control the pumped water volume. It can collect water in shallow and deep water. It is effective in the case of groundwater, which is water within a 6–8 m distance from the ground surface. Its disadvantages include possible decompression and degassing of a sample; therefore, it is not suitable for volatile substances.

In this study, water sampling by a peristaltic pump is appropriate because the main measurement object is the chloride ion, but the disadvantages (of a peristaltic pump) do not affect chloride ions. It is important to properly plan such considerations in the corresponding "water quality analysis plan".

The measurement was carried out by the silver nitrate titration method with potassium chromate as an indicator. This assay is an official analytical method prescribed in the "Standard Methods for the Examination of Water," which is one of the publicly defined water quality assays in Japan. In addition, there are "JIS K 0102: Testing Method for Industrial Wastewater" and "Standard Methods of Hygienic Chemistry," which prescribe the main testing method with regard to water pollution. The official method for chloride ion analysis includes ion chromatography and an ion electrode method (both prescribed in "JIS K 0102: Testing Method for Industrial Wastewater") in addition to the silver nitrate titration method (prescribed in the "Standard Methods for the Examination of Water"). Figure 1 shows the equipment for these assays. The ion chromatography method has an analytical accuracy of 2%–10%, and the quantitative range is 0.1–25 mg/L. It has the lowest measurable concentration (limit of quantification) among all the methods. However, because the ion chromatography analyzer is expensive (50,000–100,000 USD), it is difficult to perform the measurement. In addition, if the measurement sample is not in the low concentration range of ~0.1 mg/L, the lower limit of determination and measurements by other analytical methods are sufficient. The ion electrode method has an analytical accuracy of 5%–20%, and the analysis accuracy is slightly lower than that of ion chromatography. Its quantitative range is the widest, from 5 to 1000 mg/L, but it is not suitable for the samples with less than 5 mg/L concentration. In addition, an ion meter costs $2,000–3,000 (USD), and a chloride ion electrode costs $300–400. Although it is not as expensive as ion chromatography equipment, an appropriate budget is needed to align a set of analyzers. In the silver nitrate titration method, the analysis accuracy is ~10%, and the quantitative range is 2–170 mg/L. Analysis is performed by the burette titration operation. It is the cheapest analytical method because a set of analytical instruments can be procured for approximately $100. Furthermore, if one can master the titration operation, performing the analysis can take less time.

Since potassium chromate serves as an indicator to determine the end-point of titration, a laboratory effluent containing chromium is generated, and thus careful handling of the generated experimental effluent is required, whereas such considerations do not need to be taken into account in other methods. Because the study by Ouchi *et al.* [4] and other recent papers did not describe the experimental environment for conducting the water quality analysis, it is difficult to decide whether to use an ion chromatography analyzer or an ion meter. Ouchi *et al.* [4] suggested that the silver nitrate titration method with potassium chromate as an indicator is the easiest to perform the required experiments. All the following factors, such as the requirements, conditions, and methods for the water analysis, from the viewpoint of the presence or absence of an analyzer, the accuracy of the analysis, the range of quantitative analysis, the time required for the analysis, balance with other analysis entities, and/or the analytical methods that are believed to cause the least interference with chloride ion analysis, must be evaluated carefully in advance.

In addition to the chloride ion, the concentration of chlorophyll-a in the sample is measured by a special assay involving an ultraviolet and visible spectrophotometer. The species of the phytoplankton is identified, and the number of cells is determined by microscopic observation. Furthermore, to ascertain the environmental conditions concerning the growth of the phytoplankton, the dissolved inorganic nitrogen and phosphate phosphorus concentrations are measured on a continuous-flow analyzer for each type. As mentioned earlier, it is believed that an increase in chloride ion concentration leads to formation of a red tide due to the proliferation of cyanobacteria and dinoflagellates. This causal relation involves the phenomenon of eutrophication, wherein phosphorus and nitrogen increase in amounts. To verify this phenomenon scientifically, it is thought that these entities should be set as items to be measured simultaneously along with the quantification of the chloride ion. Ohuchi's research mainly deals with water quality analysis, but the identification of phytoplankton under a microscope is not a water quality analysis. However, microscopic examination is an important means to achieve the objective of this research. Hence, knowing the name of the species of the phytoplankton is also regarded as a piece of "basic knowledge about water quality analysis" as described above.

In this section, items to be considered for environmental analysis are introduced based on a review of recent literature. An explanation of Ohuchi *et al.*'s results [4] has been omitted because it is outside the scope of this chapter. As mentioned above, the environment in the coastal area of the Tohoku region on the Pacific coast was dramatically changed by the Great East Japan earthquake. Many research reports, such as concentration measurement of radioactive cesium in river water related to the nuclear power accident caused by the Great Earthquake [8], status of sea floor contamination due to oil flowing out with regard to the Great Earthquake [9], the change in the coastal environment [10, 11], alteration of the environment of the Amamo field [12], and the growth model of oysters [13] have been discussed. Using the "Great East Japan earthquake" as a keyword, it was found that many results have been reported on how the environmental conditions were altered via various water analysis (and environmental analysis) techniques. In these reports, "the necessity of water analysis," "the basic knowledge for water analysis," "the requirements, conditions, and methods for water analysis," a "plan for water analysis," and whether the entities to be measured satisfy the "Environmental Quality Standards" and the "Effluent Standards" are summarized. Therefore, by reading a paper in which environmental analysis is carried out, it is possible to gain new knowledge on environmental analysis.

CONSENT FOR PUBLICATION

Not applicable.

CONFLICT OF INTEREST

The author confirms that this chapter contents have no conflict of interest.

ACKNOWLEDGEMENTS

Declare none.

REFERENCES

[1]　The Basic Environment Law [Online] (Updated 13th November 1993) , 1993 [Accessed 28th November 2018]; Available from: https://www.env.go.jp/en/laws/policy/basic/ch2-1.html#section3

[2]　The Basic Environment Law [Online] (Updated 13th November 1993) , 1993 [Accessed 28th November 2018]; Available from: http://www.env.go.jp/en/laws/policy/basic/index.html

[3]　Water Pollution Control Law [Online] (Updated 1995), , 1970 [Accessed 28th November 2018]; Available from: https://www.env.go.jp/en/laws/water/wlaw/index.html

[4]　Ohuchi T, Yoshida S, Aizawa M. Change in salinity and effect of the great east japan earthquake on water quality in brackish lake hinuma. J Jpn Soc Water Environ 2018; 41(3): 55-9. [http://dx.doi.org/10.2965/jswe.41.55]

[5]　Shimadzu Corporation. ion chromatography analyzer [Online], , 1917 [Accessed 28th November 2018]; Available from: https://www.an.shimadzu.co.jp/hplc/ic/ic.htm

[6]　Horiba Ltd Ion meter [Online] , 1945 [Accessed 28th November 2018]; Available from: http://www.horiba.com/jp/application/material-property-characterization/water-analysis/water-quality-electrochemistry-instrumentation/

[7]　SHIBATA Scienct and Technology Ltd Burette for titration [Online] , 1921 [Accessed 28th November 2018]; Available from: https://www.sibata.co.jp/category/cate3/cate3-53/

[8]　Nagao S, Kanamori M, Suzuki K, Ochiai S, Inoue M. 134Cs and 137Cs radioactivity in river waters from the upper tone river. Bunseki kagaku 2017; 66(4): 243-9. [http://dx.doi.org/10.2116/bunsekikagaku.66.243]

[9]　Arakawa H, Nakamura M. For monitoring of the oil spill of Kesennuma submarine in the Great East Japan Earthquake. J Jpn Inst Mar Eng 2016; 51(6): 83-6.

[10]　Yamamoto M, Kato T, Tabata S, *et al.* Changes in the coastal environment in Kamaishi Bay after the Great East Japan Earthquake. Nippon Suisan Gakkaishi 2015; 81(2): 243-55. [http://dx.doi.org/10.2331/suisan.81.243]

[11]　Fukuda F, Nagata T. Nutrient status of Otsuchi Bay following 5 years after the Great East Japan Earthquake. Nippon Suisan Gakkaishi 2017; 83(4): 652-5. [http://dx.doi.org/10.2331/suisan.WA2432-4]

[12]　Nakaoka M, Tamaki H, Muraoka D, Tokuoka M, Komatsu T, Tanaka N. Temporal changes in seagrass beds of Sanriku Coast before and after the Great East Japan Earthquake. Nippon Suisan Gakkaishi 2017; 83(4): 659-63. [http://dx.doi.org/10.2331/suisan.WA2432-6]

[13]　Okumura Y, Nawata A, Onodera T, Ito H, Hara M. Growth Model of Oyster in Oginohama Bay near Miyagi Prefecture, Eastern Japan. J Jpn Soc Water Environ 2017; 40(4): 167-73. [http://dx.doi.org/10.2965/jswe.40.167]

CHAPTER 6

Examples of Sustainable Marine Environmental Assessments

Mitsuo Yamamoto[*]

Graduate School of Agricultural and Life Sciences, The University of Tokyo, Bunkyo-ku, Japan

Abstract: Marine environmental assessments are needed to understand and conserve the marine ecosystem. In particular, the primary production of phytoplankton, which forms the basis of ecosystem functioning, is influenced by chemical species in seawater, including nitrogen (N), phosphorus (P), and iron (Fe). Thus, previous research by the author has focused on material dynamics in rivers and coastal areas to develop marine environmental technology. The investigations in the Chikugo River, Kesennuma Bay, and Kamaishi Bay are introduced in this chapter. Fe dynamics in the Chikugo River was investigated from July 2011 to May 2012. The trend in Fe concentration was different from that of N and P. Fe distribution in the river was specific and correlated to turbidity in the estuary. Thus, Fe dynamics might have contributed to the formation of the ecosystem in the Chikugo River and the Ariake Sea. In Kesennuma Bay, Fe, N, and P concentrations were monitored to gain insights on Fe dynamics after the 2011 tsunami disaster. The dynamics of Fe concentrations was found to be different before and after one year of the disaster (March 2012). Freshwater from terrestrial areas primarily influenced Fe concentrations close to the river mouths after one year of the disaster. In Kamaishi Bay, the concentrations of nutrients (N, P, and Si) and heavy metals (Cd, Pb, Cr, Zn, As, and Se) in seawater, as well as the concentrations of radioactive materials (^{134}Cs and ^{137}Cs) in sediments, were monitored from March 2012 to November 2013. Nutrient concentrations varied differently with seasons before and after the tsunami disaster. Heavy metal concentrations were below or equal to the accepted environmental standards in Japan after the disaster. Although radioactive materials produced by the Fukushima nuclear accident were detected in the sediments, they were only found in low concentrations. These results show that the coastal environment in Kamaishi Bay has recovered from the disaster.

Keywords: Coastal environment, Iron, Linkage of forest-river-sea, Material dynamics, The Great East Japan Earthquake.

[*] **Corresponding author Mitsuo Yamamoto:** Graduate School of Agricultural and Life Sciences, The University of Tokyo, 1-1-1 Yayoi, Bunkyo-ku, Tokyo 113-8657, Japan; Tel: +81-3-5841-8887; Fax: +81-3-5841-7575; E-mail: a-myamamoto@mail.ecc.u- tokyo.ac.jp

Toshiyuki Takahashi (Ed.)

INTRODUCTION

Marine environmental assessments are important for understanding and conserving the ecosystems of marine and coastal areas. Phytoplankton influence the abundance and diversity of marine organisms, thus driving the functioning of marine ecosystems [1]. The primary production of phytoplankton is influenced by dissolved chemical species, such as nitrogen (N) and phosphorus (P), along with physical conditions, including seawater temperature, salinity, and light intensity [2]. In addition to N and P, iron (Fe) strongly regulates phytoplankton production in nutrient-rich areas of the open sea, such as the northeastern part of the Pacific Subarctic Ocean [3 - 5]. In comparison, the biological richness of coastal ecosystems is recognized globally [6], arising because terrestrial materials, such as nutrients and organic matter, are transported through rivers to coastal and estuarine environments, impacting their structure and composition. The linkage between forests, rivers, and sea (hereafter referred to as forest-river-sea linkage) is important for conserving coastal and estuarine ecosystems and has been a key subject of research in recent years [7 - 9]. Specifically, Fe dynamics has been evaluated in several coastal areas because it is an important variable in the land-ocean linkage [7, 9 - 12].

The author has been conducting research related to material dynamics in river basins and coastal areas within the framework of developing marine environmental technology focusing on Fe dynamics between terrestrial and coastal areas. The loss of seaweed beds leaving barren grounds is a serious problem along the coast of Japan, as well as other areas globally [13]. Several explanations account for the formation of barren grounds, including elevating seawater temperatures, grazing by herbivorous animals, and a shortage of iron and other nutrients, such as nitrogen and phosphorus [14, 15]. Many approaches have been trialed to restore seaweed beds, including the removal of sea urchins and seaweed-eating fishes, as well as fertilization with iron, nitrogen, and phosphate. The author's study group has focused on exploring the shortage of dissolved iron in coastal areas. Within this framework, a method to restore seaweed beds has been developed using steelmaking slag and compost containing humic substances [14, 15].

The effectiveness of this method in coastal areas was confirmed using a field test on the coast of Hokkaido adjoining the Sea of Japan [14, 15], as described in Chapter 22. In addition to this field test, many fundamental studies have been conducted in the present study to understand the characteristics of iron elution from mixtures of steelmaking slag and compost (hereafter, referred to as slag-compost fertilizer) [16], complexation between iron and humic substances together with their structural features [17], and the effect of iron on seaweed growth [18]. However, slag-compost fertilizer must be installed at suitable coastal locations to restore seaweed beds effectively. Thus, it is necessary to develop a method to assess the coastal environment before introducing the fertilizer. To achieve this goal and advance our understanding of the forest-river-sea linkage as a point of material dynamics, the author has been conducting river and marine environmental assessments in some river and coastal areas of Japan [7, 19 - 22].

To assess the forest-river-sea linkage, Fe dynamics in the Chikugo River was investigated to understand the relationship between the characteristics of Fe concentrations in the basin and the river estuarine ecosystem with that of the inner area of the Ariake Sea [7]. Within the framework of this assessment, the effects of the Great East Japan Earthquake on the coastal environment and its restoration in Kesennuma Bay [21] and Kamaishi Bay [22] in the Tohoku region of Japan were evaluated. This chapter presents the investigations in the Chikugo River basin, along with the monitoring of the coastal environments in Kesennuma Bay and Kamaishi Bay after the 2011 tsunami disaster as examples of marine environmental assessments.

IRON DYNAMICS IN THE CHIKUGO RIVER BASIN [7]

The Ariake Sea is located on the western part of Kyushu Island in Japan and is characterized by extensive tidal flats and an estuarine turbidity maximum. The Chikugo River flows into the inner Ariake Sea, where mineral particles intensively mix with organic matter through the action of large tides and waves in the estuary. These particle materials are absorbed onto nutrients and contribute towards forming the ecosystem of the inner Ariake Sea. However, the influence of Fe dynamics on the ecosystem has not been investigated. Surface water sampling and analyses were conducted once a month from the beginning of July 2011 to the second half of May 2012. These samples were used to investigate Fe dynamics and its relationship with other factors [7]. The ultimate goal was to understand how Fe influences the formation of the ecosystem in the Chikugo River and the Ariake Sea.

METHODS

Fig. (1) shows the sampling sites in the Chikugo River. Six sites from R2 to R7 are in the estuary. R7 is located 23 km from the river mouth, at the sluice gate. Twenty-three sites were situated in the middle and upper basin of Chikugo River, as well as in the five tributaries of the Houman, Kagetsu, Taio, Kawabaru, and Kusu rivers. Fe (defined as dissolved iron, D-Fe), total nitrogen (T-N), and total phosphorus (T-P) were measured using multi water quality meters, including water temperature, salinity, and turbidity.

RESULTS ANS DISCUSSION

The detected data were classified into four seasons based on the difference of averaged temperatures: summer (from July 2011 to September 2011), autumn (from October 2011 to November 2011), winter (from December 2011 to February 2012), and spring (from March 2012 to May 2012).

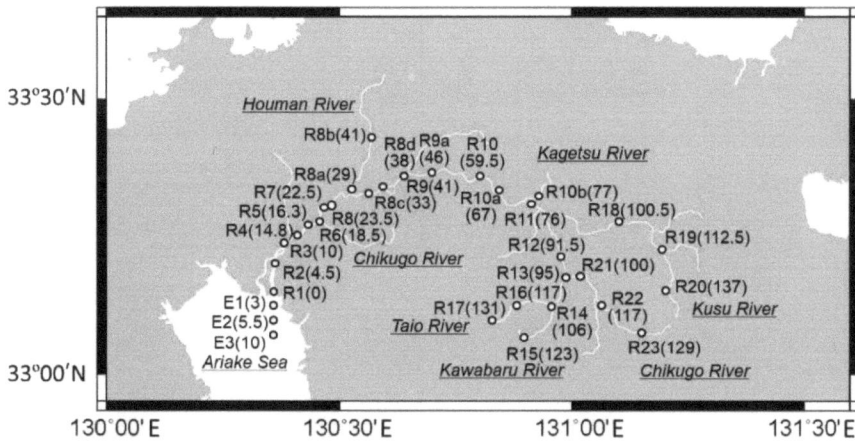

Fig. (1). Sampling sites in Chikugo River and the Ariake Sea, Japan. Numeric characters in parenthesis present the distance from the river mouth (km) [7].

Fig. (**2**) shows the seasonal changes in D-Fe concentrations at the sampling sites. The dynamics of D-Fe in the river did not present any marked seasonal variation, even though the highest D-Fe concentrations were recorded in the summer. However, it was confirmed that D-Fe concentrations increased gradually from upstream to the middle basin. In comparison, D-Fe concentrations increased noticeably in the estuary, while D-Fe concentrations at R3 or R4 were much higher than those in the middle basin and upstream sites.

This trend in D-Fe concentration differed to that obtained for T-N and T-P. T-N and T-P concentrations were relatively high in the estuary, whereas D-Fe concentrations were extremely high, peaking at R3 or R4. These D-Fe dynamics in the Chikugo River might exist because the supplied D-Fe accumulates through the effect of particles, such as particulate organic matter, in the estuary. Furthermore, D-Fe concentration in the estuary might correspond to that of the turbidity. To confirm this prediction, water samples were taken from both the surface and bottom layers at sites R7 to R2, as well as the mouth of the river (R1), and at 3 km (E1), 5.5 km (E2), and 20 km (E3) distance from the river mouth in the Ariake Sea during October 2012. Fig. (**3**) shows D-Fe concentrations

(Fig. **3a**) documented at the surface and bottom layers of estuary, together with turbidity (Fig. **3b**). D-Fe distribution was similar to turbidity distribution. The relationship between Fe concentration and turbidity was evaluated, and a high correlation coefficient was obtained (0.767). Thus, the distribution of Fe strongly correlated with the distribution of turbidity.

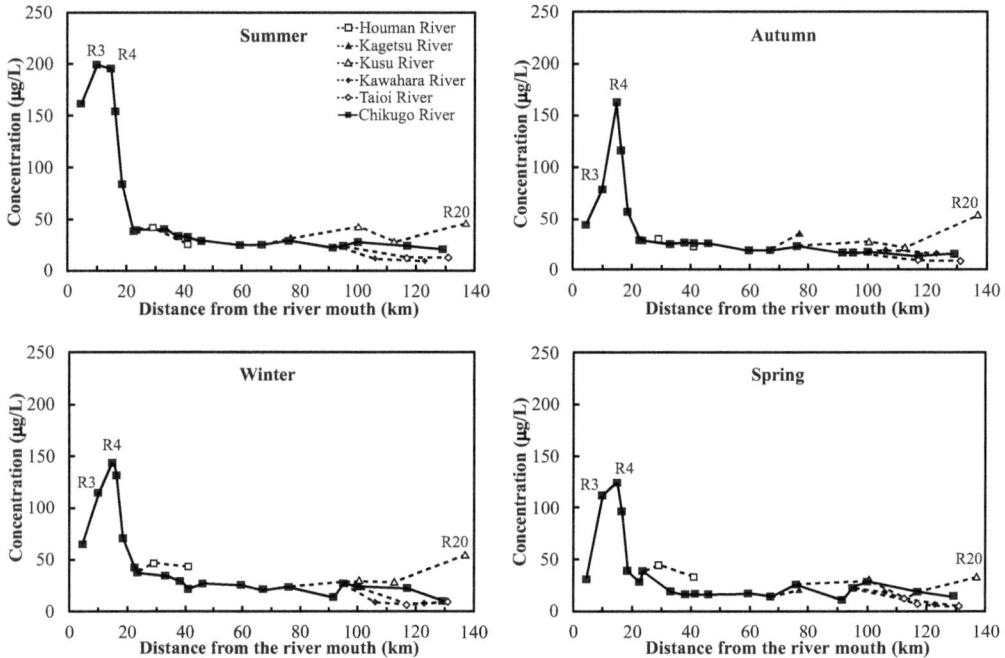

Fig. (2). Seasonal changes in Fe concentrations in the Chikugo River and its tributaries [7].

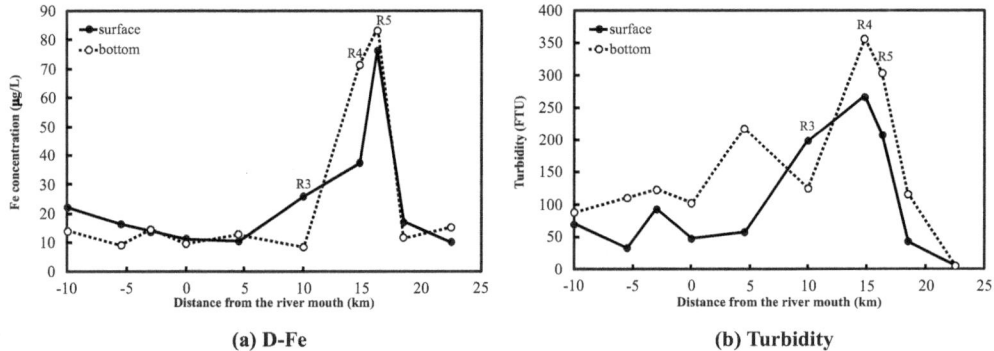

(a) D-Fe

(b) Turbidity

Fig. (3). Fe concentration and turbidity in the estuary of the Chikugo River in October 2012 (according to the distance from the river mouth and surface/bottom layer of the river). (**a**) D-Fe, (**b**) Turbidity [7].

Thus, D-Fe distribution in the river was specific and clearly differed to that of N and P throughout the year by Fe accumulating from the effect of particulate organic matter in the estuary. Because particle materials are important for forming the ecosystem [7], Fe dynamics is likely correlated to the ecosystem characteristics of the Ariake Sea.

CHANGES TO THE COASTAL ENVIRONMENT AFTER THE 2011 TSUNAMI

The earthquake and tsunami that occurred off the Pacific Coast of Tohoku and Kanto regions of Japan in 2011 caused serious damage to the coastal areas. Many studies have examined how the coastal environment and ecosystem were altered by this disaster, as well as the subsequent recovery [23 - 26]. The author has also investigated how the coastal environment changed after the disaster by focusing on changes to water quality in Kesennuma Bay [21] and Kamaishi Bay [22]. Here, the results obtained in Kesennuma Bay are briefly introduced, while those obtained in Kamaishi Bay are described in detail.

KESENNUMA BAY

In Kesennuma Bay, Miyagi Prefecture, Japan, a monthly to bi-monthly field investigation was started soon after the tsunami, to gain insights on how the tsunami affected the coastal environment [21]. Many parameters were evaluated, including changes to water quality, phytoplankton, zooplankton, and fishes; however, the author focused on assessing water quality. The sampling sites are shown in Fig. (**4**). Iron (Fe), nitrogen (N), and phosphorus (P) concentrations were monitored at four sites (Stns. 3, 8, 11, and 15). Various water quality variables, such as water temperature, salinity, and chlorophyll, were also measured with a multi-parameter water quality meter at 15 sites, including Stns. 3, 8, 11, and 15. The monitoring results showed that the characteristics of Fe concentration differed before and after one year of the disaster (March 2012). These differences were attributed to the tsunami. Furthermore, the influence of freshwater input was confirmed after one year of the tsunami. In particular, total Fe in the surface layer significantly correlated with precipitation and salinity at the river mouths (Stns. 3 and 11). Freshwater from terrestrial areas primarily influenced Fe concentrations close to the river mouths. Thus, Fe concentration was influenced by the balance between freshwater inputs and ocean water, as well as water depth in Kesennuma Bay.

KAMAISHI BAY [22]

The influences of the tsunami were clarified based on changes to the coastal ecosystem and marine pollution. In Kamaishi Bay, Iwate Prefecture, Japan, a large breakwater at the mouth of the bay was destroyed by the tsunami. Consequently, the coastal environment, including water quality, might have been noticeably altered. In addition, heavy metals from terrestrial areas and radioactive materials from the Fukushima No. 1 Nuclear Power Plant might have caused marine pollution. Thus, nutrient concentrations (NH_4-N, NO_2-N, NO_3-N, PO_4-P, and SiO_2-Si) in seawater were monitored to estimate changes to the coastal environment from March 2012 to November 2013. Furthermore, six heavy metals (Cd, Pb, Cr, Zn, As, and Se) in seawater and radioactive materials (^{134}Cs and ^{137}Cs) in sediments were evaluated to quantify the extent of marine pollution. The seawater sampling sites in Kamaishi Bay are shown in Fig. (**5**). The sediments were sampled at representative sites in Fig. (**5**).

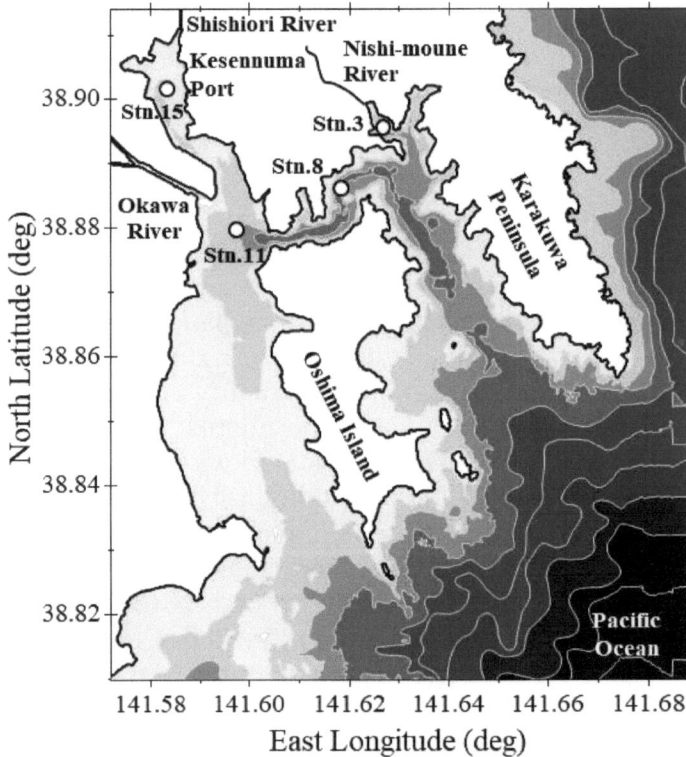

Fig. (4). Sampling sites in Kesennuma Bay, Japan.

All of the heavy metal concentrations were below or equal to the accepted environmental standards in Japan during the period of investigation. Although studies on the distribution of macrobenthos indicated disturbance to the seabed, any negative effect of the tsunami on heavy metal concentrations was not confirmed one year after the disaster. However, ^{134}Cs and ^{137}Cs produced by the Fukushima nuclear accident were detected in the sediments, but their concentrations were low. Furthermore, ^{134}Cs and ^{137}Cs concentrations differed across sampling sites. As a rule, the concentrations increased gradually during the first half of the investigation, then decreased and remained constant during the latter half of the observation. These results show that by November 2013, 2.5 years after the earthquake, the coastal environments in Kamaishi Bay had recovered from the tsunami.

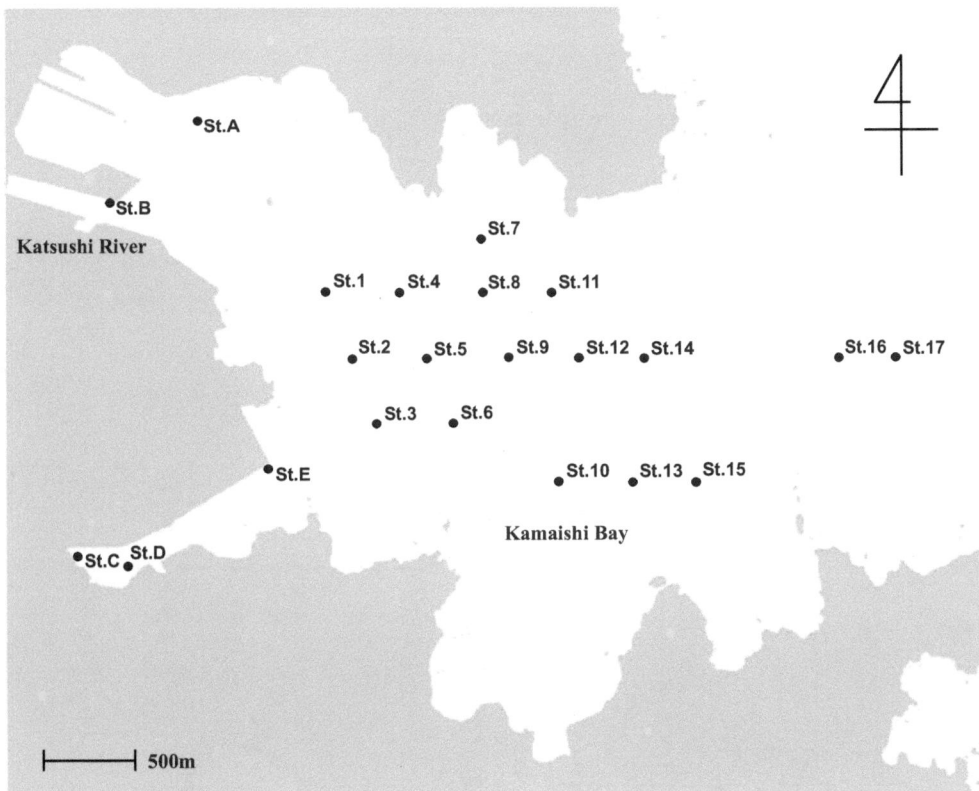

Fig. (5). Sampling sites in Kamaishi Bay, Japan [22].

The characteristics of nutrient dynamics were influenced by the breaking of the large breakwater in the mouth of the bay. Fig. (**6**) shows the seasonal changes of dissolved inorganic nitrogen (DIN: NH_4-N, NO_2-N, NO_3-N) and phosphate (PO_4-P) after 1 year of the tsunami at the representative sites (St. 1, St. 3, and St. 7) in Kamaishi Bay. A similar trend was observed at these sites. Both DIN and PO_4-P concentrations in the surface and bottom layers increased during spring

(March and May), decreased until summer and increased after autumn in 2012 and 2013. DIN and PO_4-P concentrations in the surface layer at St. 1, which was situated close to the river mouth, were higher than those in the bottom layer throughout the study. DIN and PO_4-P concentrations in the surface layer at St. 7, which was relatively close to the mouth of the bay, were lower than those in the bottom layer. At St. 3 (which was located between St. 1 and St. 7), the surface concentrations were similar to the bottom concentrations. Thus, nutrient concentrations were influenced by freshwater input from the river.

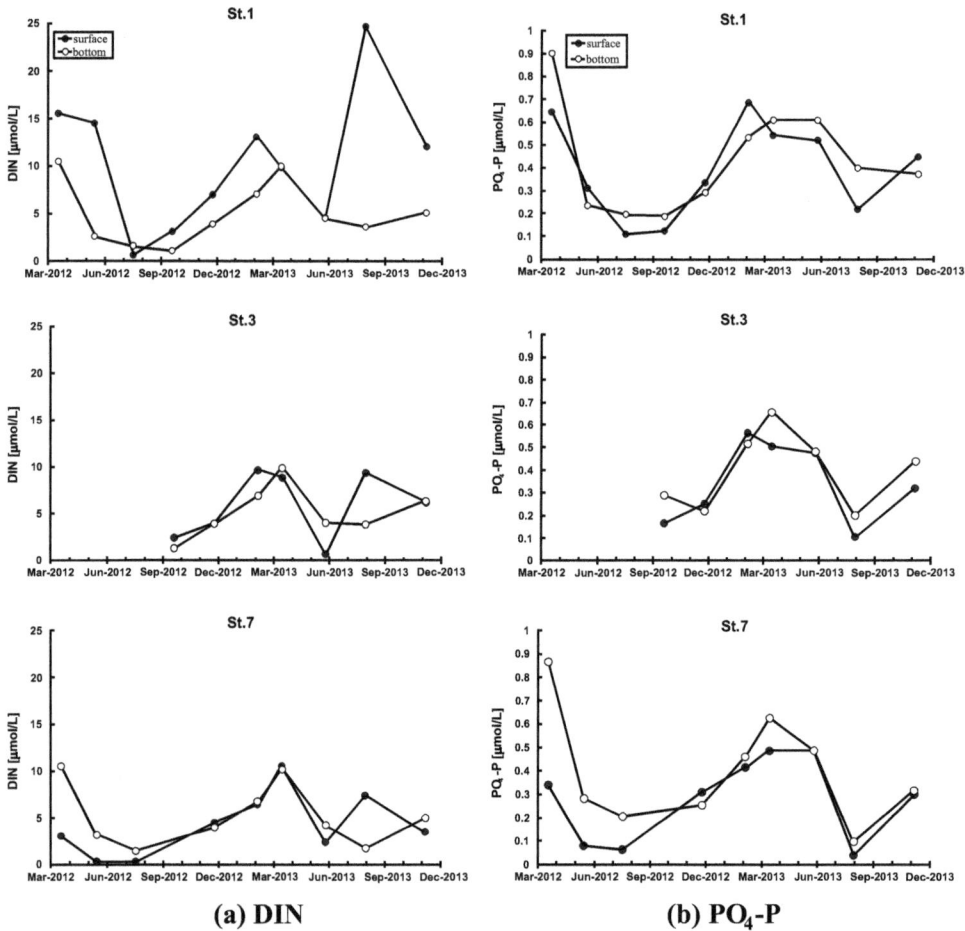

(a) DIN   **(b) PO_4-P**

Fig. (6). Changes to DIN and PO_4-P concentrations in the surface and bottom layers of Kamaishi Bay, Japan (from March 2012 to December 2013) [22].

The seasonal changes of nutrients in Kamaishi Bay after the disaster were similar to those recorded before the disaster in the surrounding coastal areas, such as

Kesennuma Bay and Shizugawa Bay [22]. Fig. (**7**) shows DIN and PO_4-P concentrations in Kamaishi Bay before and after the disaster. The investigations before the disaster were conducted during spring (April and May) of 2007, the summer (July and August) of 2007, and the autumn (October and November) of 2007 at six sites in the bay [22]. Two sites, Stn. 2 and Stn. 3 (shown in Fig. **7**) were situated close to St. 2 and St. 3 (shown in Fig. **5**), respectively. Nutrient concentrations in the bottom layer during the summer before the disaster were not low, contrasting with those documented in Kesennuma Bay and Shizugawa Bay. Seasonal changes in DIN concentrations at Kamaishi Bay were also different before the disaster. Specifically, after the disaster, nutrient concentrations were lower in summer compared to spring and autumn, which differed to that recorded before the disaster. This difference might be explained by the fact that oceanic water could not mix with seawater in the bay before the disaster because the break water was present. Overall, this study confirmed that nutrient dynamics in the bay were influenced by the tsunami, with clear changes being documented.

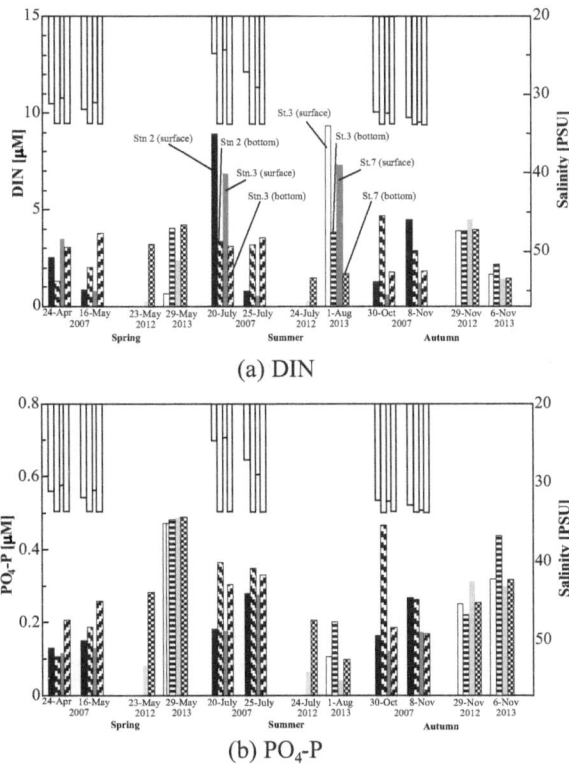

(a) DIN

(b) PO_4-P

Fig. (7). Concentrations of (a) DIN and (b) PO_4-P in Kamaishi Bay, Japan, before and after the disaster. Pre-disaster salinity is indicated as a hanging graph with white bars [22].

Thus, this study confirmed that the coastal environment of Kamaishi Bay changed with respect to the coastal ecosystem and marine pollution. This investigation was important because it advanced our understanding of the coastal environment and facilitated the revival of fisheries following the Great East Japan Earthquake. Specifically, the data assimilated on heavy metals and radioactive materials in this study were used to quantify any influence on marine produce.

CONSENT FOR PUBLICATION

Not applicable.

CONFLICT OF INTEREST

The author confirms that this chapter contents have no conflict of interest.

ACKNOWLEDGEMENTS

The river and coastal environment assessments described in this chapter were conducted in collaboration with many researchers from diverse specialized fields. In particular, iron analyses, which are essential for assessing Fe dynamics, were conducted by Professor Dan Liu (National Institute of Technology, Ariake College). The author thanks all collaborators and co-operators.

REFERENCES

[1]   Boyce DG, Lewis MR, Worm B. Global phytoplankton decline over the past century. Nature 2010; 466(7306): 591-6.
      [http://dx.doi.org/10.1038/nature09268] [PMID: 20671703]

[2]   Nishitani G, Yamamoto M, Natsuike M, Liu D. Dynamics of Phytoplankton in Kesennuma Bay and Moune Bay after the disaster 3.11. Aquabiology 2012; 203: 545-55. [in Japanese].

[3]   Martin JH, Fitzwater SE. Northeast Pacific iron distribution in relation to phytoplankton productivity. Deep-Sea Res 1988; 35: 177-96.
      [http://dx.doi.org/10.1038/371123a0]

[4]   Martin JH, Fitzwater SE. Iron deficiency limits phytoplankton growth in the north-east Pacific subarctic. Nature 1988; 331: 341-3.
      [http://dx.doi.org/10.1038/331341a0]

[5]   Martin J H, Gordon R M, Fitzwater S E, Broenkow W W. VERTEX: Phytoplankton/iron studies in the Gulf of Alaska. Deep-Sea Res 1989; 36: 649-80.

[6]   Herrera-Silveira JA, Morales-Ojeda SM. Evaluation of the health status of a coastal ecosystem in southeast Mexico: Assessment of water quality, phytoplankton and submerged aquatic vegetation. Mar Pollut Bull 2009; 59(1-3): 72-86.
      [http://dx.doi.org/10.1016/j.marpolbul.2008.11.017] [PMID: 19157464]

[7]   Yamamoto M, Liu D, Kasai A, Okubo K, Tanaka M. Dynamics of iron in the Chikugo River Basin: Comparison of iron with nitrogen and phosphate input to the estuary. Reg Stud Mar Sci 2016; 8: 89-98.

[8]   Yamashita Y, Tanaka M, Eds. Linkage of Forests. Kouseisha Kouseikaku, Tokyo: Rivers and coasts and biological production in estuarine and coastal waters 2008. [in Japanese]

[9]   Shiraiwa T, Ed. Final reports of the Amur Okhotsk project 2005–2009. Report on Amur-Okhotsk Project. 2010; pp. (6): 3-19.

[10]   Matsunaga K, Igarashi K, Fukase S, Tsubota H. Behavior of organically bound iron in seawater of estuaries. Estuar Coast Shelf Sci 1984; 18(6): 615-22.
[http://dx.doi.org/10.1016/0272-7714(84)90034-9]

[11]   Matsunaga K, Nishioka J, Kuma K, Toya K, Suzuki Y. Riverine input of bioavailable iron supporting phytoplankton growth in Kesennuma Bay (Japan). Water Res 1998; 32(11): 3436-42.
[http://dx.doi.org/10.1016/S0043-1354(98)00113-4]

[12]   Öztürk M, Bizsel N. Iron speciation and biogeochemistry in different nearshore waters. Mar Chem 2003; 83: 145-56.
[http://dx.doi.org/10.1016/S0304-4203(03)00108-7]

[13]   Fujita D. Barren ground.Current state of phycology in the 21st Century.. Yamagata: The Japanese Society of Phycology 2002; pp. 102-5.

[14]   Yamamoto M, Fukushima M, Kiso E, *et al*. Application of iron humates to barren ground in a coastal area for restoring seaweed beds. J Chem Eng of Jpn 2010; 43(7): 627-34.
[http://dx.doi.org/10.1252/jcej.43.627]

[15]   Yamamoto M, Kato T, Kanayama S, Nakase K, Tsutsumi N. Effectiveness of iron fertilization for seaweed bed restoration in coastal areas. J Water Environ Technol 2017; 15(5): 186-97.
[http://dx.doi.org/10.2965/jwet.16-080]

[16]   Yamamoto M, Fukushima M, Liu D. The effect of humic substances on iron elution in the method of restoration of seaweed beds using steelmaking slag. ISIJ Int 2012; 52(10): 1909-13.
[http://dx.doi.org/10.2355/isijinternational.52.1909]

[17]   Yamamoto M, Nishida A, Otsuka K, Komai T, Fukushima M. Evaluation of the binding of iron(II) to humic substances derived from a compost sample by a colorimetric method using ferrozine. Bioresour Technol 2010; 101(12): 4456-60.
[http://dx.doi.org/10.1016/j.biortech.2010.01.050] [PMID: 20163958]

[18]   Iwai H, Fukushima M, Yamamoto M, Motomura T. Seawater extractable organic matter (SWEOM) derived from a compost sample and its effect on the serving bioavailable fe to the brown macroalga, saccahrina japonica. Humic Subs Res 2015; 12(1): 5-20.

[19]   Yamamoto M, Liu D. Assessment of coastal environment for introducing the method of seaweed bed restoration using a mixture of steelmaking slag and compost. Proceedings of 18th International Conference of International Humic Substances Society, 2016. 0-5.

[20]   Yamamoto M, Iwai H, Yamaguchi A, Liu D. Seasonal changes of iron in rivers and relation to structural features of humic acids in their sediments of the northwest of Hokkaido, Japan. Proceedings of 19th International Conference of International Humic Substances Society, 2018. 03-6.

[21]   Yamamoto M, Liu D. Change of coastal environment based on water quality in kesennuma bay and moune bay after the disaster 3.11. Aquabiology 2013; 35(6): 547-53. [in Japanese].

[22]   Yamamoto M, Kato T, Tabeta S, *et al*. Changes in the coastal environment in Kamaishi Bay after the Great East Japan Earthquake. Bull Jpn Soc Sci Fish 2015; 81(2): 243-55.

[23]   Masuda R, Hatakeyama M, Yokoyama K, Tanaka M. Recovery of coastal fauna after the 2011 tsunami in japan as determined by bimonthly underwater visual censuses conducted over five years. PLoS One 2016; 11(12): e0168261.
[http://dx.doi.org/10.1371/journal.pone.0168261] [PMID: 27942028]

[24]   Natsuike M, Yokoyama K, Nishitani G, Yamada Y, Yoshinaga I, Ishikawa A. Germination fluctuation of toxic Alexandrium fundyense and A. pacificum cysts and the relationship with bloom occurrences in Kesennuma Bay, Japan. Harmful Algae 2017; 62: 52-9.
[http://dx.doi.org/10.1016/j.hal.2016.11.018] [PMID: 28118892]

[25]   Tachibana A, Nishibe Y, Fukuda H, Kawanobe K, Tsuda A. Phytoplankton community structure in Otsuchi Bay, northeastern Japan, after the 2011 off the Pacific coast of Tohoku Earthquake and tsunami. J Oceanogr 2017; 73: 55-65.
[http://dx.doi.org/10.1007/s10872-016-0355-3]

[26]   Yamada Y, Kaga S, Kaga Y, Naiki K, Watanabe S. Change of seawater quality in Ofunto Bay, Iwate, after the 2011 off the Pacific coast of Tohoku Earthquake. J Oceanogr 2017; 73: 11-24.
[http://dx.doi.org/10.1007/s10872-015-0336-y]

CHAPTER 7

A Sequence Between Microfouling and Macrofouling in Marine Biofouling

Hideyuki Kanematsu[1,*] and **Dana M. Barry**[2]

[1] *Department of Materials Science and Engineering, National Institute of Technology (KOSEN), Suzuka College, Japan*

[2] *Department of Electrical and Computer Engineering, Clarkson University, Potsdam, New York State, 13699, the USA/SUNY Canton, Canton, New York State, 13617, the USA*

Abstract: Biofouling is described in this chapter. Biofouling is the overall phenomenon and processes related to the attachment of organisms to surfaces. It is classified into two main categories: microfouling and macrofouling. Microfouling is the phenomenon induced by bacterial activity. Bacteria, microalgae, *etc.* attach to materials' surfaces, This induces the phenomenon. On the other hand, macrofouling occurs by the attachment of larger organisms such as barnacles, oysters, *etc.* Both types have a close relationship to each other. Actually, it occurs in a series. In this chapter, we provide an overview of each type and discuss how to effectively control them.

Keywords: Bacteria, Biofilm, Biofouling, Macrofouling, Microfouling.

WHAT IS BIOFOULING?

We would like to begin this chapter by defining the important technical term, biofouling [1 - 3]. Biofouling is a process where organisms are attached to materials' surfaces. This causes materials to deteriorate in various ways. As a result, ships in marine environments will sail slower and increase the traveling time of the journey. Also, the materials used to build the ships may rust [4 - 10]. These phenomena could be attributed to biofouling in varying degrees.

Biofouling is classified into two main categories: microfouling and macrofouling [1, 11 - 19].

Microfouling refers to a material's degradation by the attachment of bacteria to them. Bacteria tend to attach to materials' surfaces, because carbon compounds (a form of needed nutrition) generally exist on them. Various carbon compounds

* **Corresponding author Hideyuki Kanematsu:** Department of Materials Science and Engineering, NIT Suzuka College, Suzuka, Japan; Tel/Fax: +81-59-368-1848; E-mail: kanemats@mse.suzuka-ct.ac.jp

Toshiyuki Takahashi (Ed.)

form very thin, film-like matters on materials' surfaces. They are called "conditioning films". Many kinds of carbon compounds exist in conditioning films. As for marine environments, the main components are accumulated carbohydrates. Fig. (**1**) shows conditioning films.

(a) The attachment of carbon compounds.

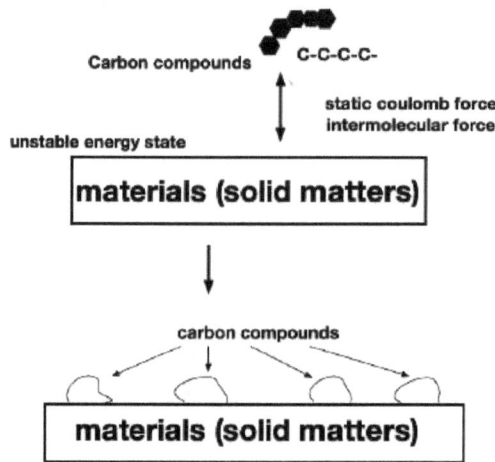

Carbon compounds   C-C-C-C-

static coulomb force
intermolecular force

unstable energy state

materials (solid matters)

carbon compounds

materials (solid matters)

(b) The bacteria attracted by carbon compounds through chemotaxes.

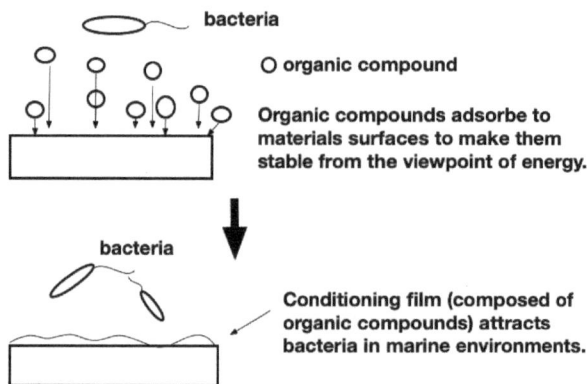

bacteria

O organic compound

Organic compounds adsorbe to
materials surfaces to make them
stable from the viewpoint of energy.

bacteria

Conditioning film (composed of
organic compounds) attracts
bacteria in marine environments.

Fig. (1). Schematic image of conditioning films and bacteria.

Bacteria attach to materials' surfaces and obtain nutrition (conditioning film) in order to survive [20 - 29]. The driving force for the attachment process must be the tendency for bacteria to seek areas of concentrated nutrition. This tendency is called chemotaxis. When the number of bacteria (attaching to materials' surfaces) increases to a certain threshold value, then the bacteria excrete organic polymers outside of their cells (Fig. **2**).

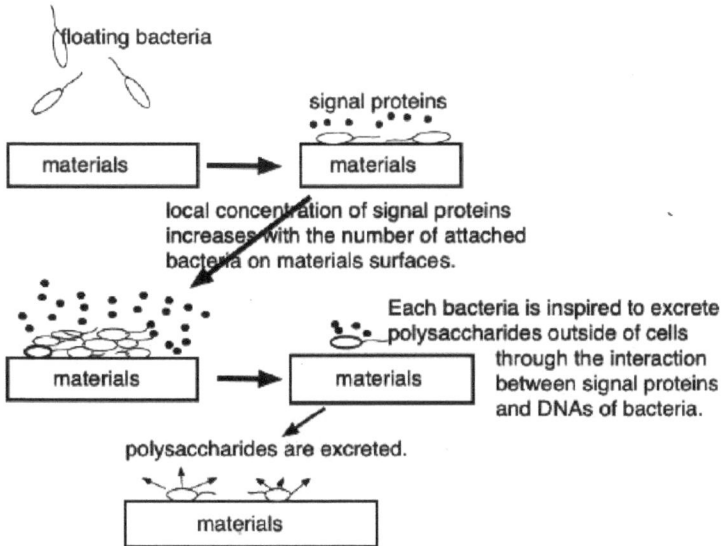

Fig. (2). Excretion of Polysaccharides through quorum sensing process.

This phenomenon depends on the localized number of bacteria. As a result, bacteria become encased by the polymers, which have a sticky surface. These film-like polymers (enclosing bacteria and materials' surfaces) are called "biofilm". The process is schematically shown in Fig. (**3**). Therefore, biofilm could be considered a product of microfouling [30 - 52].

On the other hand, macrofouling is the attachment of larger organisms to materials' surfaces. Examples for the marine environment include oysters and barnacles. These organisms generally live on solid surfaces like rocks and artificial ones such as ships, marine structures, *etc*. Just like bacteria on materials' surfaces, those organisms also attach to materials' surfaces to get nutrition. Since barnacles and oysters use plankton as nutrition, it is very natural for them to get plankton in biofilms formed on solid surfaces. Even though there not so many research papers about this type of relationship, we can say that microfouling must be followed by macrofouling. Therefore, the two phenomena - microfouling and macrofouling are usually connected in a series, as shown in Fig. (**4**).

In this chapter, we describe biofouling in more detail and propose some tips on how to control it for both materials' science and industrial viewpoints.

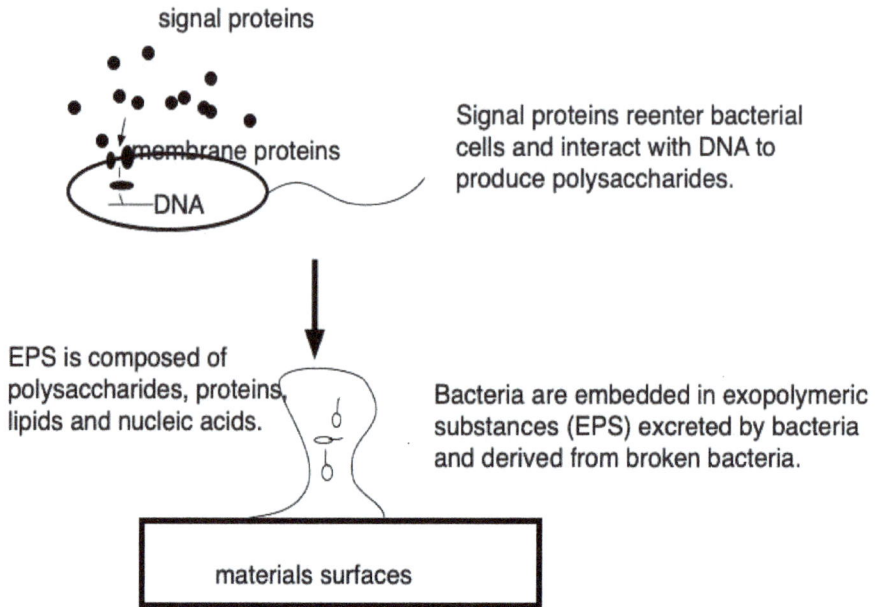

signal proteins

membrane proteins

DNA

Signal proteins reenter bacterial cells and interact with DNA to produce polysaccharides.

EPS is composed of polysaccharides, proteins, lipids and nucleic acids.

Bacteria are embedded in exopolymeric substances (EPS) excreted by bacteria and derived from broken bacteria.

materials surfaces

Fig. (3). Schematic illustration of Biofilms' formation.

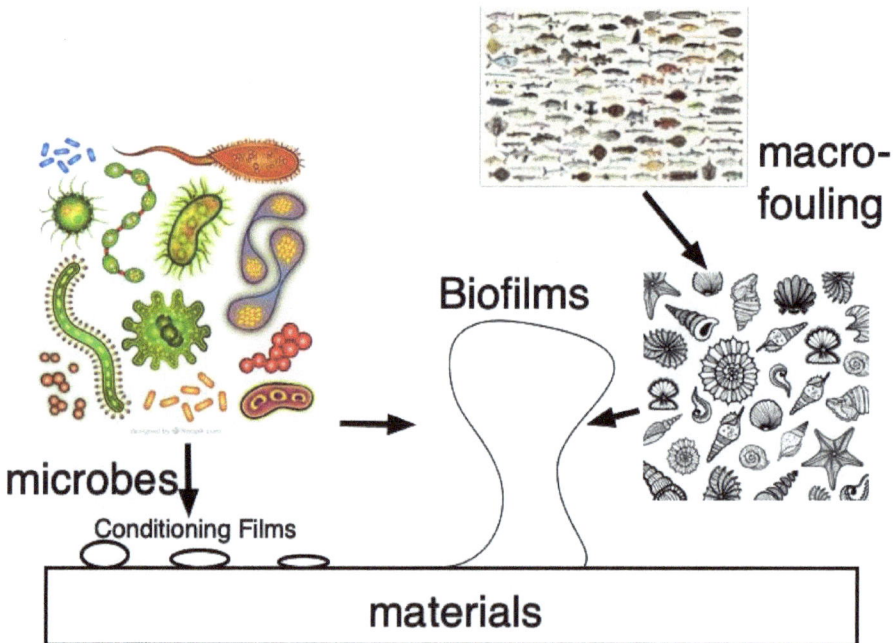

macro-fouling

Biofilms

microbes

Conditioning Films

materials

Fig. (4). Biofouling, microfouling and macrofouling.

Microfouling in Marine Environments

As shown in the previous section, biofilms are the main player for microfouling. This is also true in marine environments. As a premise for biofilm formation, the existence of conditioning film has already been hypothesized. Even though it would be very hard to demonstrate the existence (conditioning films) by experimental results, it is very natural for us to postulate this process. Biofilms are produced on solid surfaces by bacterial activities after a threshold number of bacteria exist. Therefore, we need to explain why a number of bacteria could exist on a solid surface. One reason is that organic compounds (a form of nutrition) already present on the surface. Thus, chemotaxis is the tendency for bacteria to attach to a surface. It is difficult to carry out experiments to prove the existence of condition films for two main reasons. One of them could be attributed to the problem of preparing control specimens without conditioning films. The other one could be attributed to the lack of analytical methods. Both difficulties will be solved with the development of analytical methods in the future.

Even though the analyses of conditioning films might be difficult, we have many indirect signs for conditioning films and there have been quite a few investigations for conditioning film in marine environments. Compere, Poleunis, and their collaborators suggest in their research papers that stainless steels immersed in marine environments formed conditioning films (on them). These films were mainly composed of proteins and carbohydrates. According to their ideas, protein is adsorbed to solid surfaces first and then carbohydrates form. Finally, the conditioning film would mainly be composed of carbohydrate. Carbohydrates generally exist in every path of metabolism and are stored as ATP in organisms' cells. The carbohydrate is called storage carbohydrate. Some researchers point out that half of all dissolved carbon compounds in seawater might be composed of storage carbohydrates.

Therefore, once certain solids have contact with seawater, their surfaces would soon allow carbon compounds to be adsorbed. This will allow the unstable energy state of solid surfaces to be stabilized. Therefore, as previously described, marine bacteria are attracted to solid surfaces and form biofilms.

The characteristics of marine biofilms must be characterized by marine bacteria. The distribution and the number of species must be huge. The flora would change according to variable like location, the level under water, weather, temperatures, the oxygen content, *etc.* The species of bacteria present might depend on the substrates.

Table **1** and Table **2** show our examples [53]. Several kinds of metallic specimens were simultaneously immersed in Ise Bay, Japan. The result show that the constitution of flora changed with time (Table **1**). We understand how and why the changes occurred. Obviously, a species of marine bacteria became attached to the materials' surfaces. The species must have the highest capability to attach there and to form biofilms. However, other bacteria would attach and enter the biofilms. The second, the third and the following bacteria would be easier to attach to materials' surfaces as compared to the primary one, since biofilms already exist in advance to some extent. In addition, within biofilms, the portion of each bacteria must change depending on the components, temperature, *etc.* Such a change of flora would happen in many aspects.

Table 1. The change of microbiota on tin plated carbon steels.

Immersion Time	Bacteria	Percentages
One week	α-Proteobacteria	24%
	γ-Proteobacteria	67%
	Actinobacteria	3%
	ε-Proteobacteria	3%
	Firmicutes	3%
One month	ε-proteobacteria	71%
	Actinobacteria	29%

However, Table **2** shows a quite different aspect. It clearly shows that the primary biota within biofilms would change from substrate to substrate. We presume the correlation between metallic factor and biofilm's one would contribute to biofilm formation.

When biofilms form on materials' surfaces, geographical profiles would change there. In such a case, the substrate would be covered with biofilms. However, the extent of coverage is inhomogeneous, as shown in Fig. (**5**). Then dissolved oxygen in the environment has contacts to substrate in different ways, depending on places. If the substrate would be metallic materials, the insufficient oxygen part would become cathode, while the sufficient oxygen part would become anode. Each nearest anode/cathode cite would be coupled to form a local electrochemical cell and the substrate metal would dissolve at the anode. At this point, the metallic surface is composed of many local oxygen concentration cells and the dissolution would be activated at those local anodic cites. Therefore, the microfouling would induce corrosion on metallic surface in this way.

As described in the previous section, the microfouling would lead to the following macrofouling. One effective protection measure to control macrofouling is to control microfouling. In the following section, we will describe macrofouling.

Table 2. The difference of microbiota on various specimens.

Substrates	Bacteria	Percentages
Carbon steel	γ-Proteobacteria	9%
	Firmicutes	55%
	Actinobacteria	36%
Stainless steel (18Cr-8Ni)	α-Proteobacteria	19%
	γ-Proteobacteria	16%
	δ-Proteobacteria	4%
	ε-Proteobacteria	6%
	Firmicutes	13%
	SRB	6%
	Others	36%
Tin plates carbon steel	α-Proteobacteria	24%
	γ-Proteobacteria	67%
	Actinobacteria	3%
	ε-Proteobacteria	3%
	Firmicutes	3%

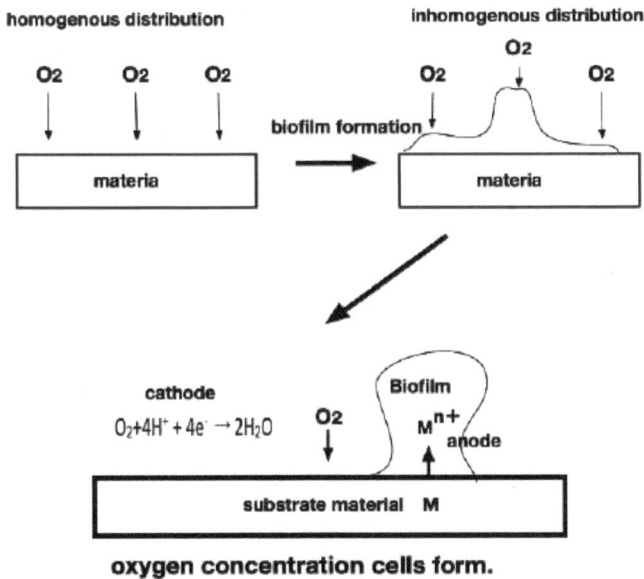

homogenous distribution　　　　inhomogenous distribution

O_2　O_2　O_2　　　　O_2　O_2　O_2

biofilm formation

materia　　　→　　　materia

cathode

$O_2 + 4H^+ + 4e^- \rightarrow 2H_2O$　O_2　M^{n+}　Biofilm　anode

substrate material　M

oxygen concentration cells form.

Fig. (5). Biofilm formation and oxygen concentration cell.

Macrofouling in Marine Environments

While microfouling's center players are bacteria and microalgae, *etc.*, macrofouling's would be larger organisms. The representative organisms for macrofouling are barnacles, oysters, blue sea mussels, *etc*. They attach to materials' surfaces just like bacteria. What kinds of organisms would attach to the materials' surfaces? The answer depends on the combination of environments and materials, since the dominant species differ from environment to environment. However, the species would be affected by the predecessor biofilms (rather than by materials), since biofilms must exist in advance for the attachment of those larger organisms. Fig. (**6**) shows the schematic illustration for the relationship between macrofouling and microfouling. As shown in this figure, they occur in a series.

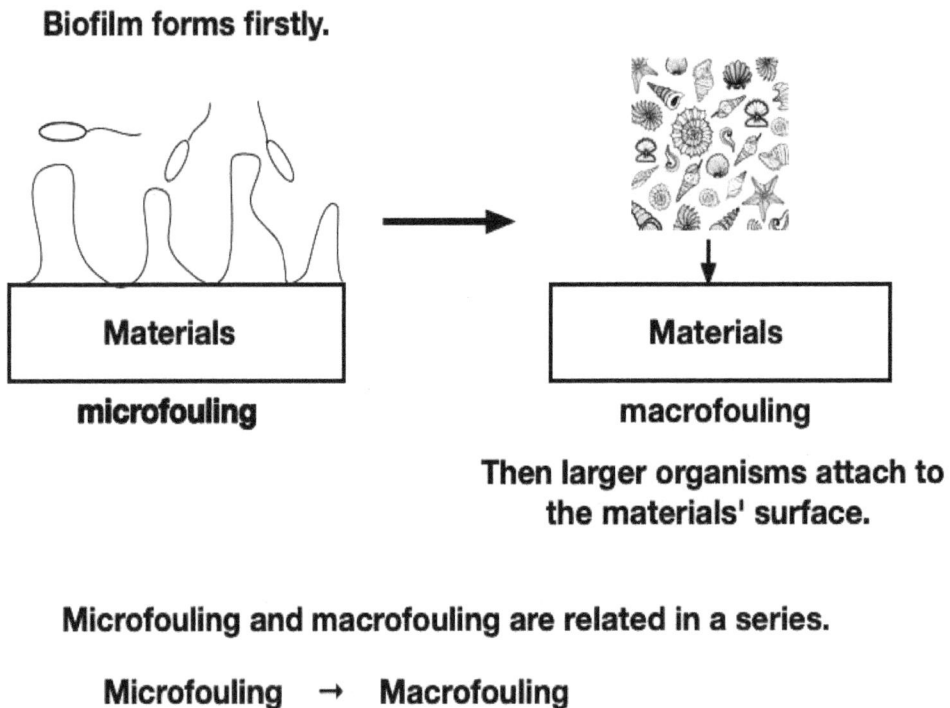

Biofilm forms firstly.

microfouling

macrofouling

Then larger organisms attach to the materials' surface.

Microfouling and macrofouling are related in a series.

Microfouling → Macrofouling

Fig. (6). The relation between microfouling and macrofouling.

Bacteria attach to materials' surfaces, as explained in the previous section. Then larger organisms approach the materials' surfaces, because they contain biofilms composed of polymers, microorganisms, *etc.* which serve as nutrition for them. Therefore, they attach to materials' surfaces. The larger organisms such as barnacles, oysters are attached to materials' surfaces well, since their biology is sedentary.

Barnacles look like soft-bodied creatures at a glance. However today, they basically belong to crustaceans such as shrimp, crab, *etc*. They live in groups and tend to select places where other barnacles already exist attached to materials' surfaces. The reason is that they would have difficulty to find nutrition by moving. For the same reason, we presume that they tend to get nutrition from biofilms. They exist in most marine environments all over the world. However, they do not live in fresh water environments. Their main source for nutrition is said to be various kinds of planktons.

Oysters contain a wide variety of clams. They have hard shells composed of calcites. They also tend to attach solid matters. However, they could live also in clay at the bottom of sea and attach to the bottom of ships. From the moment of fertilization, an oyster takes just one day to form a hard shell. Generally, we could observe an oyster's existence only one day after the immersion of materials into the sea.

Macrofouling leads to serious corrosion and deterioration of materials. As for corrosion, materials' surfaces are clearly changed as compared with the case of microfouling. Therefore, it would lead to differential aeration corrosion very easily. In addition, the excreted chemicals from the larger organisms might be the source to induce corrosion. When the acidic matter would be excreted, the local pH at the vicinity of surfaces would decrease the pH and become acidic. Then the iron and steel might be vulnerable to corrosion. The accumulation of organisms on materials' surfaces also leads to the deterioration of materials or structural functions. When they attach to ship bottoms, the power of propulsion would decrease seriously. Or, when they attach to various nets at the inlet of the sea, conduits, motors, *etc*. (for the introduction or discharge of sea water or fresh water), the macrofouling might decrease the original function of those structures. Therefore, macrofouling (which is an important and difficult problem for industry) should be properly controlled.

How Should We Control Biofouling Effectively?

We provided an overview about the biofouling in the previous sections. And, we already know that the biofouling leads to huge economic loss due to materials' deterioration including corrosion. Therefore, it is obviously very important for us to control biofouling. So how should we control biofouling?

As described above, macrofouling causes industrial problems and tremendous economic loss, since the phenomenon is visible at the macroscale. However, it is closely related to the biofilm phenomenon, *i.e.*, caused by bacteria on the microscale. In such a situation, the countermeasure should be considered from both the anti-macrofouling approach and from the anti-microfouling, since they occur in a series. Fig. (**7**) shows the general concept for controlling biofouling.

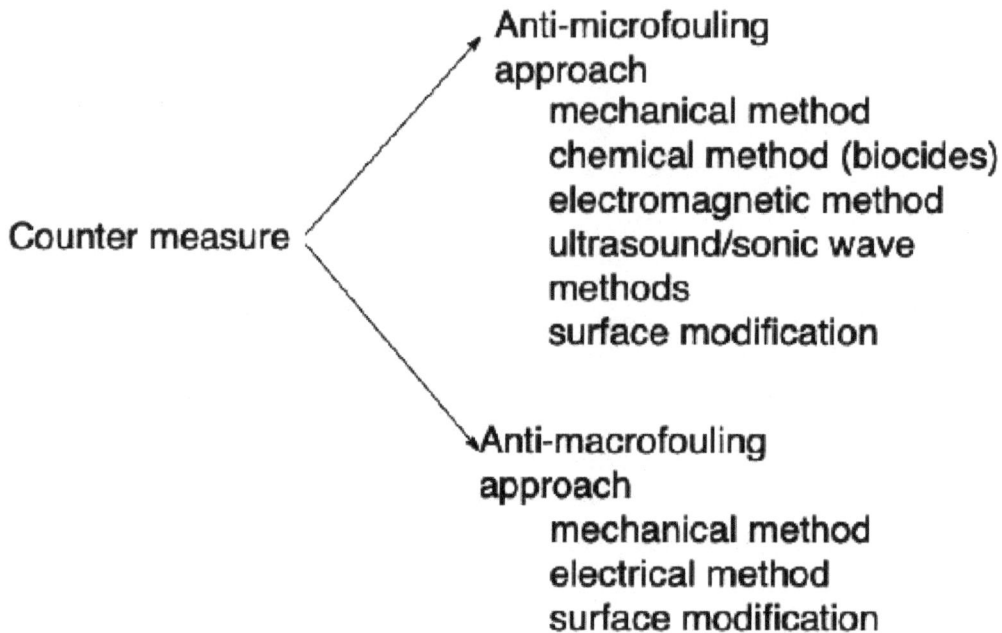

Anti-microfouling
approach
    mechanical method
    chemical method (biocides)
    electromagnetic method
    ultrasound/sonic wave
    methods
    surface modification

Counter measure

Anti-macrofouling
approach
    mechanical method
    electrical method
    surface modification

Since macrofouling is more versatile, it is harder to control it.

Fig. (7). The concept of counter measures to control biofilms.

The countermeasure is classified into two main types: the macrofouling approach and the microfouling one. The former is also classified into physics and chemical methods. The former might provide an electrical shock to the organisms and the use of mechanical machines to properly remove them. It is possible that the geographical surface profiles will control the attachment of organisms effectively.

On the other hand, the main measure of the latter is biocides. In the past, organic tin compounds (tributyl tin) dominated the world's antifouling markets. It could control biofouling very effectively. However, the problem was the environmental negative burden of that product. Therefore, it has been banned to use for such a purpose. On the other hand, researchers found that cupric oxide (CuO) has a strong resistance to macrofouling. As a result, it has already had some practical uses.

CONSENT FOR PUBLICATION

Not applicable.

CONFLICT OF INTEREST

The author confirms that this chapter contents have no conflict of interest.

ACKNOWLEDGEMENT

A part of the Figs. (**4 & 6**) was designed by Freepik (https://www.freepik.com). We highly appreciate them for it.

REFFERENCES

[1]   Sanders P, Maxwell S. Microfouling, macrofouling and corrosion of metal test specimens in seawater. Microbial corrosion 1983; p. 74-83.

[2]   Baker J, Dudley L. Biofouling in membrane systems-a review. Desalination 1998; 118: 81-9.
        [http://dx.doi.org/10.1016/S0011-9164(98)00091-5]

[3]   Nguyen T, Roddick FA, Fan L. Biofouling of water treatment membranes: a review of the underlying causes, monitoring techniques and control measures. Membranes (Basel) 2012; 2(4): 804-40.
        [http://dx.doi.org/10.3390/membranes2040804] [PMID: 24958430]

[4]   Schultz MP, Bendick JA, Holm ER, Hertel WM. Economic impact of biofouling on a naval surface ship. Biofouling 2011; 27(1): 87-98.
        [http://dx.doi.org/10.1080/08927014.2010.542809] [PMID: 21161774]

[5]   Schultz MP. Effects of coating roughness and biofouling on ship resistance and powering. Biofouling 2007; 23(5-6): 331-41.
        [http://dx.doi.org/10.1080/08927010701461974] [PMID: 17852068]

[6]   Davidson I, Scianni C, Hewitt C, *et al.* Mini-review: Assessing the drivers of ship biofouling management--aligning industry and biosecurity goals. Biofouling 2016; 32(4): 411-28.
        [http://dx.doi.org/10.1080/08927014.2016.1149572] [PMID: 26930397]

[7]   Zargiel KA, Coogan JS, Swain GW. Diatom community structure on commercially available ship hull coatings. Biofouling 2011; 27(9): 955-65.
        [http://dx.doi.org/10.1080/08927014.2011.618268] [PMID: 21932984]

[8]   Dhanasekaran D, Thajuddin N, Rashmi M, Deepika T, Gunasekaran M. Screening of biofouling activity in marine bacterial isolate from ship hull. Int J Environ Sci Technol 2009; 6: 197-202.
        [http://dx.doi.org/10.1007/BF03327622]

[9]   Evans L. Marine algae and fouling: a review, with particular reference to ship-fouling. Bot Mar 1981; 24: 167-72.
        [http://dx.doi.org/10.1515/botm.1981.24.4.167]

[10]   Zargiel KA, Swain GW. Static *vs* dynamic settlement and adhesion of diatoms to ship hull coatings. Biofouling 2014; 30(1): 115-29.
        [http://dx.doi.org/10.1080/08927014.2013.847927] [PMID: 24279838]

[11]   Cowie PR, Smith MJ, Hannah F, Cowling MJ, Hodgkeiss T. The prevention of microfouling and macrofouling on hydrogels impregnated with either Arquad 2C-75 or benzalkonium chloride. Biofouling 2006; 22(3-4): 173-85.
        [http://dx.doi.org/10.1080/08927010600783296] [PMID: 17290862]

[12]   Little BJ, Wagner PA. Succession in microfouling.Fouling Organisms of the Indian Ocean: Biology and Control Technology. New Delhi: Oxford and IBH 1997; pp. 105-34.

[13]   Dobretsov S, Railkin A. Correlative connections between marine maicro-and macrofouling. Biol Mora (Vladivost) 1994; 20: 115-9.

[14]   Dobretsov S, Abed RM, Voolstra CR. The effect of surface colour on the formation of marine micro and macrofouling communities. Biofouling 2013; 29(6): 617-27.
[http://dx.doi.org/10.1080/08927014.2013.784279] [PMID: 23697809]

[15]   Winters H, Isquith IR. In-plant microfouling in desalination. Desalination 1979; 30: 387-99.
[http://dx.doi.org/10.1016/S0011-9164(00)88468-4]

[16]   Jenner HA, Taylor C, Van Donk M, Khalanski M. Chlorination by-products in chlorinated cooling water of some European coastal power stations. Mar Environ Res 1997; 43: 279-93.
[http://dx.doi.org/10.1016/S0141-1136(96)00091-8]

[17]   Muthukrishnan T, Abed RM, Dobretsov S, Kidd B, Finnie AA. Long-term microfouling on commercial biocidal fouling control coatings. Biofouling 2014; 30(10): 1155-64.
[http://dx.doi.org/10.1080/08927014.2014.972951] [PMID: 25390938]

[18]   Sokolova A, Cilz N, Daniels J, *et al.* A comparison of the antifouling/foul-release characteristics of non-biocidal xerogel and commercial coatings toward micro- and macrofouling organisms. Biofouling 2012; 28(5): 511-23.
[http://dx.doi.org/10.1080/08927014.2012.690197] [PMID: 22616756]

[19]   Dobretsov S, Dahms H-U, Tsoi MY, Qian P-Y. Chemical control of epibiosis by Hong Kong sponges: the effect of sponge extracts on micro-and macrofouling communities. Mar Ecol Prog Ser 2005; 297: 119-29.
[http://dx.doi.org/10.3354/meps297119]

[20]   Donlan RM. Biofilms: microbial life on surfaces. Emerg Infect Dis 2002; 8(9): 881-90.
[http://dx.doi.org/10.3201/eid0809.020063] [PMID: 12194761]

[21]   Murga R, Miller J, Donlan R. Biofilm formation by gram-negative bacteria on central venous catheter connectors: effect of conditioning films in a laboratory model. J clinical microbial 2001; 39: 2294-2297.

[22]   Ramage G, Tomsett K, Wickes BL, López-Ribot JL, Redding SW. Denture stomatitis: a role for Candida biofilms. Oral Surg Oral Med Oral Pathol Oral Radiol Endod 2004; 98(1): 53-9.
[http://dx.doi.org/10.1016/j.tripleo.2003.04.002] [PMID: 15243471]

[23]   Ramage G, Vandewalle K, Wickes BL, López-Ribot JL. Characteristics of biofilm formation by Candida albicans. Rev Iberoam Micol 2001; 18(4): 163-70.
[PMID: 15496122]

[24]   Lorite GS, Rodrigues CM, de Souza AA, Kranz C, Mizaikoff B, Cotta MA. The role of conditioning film formation and surface chemical changes on *Xylella fastidiosa* adhesion and biofilm evolution. J Colloid Interface Sci 2011; 359(1): 289-95.
[http://dx.doi.org/10.1016/j.jcis.2011.03.066] [PMID: 21486669]

[25]   Busscher HJ, Bos R, van der Mei HC. Initial microbial adhesion is a determinant for the strength of biofilm adhesion. FEMS Microbiol Lett 1995; 128(3): 229-34.
[http://dx.doi.org/10.1111/j.1574-6968.1995.tb07529.x] [PMID: 7781968]

[26]   Characklis WG, Cooksey KE. Biofilms and microbial fouling.Advances in applied microbiology. Elsevier 1983; 29: pp. 93-138.
[http://dx.doi.org/10.1016/S0065-2164(08)70355-1]

[27]   Bradshaw D, Marsh P, Watson G, Allison C. Effect of conditioning films on oral microbial biofilm development. Biofouling 1997; 11: 217-26.
[http://dx.doi.org/10.1080/08927019709378332]

[28]   Busscher HJ, van der Mei HC. Physico-chemical interactions in initial microbial adhesion and relevance for biofilm formation. Adv Dent Res 1997; 11(1): 24-32.
        [http://dx.doi.org/10.1177/08959374970110011301] [PMID: 9524439]

[29]   Marsh P D, Bradshaw D J. Dental plaque as a biofilm. J Ind Microbial 1995; 15: 169-175..

[30]   Solano C, Echeverz M, Lasa I. Biofilm dispersion and quorum sensing. Curr Opin Microbiol 2014; 18: 96-104.
        [http://dx.doi.org/10.1016/j.mib.2014.02.008] [PMID: 24657330]

[31]   Ikegai H. Genomics Approach. Biofilm and Material Science. New York, the USA: Springer 2015; pp. 53-60.

[32]   Whiteley M, Diggle SP, Greenberg EP. Progress in and promise of bacterial quorum sensing research. Nature 2017; 551(7680): 313-20.
        [http://dx.doi.org/10.1038/nature24624] [PMID: 29144467]

[33]   Kim MK, Zhao A, Wang A, *et al.* Surface-attached molecules control *Staphylococcus aureus* quorum sensing and biofilm development. Nat Microbiol 2017; 2: 17080.
        [http://dx.doi.org/10.1038/nmicrobiol.2017.80] [PMID: 28530651]

[34]   Jemielita M, Wingreen NS, Bassler BL. Quorum sensing controls *Vibrio cholerae* multicellular aggregate formation. eLife 2018; 7e42057
        [http://dx.doi.org/10.7554/eLife.42057.] [PMID: 30582742]

[35]   Morales-Soto N, Cao T, Baig NF, Kramer KM, Bohn PW, Shrout JD. Surface-growing communities of *Pseudomonas aeruginosa* exhibit distinct alkyl quinolone signatures. Microbiol Insights 2018; 11,1178636118817738.
        [http://dx.doi.org/10.1177/1178636118817738] [PMID: 30573968]

[36]   Singh N, Rajwade J, Paknikar KM. Transcriptome analysis of silver nanoparticles treated Staphylococcus aureus reveals potential targets for biofilm inhibition. Coll Surf B Biointer 2019; 175: 487-97.
        [http://dx.doi.org/10.1016/j.colsurfb.2018.12.032] [PMID: 30572157]

[37]   Heo YM, Kim K, Ryu SM, *et al.* Diversity and ecology of marine algicolous arthrinium species as a source of bioactive natural products. Mar Drugs 2018; 16(12): 508.
        [http://dx.doi.org/10.3390/md16120508] [PMID: 30558255]

[38]   Sully EK, Malachowa N, Elmore BO, *et al.* Selective chemical inhibition of agr quorum sensing in *Staphylococcus aureus* promotes host defense with minimal impact on resistance. PLoS Pathog 2014; 10(6)e1004174
        [http://dx.doi.org/10.1371/journal.ppat.1004174] [PMID: 24945495]

[39]   Singh R, Ray P. Quorum sensing-mediated regulation of staphylococcal virulence and antibiotic resistance. Future Microbiol 2014; 9(5): 669-81.
        [http://dx.doi.org/10.2217/fmb.14.31] [PMID: 24957093]

[40]   Zhang X, Crippen TL, Coates CJ, Wood TK, Tomberlin JK. Effect of quorum sensing by *Staphylococcus epidermidis* on the attraction response of female adult yellow fever mosquitoes, *Aedes aegypti aegypti* (linnaeus) (diptera: culicidae), to a blood-feeding source. PLoS One 2015; 10(12)e0143950
        [http://dx.doi.org/10.1371/journal.pone.0143950] [PMID: 26674802]

[41]   Le KY, Otto M. Quorum-sensing regulation in staphylococci-an overview. Front Microbiol 2015; 6: 1174.
        [http://dx.doi.org/10.3389/fmicb.2015.01174] [PMID: 26579084]

[42]   Harris LG, Dudley E, Rohde H, *et al.* Limitations in the use of PSMγ, agr, RNAIII, and biofilm formation as biomarkers to define invasive Staphylococcus epidermidis from chronic biomedical device-associated infections. Int J Med Microbiol 2017; 307(7): 382-7.
        [http://dx.doi.org/10.1016/j.ijmm.2017.08.003] [PMID: 28826573]

[43]   Kolari M. 2003. Attachment mechanisms and properties of bacterial biofilms on non-living surfaces

[44]   Kanematsu H, Barry DM. Introduction. Biofilm and Material Science. 1st ed. New York, the USA: Springer 2015; pp. 3-8.

[45]   Flemming HC, Wingender J. The biofilm matrix. Nat Rev Microbiol 2010; 8(9): 623-33.
[http://dx.doi.org/10.1038/nrmicro2415] [PMID: 20676145]

[46]   Percival S L, Malic S, Cruz H, Williams D W. Introduction to Biofilms. Biofilms veterin med 2011; 41-68.
[http://dx.doi.org/10.1007/978-3-642-21289-5_2]

[47]   Fang F, Lu WT, Shan Q, Cao JS. Characteristics of extracellular polymeric substances of phototrophic biofilms at different aquatic habitats. Carbohydr Polym 2014; 106: 1-6.
[http://dx.doi.org/10.1016/j.carbpol.2014.02.010] [PMID: 24721043]

[48]   Flemming HC. EPS-Then and Now. Microorganisms 2016; 4(4): 41.
[http://dx.doi.org/10.3390/microorganisms4040041] [PMID: 27869702]

[49]   Brian F, Molmeret M, Fahs A, *et al.* Characterization and anti-biofilm activity of extracellular polymeric substances produced by the marine biofilm-forming bacterium *Pseudoalteromonas ulvae* strain TC14. Biofouling 2016; 32: 547-60.
[http://dx.doi.org/10.1080/08927014.2016.1164845]

[50]   Ding Y, Zhou Y, Yao J, *et al. In Situ* molecular imaging of the biofilm and its matrix. Anal Chem 2016; 88(22): 11244-52.
[http://dx.doi.org/10.1021/acs.analchem.6b03909] [PMID: 27709903]

[51]   Gunn JS, Bakaletz LO, Wozniak DJ. What's on the outside matters: The role of the extracellular polymeric substance of gram-negative biofilms in evading host immunity and as a target for therapeutic intervention. J Biol Chem 2016; 291(24): 12538-46.
[http://dx.doi.org/10.1074/jbc.R115.707547] [PMID: 27129225]

[52]   Decho AW, Gutierrez T. Microbial extracellular polymeric substances (EPSs) in ocean systems. Front Microbiol 2017; 8: 922.
[http://dx.doi.org/10.3389/fmicb.2017.00922] [PMID: 28603518]

[53]   Maseda H, Ikigai H, Kuroda D, Ogawa A, Kanematsu H. Immersion of iron and steel materials into marine environment at ise gulf and gene analysis of attached microorganism. CAMP-ISIJ 2010; 23: 668-9.

Part 2: Fundamentals to Detect and Evaluate Microorganisms and Pathogenic Viruses from Seawater Environments

CHAPTER 8

Introduction to Methods for Collection, Detection, and Evaluation of Microbes

Toshiyuki Takahashi[*]

Department of Chemical Science and Engineering, National Institute of Technology, Miyakonojo College, Miyakonojo, Japan

Abstract: Microbes, including bacteria and plankton, play important roles in aquatic ecosystems, particularly marine ecosystems. Techniques to detect and evaluate marine microbes are necessary for the understanding and conservation of the ecosystems, and to better understand microbial corrosion of artificial structures built in the marine environment. Fundamental techniques to detect and evaluate microbes have been developed from laboratory experiments using cultured model microorganisms, such as *Escherichia coli*. Such laboratory-based techniques are often insufficient in the marine environment because of interference from organisms other than target microbes of each research, and from organic and inorganic compounds in the environment. Precise evaluation of microbial communities under such circumstances demands techniques with greater sensitivity and specificity than those of the laboratory-based approaches. This chapter considers more precise and sensitive techniques than the conventional techniques for the collection, detection, and evaluation of microbes. These techniques can also help experts and professional readers achieve practical evaluation and control of microbes in complex conditions such as marine environments.

Keywords: Aqueous ecosystems, Marine ecosystems, Microorganisms.

INTRODUCTION

Compared to terrestrial ecosystems, which are supported mainly by grassland and geographically extensive forest ecosystems, marine ecosystems are supported by seaweed and microscopic-scale organisms [1 - 3]. Plankton are a crucial component of marine ecosystems, both as primary producers (phytoplankton) and primary consumers (zooplankton). In addition, as in terrestrial ecosystems, microbes such as bacteria are important as decomposers [1 - 4]. Cyanobacteria and chemosynthetic bacteria are also recognized as primary producers in marine ecosystems. Although certain organisms are central targets in marine ecology, we

[*] **Corresponding author Toshiyuki Takahashi:** Department of Chemical Science and Engineering, National Institute of Technology, Miyakonojo College, Miyakonojo, Japan; Tel; +81-986-47-1219; Fax; +81-986-47-1231; E-mail: mttaka@cc.miyakonojo-nct.ac.jp

must also understand the diversity of microbes in the marine ecosystem as the same as that of macro-organisms.

Microbes cause microbiologically influenced corrosion (MIC) of artificial structures built in marine environments [5, 6]. Thermal and nuclear power plants built along marine coasts use sea water as the cooling water. The sea water naturally contains microorganisms that are capable of causing MIC, which could be problematic. In addition to understanding the diversity of microbes in marine ecosystem, developing techniques to detect and evaluate microbes will increase the knowledge of interactions of marine microbes with artificial materials.

From these perspectives, marine ecologists and industrial engineers concerned with marine systems have an obvious interest in marine ecology. Molecular microbiological methods such as polymerase chain reaction (PCR) and next-generation sequencing (NGS) technologies have been used recently to evaluate microorganisms [7, 8]. However, these methods are often complicated to perform, and require microbiology specialists. This sometimes excludes marine ecologists and industrial engineers working on ocean development who lack microbiology experience.

The second part of this book focuses on detection, collection, and evaluation methods for microorganisms. This part presents several unique methods to put them into practice (Fig. 1). Readers can acquire practical knowledge and learn the necessary techniques to evaluate and control microbes including those in the marine environment.

Unique Enrichment Methods for Sparse Target Microbes

This section focuses on collection or isolation methods for microorganisms. Fundamental techniques to detect and evaluate microbes have been developed from experiments using cultured model microorganisms such as *Escherichia coli*.

These laboratory-based techniques do not often work sufficiently because of interference from organisms other than the target microbes for each research, and from organic and inorganic compounds in the environment. Precise evaluation of microbial communities under such circumstances demands techniques with greater sensitivity and specificity than those performed in laboratory conditions. These associated techniques can also contribute to the improvement of public health and aqueous environments. The latter can be achieved through aquatic conservation because some microbes are pathogenic for fish, aquarium residents, and humans.

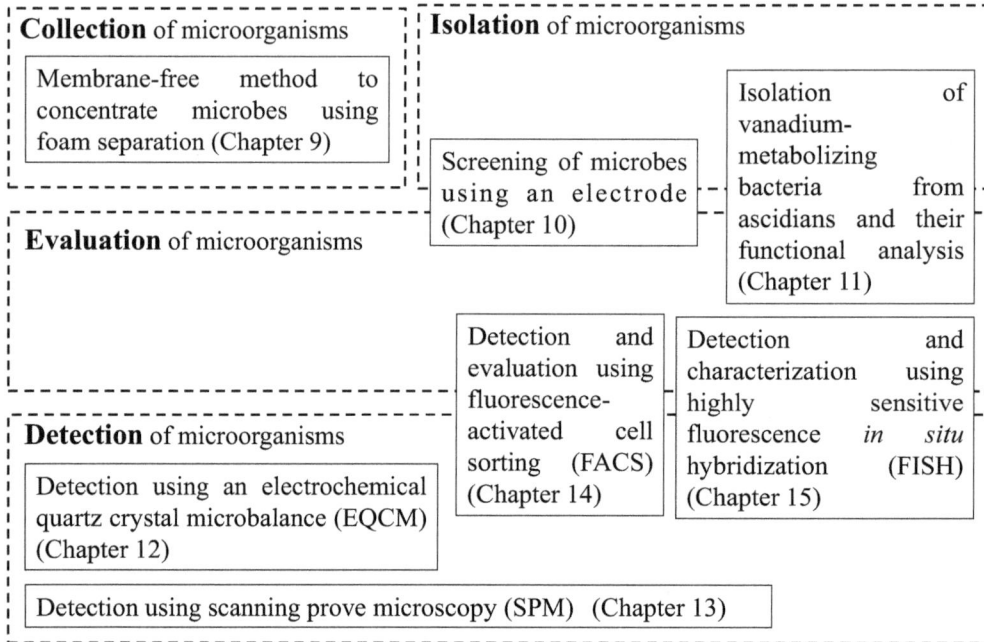

Fig. (1). Techniques presented in this chapter.

This section explains how to isolate and collect microbes using foam separation to concentrate living microorganisms and using a potential-controlled electrode as a selective method for difficult-to-cultivate microbes. This section also presents an experimental example of the isolation of vanadium-accumulating or -reducing microbes from ascidians and their functional analysis.

Instead of a conventional membrane filter method to remove microbes, the foam separation method uses dispersed bubbles and surfactants to remove and concentrate microbes [9]. Chapter 9 presents some examples of the removal of bacteria (*Vibrio*) and fungus (*Fusarium*).

Chapter 10 presents an electrical retrieval method for screening environmental microbes using a potential-controlled electrode [10, 11]. The method takes advantage of microbial characteristics to selectively attract living microbes to the potential-controlled electrode. This section provides some experimental examples of the isolation of microbes from deep-sea sediment core samples and 16S ribosomal RNA gene sequencing analysis.

Blood cells of ascidians accumulate higher levels of vanadium than that found in seawater [12, 13]. Vanadium is a rare metal in nature. It is generally used for production of steel, alloy components, catalyst, and pigments [14]. The accumulation of vanadium in ascidians is also related to vanadium-accumulating or -reducing bacteria, which live symbiotically with the host ascidian [15]. Chapter 11 provides an overview of the isolation and functional analyses of vanadium-metabolizing bacteria in ascidians.

Detection and Evaluation Methods for Marine Microorganisms

This section focuses on detection and evaluation methods for microorganisms. These goals require an understanding of marine environments, which include interactions between organisms and also between microbes and artificial materials used in the marine environment [5, 6, 16]. This section focuses on how to detect and evaluate microbes using an electrochemical method (electrochemical quartz crystal microbalance, EQCM) and scanning probe microscopy (SPM). Furthermore, this section presents microbial detection by fluorescence-activated cell sorting (FACS) [17], and also shows both single cell imaging- and DNA sequencing-based techniques of detecting microorganisms by fluorescence *in situ* hybridization (FISH) [18].

EQCM is the focus of Chapter 12. EQCM detects micro-weight changes from the order of ng to µg on an electrode surface using electrochemical measurement [19]. It is challenging to precisely detect microbes invisible to the naked eye. A biofilm [20], which is a sticky and gel-like clump produced by bacteria, is the primary cause of microbial corrosion of materials built in aqueous environments [21]. The section introduces the application of EQCM for detection and evaluation of biofilm formation.

SPM is the focus of Chapter 13. SPM detects surface microgeometry and physical properties from changes of a measurable physical quantity between the sample surface and a micro-probe positioned close to the sample surface. SPM enables the observation of specimens with a resolution of nm in aqueous media. The resolution using SPM is comparable to that of conventional transmission electron microscopy (TEM) [22]. SPM can also provide 3-demential (3D) imaging and visualization of concavo-convex shape on a sample surface [23]. The 3D resolution using SPM is generally superior to that using scanning electron microscopy (SEM). This section introduces the application of SPM for detection and evaluation of microorganism's adhesion to the material surface.

Chapter 14 presents microbial detection and evaluation using FACS [17]. FACS, which is often used almost interchangeably with flow cytometry (FCM), performs multiparametric analysis of physical characteristics of an individual particle and cell [24 - 26]. By detecting optical signals from target particles irradiated by light with an appropriate excitation wavelength, FACS systems screen the characteristics of targets. As several studies have already used unicellular microalgae [26 - 30], FACS (or FCM) can detect an individual microalga and evaluate its physiological characteristics. This section presents some experimental examples of detection of microalgae using FACS. Furthermore, life cycle analysis based on life cycle parameters such as cell size, DNA content, and chlorophyll content of each cell is detailed.

Chapter 15 presents microbial detection using FISH [18], which is one of the standard techniques used in molecular biology. It generally enables visual detection of microbes using genetic sequence-based information, helps in the understanding of genome organization such as gene mapping and genetic rearrangements in each target organism, and provides phylogenetic information [31 - 34]. The original FISH method, however, is not always sufficient for microbial detection from environmental samples. In addition to the standard FISH, this section also presents several highly sensitive FISH methods. These include catalyzed reporter deposition (CARD)-FISH, two-pass tyramide signal amplification (TSA)-FISH and *in situ* DNA-hybridization chain reaction (HCR) FISH [35 - 37].

CONSENT FOR PUBLICATION

Not applicable.

CONFLICT OF INTEREST

The author confirms that this chapter contents have no conflict of interest.

ACKNOWLEDGEMENTS

Declare none.

REFFERENCES

[1]   Glöckner FO, Stal LJ, Sandaa R-A, *et al.* Marine Microbial Diversity and its role in Ecosystem Functioning and Environmental Change. Marine Bord Position Paper 17. In: Calewaert JB, McDonough N, Eds. Ostend, Belgium: Marine Board-ESF 2012; pp. 1-80.

[2]   Herbert RA. Nitrogen cycling in coastal marine ecosystems. FEMS Microbiol Rev 1999; 23(5): 563-90.
[http://dx.doi.org/10.1111/j.1574-6976.1999.tb00414.x] [PMID: 10525167]

[3]   Azam F, Malfatti F. Microbial structuring of marine ecosystems. Nat Rev Microbiol 2007; 5(10): 782-91.
[http://dx.doi.org/10.1038/nrmicro1747] [PMID: 17853906]

[4]   Das S, Mangwani N. Ocean acidification and marine microorganisms: responses and consequences. Oceanologia 2015; 57: 349-61.
[http://dx.doi.org/10.1016/j.oceano.2015.07.003]

[5]   Eashwar M, Chandrasekaran P, Subramanian G. Marine microbial films and the corrosion of steel.

Corrosion Sci Eng 1988; 4(2): 115-9.

[6]   Ramírez GA, Hoffman CL, Lee MD, *et al.* Assessing marine microbial induced corrosion at Santa Catalina Island, California. Front Microbiol 2016; 7: 1679.
[http://dx.doi.org/10.3389/fmicb.2016.01679] [PMID: 27826293]

[7]   Agrawal PK, Agrawal S, Shrivastava R. Modern molecular approaches for analyzing microbial diversity from mushroom compost ecosystem. Biotech 2015; 5(6): 853-66.
[http://dx.doi.org/10.1007/s13205-015-0289-2] [PMID: 28324393]

[8]   Tan B, Ng C, Nshimyimana JP, Loh LL, Gin KY, Thompson JR. Next-generation sequencing (NGS) for assessment of microbial water quality: Current progress, challenges, and future opportunities. Front Microbiol 2015; 6: 1027.
[http://dx.doi.org/10.3389/fmicb.2015.01027] [PMID: 26441948]

[9]   Suzuki Y, Hanagasaki N, Furukawa T, Yoshida T. Removal of bacteria from coastal seawater by foam separation using dispersed bubbles and surface-active substances. J Biosci Bioeng 2008; 105(4): 383-8.
[http://dx.doi.org/10.1263/jbb.105.383] [PMID: 18499055]

[10] Koyama S, Konishi MA, Ohta Y, *et al.* Attachment and detachment of living microorganisms using a potential-controlled electrode. Mar Biotechnol (NY) 2013; 15(4): 461-75.
[http://dx.doi.org/10.1007/s10126-013-9495-2] [PMID: 23420537]

[11] Koyama S, Nishi S, Tokuda M, *et al.* Electrical retrieval of living microorganisms from cryopreserved marine sponges using a potential-controlled electrode. Mar Biotechnol (NY) 2015; 17(5): 678-92.
[http://dx.doi.org/10.1007/s10126-015-9651-y] [PMID: 26242755]

[12] Michibata H, Iwata Y, Hirata J. Isolation of highly acidic and vanadium-containing blood cells from among several types of blood cell from ascidiidae species by density-gradient centrifugation. J Exp Zool 1991; 257: 306-13.
[http://dx.doi.org/10.1002/jez.1402570304]

[13] Michibata H, Terada T, Anada N, Yamakawa K, Numakunai T. The accumulation and distribution of vanadium, iron, and manganese in some solitary ascidians. Biol Bull 1986; 171(3): 672-81.
[http://dx.doi.org/10.2307/1541632] [PMID: 29314894]

[14] Miyauchi A, Okabe TH. Production of metallic vanadium by preform reduction process. Mater Trans 2010; 51(6): 1102-8.
[http://dx.doi.org/10.2320/matertrans.M2010027]

[15] Ueki T, Fujie M, Romaidi , Satoh N. Symbiotic bacteria associated with ascidian vanadium accumulation identified by 16S rRNA amplicon sequencing. Mar Genomics 2019; 43: 33-42.
[http://dx.doi.org/10.1016/j.margen.2018.10.006] [PMID: 30420273]

[16] Eashwar M, Chandrasekaran P, Subramanian G. Marine microbial films and the corrosion of steel. B Electrochem 1988; 4(2): 115-9.

[17] Gerashchenko B I, Takahashi T, Kosaka T, Hosoya H. Life cycle analysis of unicellular algae. Curr Protoc Cytom 2010; 11: 11.19.1-6.
[http://dx.doi.org/10.1002/0471142956.cy1119s52]

[18] Yamaguchi T, Kawakami S, Hatamoto M, *et al. In situ* DNA-hybridization chain reaction (HCR): A facilitated *in situ* HCR system for the detection of environmental microorganisms. Environ Microbiol 2015; 17(7): 2532-41.
[http://dx.doi.org/10.1111/1462-2920.12745] [PMID: 25523128]

[19] Shen Z, Huang M, Xiao C, Zhang Y, Zeng X, Wang P-G. Non-labeled qcm biosensor for bacteria detection using carbohydrate and lectin recognitions. Anal Chem 2007; 79(6): 2312-9.
[http://dx.doi.org/10.1021/ac061986j] [PMID: 17295446]

[20] Costerton JW, Lewandowski Z, Caldwell DE, Korber DR, Lappin-Scott HM. Microbial biofilms. Annu Rev Microbiol 1995; 49: 711-45.

[http://dx.doi.org/10.1146/annurev.mi.49.100195.003431] [PMID: 8561477]

[21]   de Queiroz GA, Andrade JS, Malta TBS, Vinhas G, de Andrade Lima MAG. Biofilm formation and corrosion on carbon steel api 5lx60 in clayey soil. Mater Res 2018; 21(3)e20170338
[http://dx.doi.org/10.1590/1980-5373-mr-2017-0338]

[22]   Firtel M, Beveridge TJ. Scanning probe microscopy in microbiology. Micron 1995; 26(4): 347-62.
[http://dx.doi.org/10.1016/0968-4328(95)00012-7] [PMID: 8574524]

[23]   Baer DR, Gaspar DJ, Nachimuthu P, Techane SD, Castner DG. Application of surface chemical analysis tools for characterization of nanoparticles. Anal Bioanal Chem 2010; 396(3): 983-1002.
[http://dx.doi.org/10.1007/s00216-009-3360-1] [PMID: 20052578]

[24]   Becton, Dickinson and Company. Introduction to Flow Cytometry: A Learning Guide. BD Biosciences 2002; 1-52.

[25]   Adan A, Alizada G, Kiraz Y, Baran Y, Nalbant A. Flow cytometry: basic principles and applications. Crit Rev Biotechnol 2017; 37(2): 163-76.
[http://dx.doi.org/10.3109/07388551.2015.1128876] [PMID: 26767547]

[26]   Takahashi T. Quality assessment of microalgae exposed to trace metals using flow cytometry. Superfood and functional food - development of superfood and its role in medicine (Eds Shiomi, N, Waisundara, V Y) InTechOpen. 2017; pp. 29-45.
[http://dx.doi.org/10.5772/65516]

[27]   Olson RJ, Frankel SL, Chisholm SW, Shapiro HM. An inexpensive flow cytometer for the analysis of fluorescence signals in phytoplankton: Chlorophyll and DNA distributions. J Exp Mar Biol Ecol 1983; 68: 129-44.
[http://dx.doi.org/10.1016/0022-0981(83)90155-7]

[28]   Stauber JL, Franklin NM, Adams MS. Applications of flow cytometry to ecotoxicity testing using microalgae. Trends Biotechnol 2002; 20(4): 141-3.
[http://dx.doi.org/10.1016/S0167-7799(01)01924-2] [PMID: 11906740]

[29]   Gerashchenko BI, Kosaka T, Hosoya H. Growth kinetics of algal populations exsymbiotic from *Paramecium bursaria* by flow cytometry measurements. Cytometry 2001; 44(3): 257-63.
[http://dx.doi.org/10.1002/1097-0320(20010701)44:3<257::AID-CYTO1118>3.0.CO;2-V] [PMID: 11429776]

[30]   Gerashchenko BI, Gerashchenko II, Kosaka T, Hosoya H. Stimulatory effect of aerosil on algal growth. Can J Microbiol 2002; 48(2): 170-5.
[http://dx.doi.org/10.1139/w01-143] [PMID: 11958570]

[31]   Garimberti E, Tosi S. Fluorescence *in situ* hybridization (FISH), basic principles and methodology. Methods Mol Biol 2010; 659: 3-20.
[http://dx.doi.org/10.1007/978-1-60761-789-1_1] [PMID: 20809300]

[32]   Ratan ZA, Zaman SB, Mehta V, Haidere MF, Runa NJ, Akter N. Application of fluorescence *In situ* hybridization (fish) technique for the detection of genetic aberration in medical science. Cureus 2017; 9(6)e1325
[http://dx.doi.org/10.7759/cureus.1325] [PMID: 28690958]

[33]   Hoshino T, Yilmaz LS, Noguera DR, Daims H, Wagner M. Quantification of target molecules needed to detect microorganisms by fluorescence *in situ* hybridization (FISH) and catalyzed reporter deposition-FISH. Appl Environ Microbiol 2008; 74(16): 5068-77.
[http://dx.doi.org/10.1128/AEM.00208-08] [PMID: 18552182]

[34]   Pernthaler A, Pernthaler J. Fluorescence *in situ* hybridization for the identification of environmental microbes. Methods Mol Biol 2007; 353: 153-64.
[http://dx.doi.org/10.1385/1-59745-229-7:153] [PMID: 17332640]

[35]   Ferrari BC, Tujula N, Stoner K, Kjelleberg S. Catalyzed reporter deposition-fluorescence *in situ* hybridization allows for enrichment-independent detection of microcolony-forming soil bacteria. Appl

Environ Microbiol 2006; 72(1): 918-22.
[http://dx.doi.org/10.1128/AEM.72.1.918-922.2006] [PMID: 16391135]

[36] Kawakami S, Kubota K, Imachi H, Yamaguchi T, Harada H, Ohashi A. Detection of single copy genes by two-pass tyramide signal amplification fluorescence *in situ* hybridization (Two-Pass TSA-FISH) with single oligonucleotide probes. Microbes Environ 2010; 25(1): 15-21.
[http://dx.doi.org/10.1264/jsme2.ME09180] [PMID: 21576847]

[37] Yamaguchi T, Fuchs BM, Amann R, *et al.* Rapid and sensitive identification of marine bacteria by an improved in situ DNA hybridization chain reaction (quickHCR-FISH). Syst Appl Microbiol 2015; 38(6): 400-5.
[http://dx.doi.org/10.1016/j.syapm.2015.06.007] [PMID: 26215142]

CHAPTER 9

A Membrane-Free Alternative Method for Concentration of Live Microorganisms by Foam Separation

Yoshihiro Suzuki*

Department of Civil and Environmental Engineering, University of Miyazaki, Miyazaki, Japan

Abstract: The collection and isolation of microorganisms from seawater in aquaculture systems, aquariums, fishing port facilities, and recreation areas is the most fundamental means of studying microbiological, physiological, and pathogenic properties of water to improve public health and the environment. However, there are few methods for concentration of living microorganisms. Thus, it is necessary to develop a technology for collection and concentration of microorganisms from seawater. In this study, we examined the removal and concentration of the bacterium *Vibrio* and fungus *Fusarium* from seawater by foam separation using dispersed bubbles and surfactants. After batch processing with only 1 mg/L milk casein added as a surfactant and after injection of bubbles, *Vibrio* and *Fusarium* were isolated at removal efficiency rates of more than 80% and 99.9%, respectively, and most of the microbial cells were concentrated alive in the foam water within 5 min. When the continuous foam separation unit was installed at the actual site of a fishery harbor, though the removal efficiency for viable bacteria was 49.2%, the bacteria were isolated at a huge concentration in the foam water, and the concentration factor was 18.5. Foam separation is a feasible convenient technology for not only seawater purification but also membrane-free concentration of live microorganisms.

Keywords: Concentration, Dispersed bubbles, *Fusarium*, Live state, Milk casein, Removal, Seawater, *Vibrio*.

INTRODUCTION

The coastal area is filled up by various objects such as industrial plants, fisheries, aquaculture enterprises, marine sports facilities, and recreational organizations. In the near future, the types of coastal-area use will become more and more diverse. However, sewage and drainage generated by human activities are sometimes

---

* **Corresponding author Yoshihiro Suzuki:** Department of Civil and Environmental Engineering, University of Miyazaki, Miyazaki, Japan; Tel: +81-985-58-7339; Fax: +81-985-58-7344; E-mail: ysuzuki@cc.miyazaki-u.ac.jp

Toshiyuki Takahashi (Ed.)

untreated and are released eventually into the coastal area *via* rivers or estuaries. Therefore, the most important problem for management of coast utilization is protection of human health and the pathogen risks associated with economic activities on the coast, in particular with some materials and food resources. There are various microorganisms—a mixture of useful, harmless, harmful, and pathogenic species—in seawater. It is necessary for the management of public and environmental health to compile and maintain a database and information about the microorganisms including the pathogens in seawater for safe use of coastal areas. However, most of the information on the microorganisms in seawater is limited to viruses and bacteria in relation to the diseases of cultured fish and aquarium animals. In comparison with inland water, there is still a remarkable lack of data on clinically important microorganisms in seawater. This is because the unspecified pathogenic and nonpathogenic bacteria are mixed at very low concentrations, making it difficult to study them. To obtain accurate and comprehensive microbiological information about the coastal area, it is necessary to collect bacteria from water samples. Although there are established methods for collecting bacteria from water samples using membrane filters [1] and membrane modules [2], clogging is inevitable with filtration because coastal seawater contains various suspended particles (particulate matter). Such clogging makes it difficult to filter large volumes of turbid water. As a possible means of concentrating bacteria without filtration, chemical flocculation [3] and skimmed-milk flocculation [4] have been developed for real-time PCR analysis with DNA extraction from such sources with good recovery rates. Nevertheless, to date, nobody has developed a technique for collecting live bacteria from water so that a high cell concentration is obtained in a short period.

When one turns an eye to the sea, "sea foam" containing various concentrated bacteria and undefined suspended matter is frequently observed along the water's edge in coastal zones [5]. Surface-active substances (surfactants) among biological compounds, such as polysaccharides [6] and proteins [7 - 9] appear to play an important role in the concentration of suspended solids in the foam. These surfactants not only generate foam on the water surface but also change the interface of solids from hydrophilic to hydrophobic. Materials with a hydrophobic interface adsorb onto the gas–liquid interface of bubbles. By applying the principle of concentrating suspended solids in sea foam, we have previously developed a foam separation method by means of dispersed bubbles and surfactants [10]. The primary advantage of this method is the use of milk casein or natural surfactants as chemical agents for water processing. Therefore, foam separation *via* dispersed bubbles and a surface-active protein has a high potential as a way to collect and concentrate microorganisms from water into the generated foam (Fig. **1**), because it is relatively easy to culture and isolate the microorganisms from the foam.

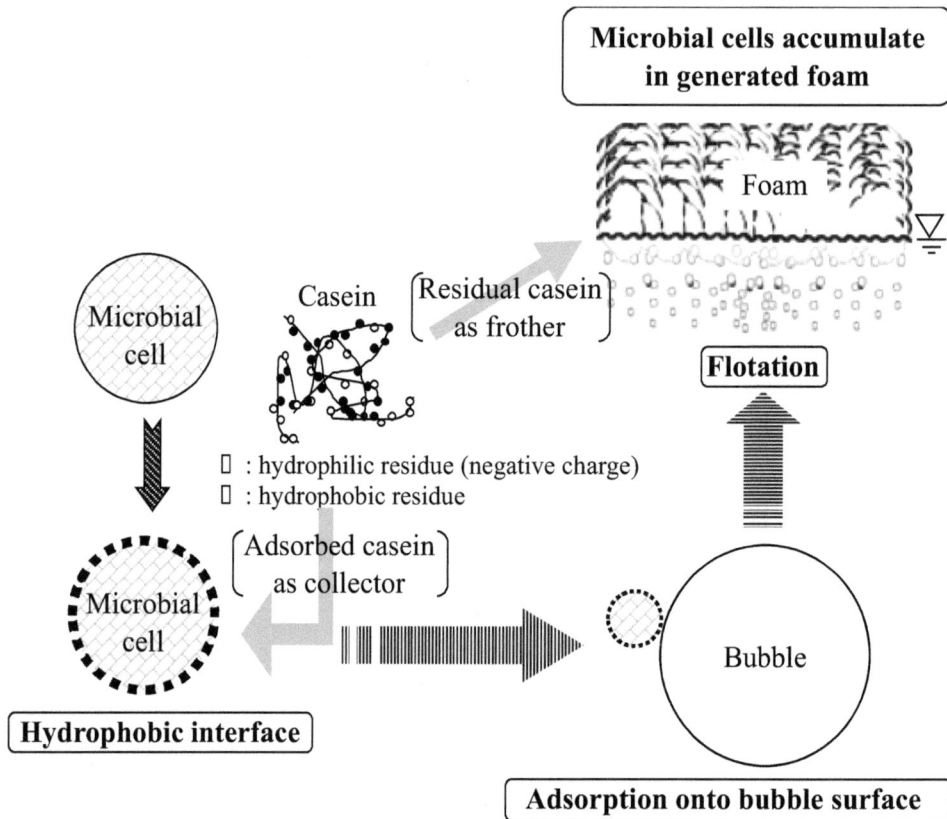

Fig. (1). A conceptual schematic of the mechanism of microbial concentration *via* foam separation.

In the present study, we examined foam-based concentration of *Vibrio* and *Fusarium* as model microorganisms of bacteria and fungi, respectively. We employed dispersed bubbles and milk casein as a surface-active protein for foam separation to test the method for concentration of live microorganisms from microbe-spiked seawater. In addition, this method was applied to actual coastal seawater by means of a continuous foam separation unit.

MATERIALS AND METHODS

Preparation of *Vibrio*-Spiked Seawater

Vibrio is one of the major indigenous bacteria existing in seawater, and pathogens such as *Vibrio cholerae* and *V. parahaemolyticus* are members of this genus. In this study, *V. nigripulchritudo*, which is considered one of the major pathogens threatening shrimp aquaculture, served as a model *Vibrio* species.

The *V. nigripulchritudo* strain used in this study was kindly provided by Kyushu Medical Co., Fukuoka, Japan. The bacterium was grown in 3 mL of Marine Broth 2216E (Becton, Dickinson and Company, MD, USA) at 25°C with continuous shaking for 16–18 h. The broth was then centrifuged at 3,000 rpm (H-103N Series, Kokusan Co., Tokyo, Japan) for 5 min and was washed three times with centrifugation at 3,000 rpm for 10 min by means of sterilized artificial seawater (NaHCO$_3$ 96 mg, CaSO$_4$·2H$_2$O 60 mg, MgSO$_4$ 60 mg, and KCl 4.0 mg in 1,000 mL of distilled deionized water; pH 7.5; autoclaved at 121°C for 15 min). Next, the supernatant was removed, and the collected *V. nigripulchritudo* cells were resuspended in 1 mL of artificial seawater. Sterilized artificial seawater (3000 mL) was inoculated with 1 mL of the *V. nigripulchritudo* suspension, whereupon the final concentration of *V. nigripulchritudo* was 10^6 colony-forming units (CFU)/mL in the *Vibrio*-spiked seawater.

Preparation of *Fusarium*-Spiked Seawater

Fusarium is the most important fungal genus widespread in the soil, human living environments, and water environments such as river water and seawater. *Fusarium* is a pathogen of humans, plants, fish, prawns, and insects, *i.e.*, causes many diseases. In this study, *Fusarium solani* which is considered one of the major pathogens of humans, plants, and prawns was tested as a model fungus.

The *F. solani* stock suspension employed in this study was kindly provided by Kyushu Medical Co. Sterilized artificial seawater (3000 mL) was inoculated with 3 mL of the *F. solani* stock suspension, whereupon the final concentration of *F. solani* was 10^3 CFU/mL in the *Fusarium*-spiked seawater.

Sampling of Actual Seawater

Actual seawater was collected from the shrimp farm pond in Miyazaki, Japan. A sterilized bucket was used to sample the surface water (to a depth of 0-0.3 cm). The collected water sample was stored in a sterile 1 L polyethylene bottle and transported immediately to the laboratory for the foam separation test. After sampling, foam separation was started within 2 h.

Foam Separation *via* Batch Equipment

In our previous study involving batch equipment [10], suspended particulate matter was removed by a 90% removal rate from polluted seawater by foam separation at 1 mg/L casein. The efficiency of removal of the suspended particulate matter did not improve even with a further increase in the casein concentration. Therefore, the casein level was fixed at 1 mg/L. Thus, the appropriate amount of milk casein (1 mg/L) as a surfactant was added to each seawater sample (180 mL). The stock solution of milk casein (Reagent grade; Wako Chemical, Osaka, Japan) was prepared in 0.01 M NaOH at a final concentration of 10,000 mg/L. Each seawater sample was rapidly stirred (150 rpm) using a jar tester for 1 min. *via* the transfer of a sample to a cylindrical column (height, 54 cm; diameter, 2.6 cm) of the batch flotation apparatus, foam separation was carried out. Dispersed air was supplied from the bottom of the column with a glass-ball filter (pore size 5-10 μm; Kinoshita Rika, Tokyo, Japan). The foam generated on the water surface was drawn into a trap bottle by a vacuum pump. The recovered foam was defoamed and designated as foam water. The processing time for foam separation was 3 min; the air supply flow rate was 0.3 L/min. The average volume of generated foam water was 20 mL. The treated water was sampled from the drain. The same operation was repeated three times. Next, raw seawater, treated seawater, and foam water were analyzed for *Vibrio* and *Fusarium*.

Enumeration of *Vibrio* and *Fusarium* cells

Marine Agar-2216 plates (Becton, Dickinson and Company) were employed to count *V. nigripulchritudo* cells. To analyze *Vibrio*-spiked seawater, treated seawater and foam water, 0.1 mL each (or 0.1 mL of a sample diluted with sterilized physiological saline), were applied to agar plates and incubated for 24 h at 25°C. The counts of *V. nigripulchritudo* were determined by the pour plate method.

To count *Vibrio* cells in actual seawater, TCBS Agar plates (Becton, Dickinson and Company) were chosen. To analyze the seawater, treated seawater, and foam water, 0.1 mL samples (or 0.1 mL of a sample diluted with sterilized physiological saline) were applied to agar plates and incubated for 24 h at 25°C. The resultant colonies were counted as the genus *Vibrio*.

Potato dextrose agar (containing 0.01% of chloramphenicol and 1% of NaCl) plates (Becton, Dickinson and Company) were used to count *F. solani*. To analyze the *Fusarium*-spiked seawater, treated seawater, and foam water, 0.1 mL

samples (or 0.1 mL of a sample diluted with sterilized physiological saline containing 0.05% of Tween 80) were applied to agar plates and incubated for 1 week at 27°C. The formed colonies were counted as *F. solani*.

A Continuous-Flow Experiment with the Foam Separation Unit [11]

The foam separation unit (Fig. **2**) was equipped with an inhalation-type aerator (KA type, Puresuka Co., Tokyo, Japan) [9]. By the rotating impeller (1700 rpm), negative pressure is generated at the back of the impeller, and air is drawn from the water in the shaft tube connected to the outside environment. Air is sheared with a blade immersed in the water, and numerous bubbles are extensively dispersed in the water. The air bubbles were vigorously mixed with the water to enable foam separation by this unit for the samples containing small amounts of surfactants such as coastal seawater with and without added casein as a frother. Surfactants in water adsorb onto bubbles, and the bubbles are carried to the water surface. Then, foam is generated on the water surface. The resulting foam is continuously removed spontaneously *via* a foam duct placed in the upper part of the cell equipped with an air exhaust. In this study, the foam separation unit was installed in a fishery harbor. Harbor seawater was introduced continuously with the addition of casein (1 mg/L) into the foam separation unit (volume, 14 L). The air supply rate of the aerator was 24 L/min. The water flow rate was adjusted to 2.4 L/min (air/water volume ratio 1:10). The suspension was kept in seawater by continuous stirring with a paddle. After adjustment of the water flow rate, samples of raw water in the tank, treated water, and foam water were collected at 5 min intervals for 15 min. The samples of raw water, treated water, and foam water were analyzed for the counts of viable bacteria. The diluted samples (0.1 mL) were plated by the spread plate method in duplicate on Marine Agar-2216 plates and incubated at 37°C for 24 h. Colonies of all sizes were counted as viable bacteria. Turbidity was determined with a turbidity meter (SEP-PT-706D; Mitsubishi Kagaku, Tokyo, Japan).

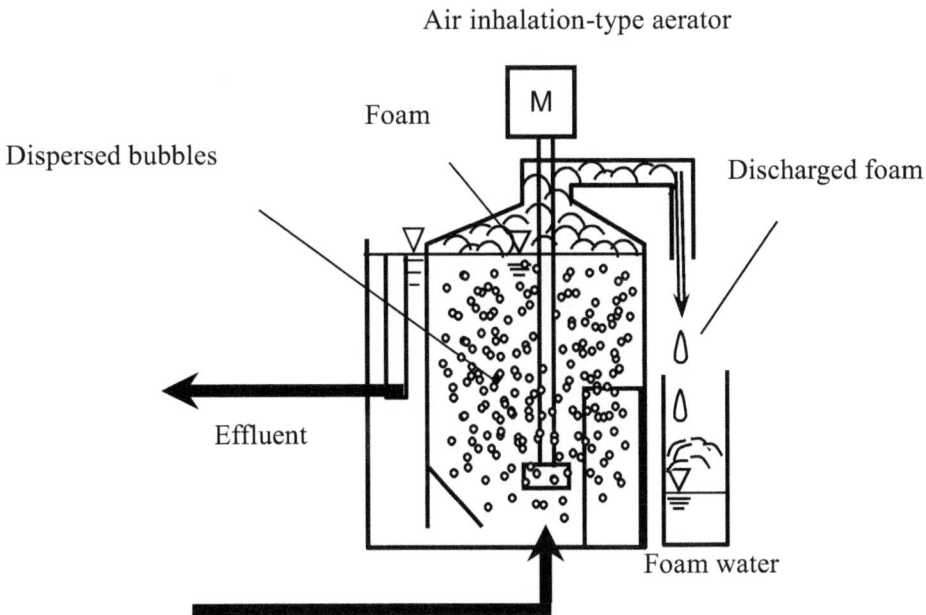

Fig. (2). A schematic diagram (not to scale) of the foam separation unit for a continuous-flow system.

RESULTS AND DISCUSSION

Removal and Concentration of *Vibrio*

The counts of *V. nigripulchritudo* cells in the *Vibrio*-spiked seawater, treated seawater, and foam water are shown in Fig. (**3**). The cells of *V. nigripulchritudo* were removed from *Vibrio*-spiked seawater by foam separation. The removal efficiency was 84.4% ± 1.3% (n = 3). In contrast, *V. nigripulchritudo* was detected in the foam water at a huge concentration. According to the mass balance of the treated water and foam water in terms of *V. nigripulchritudo* counts based on the count in *Vibrio*-spiked seawater, 55.9% of active *V. nigripulchritudo* cells were recovered in the foam water (Fig. **4**). The added *V. nigripulchritudo* cells were removed from the raw seawater and concentrated effectively in the foam water as live cells.

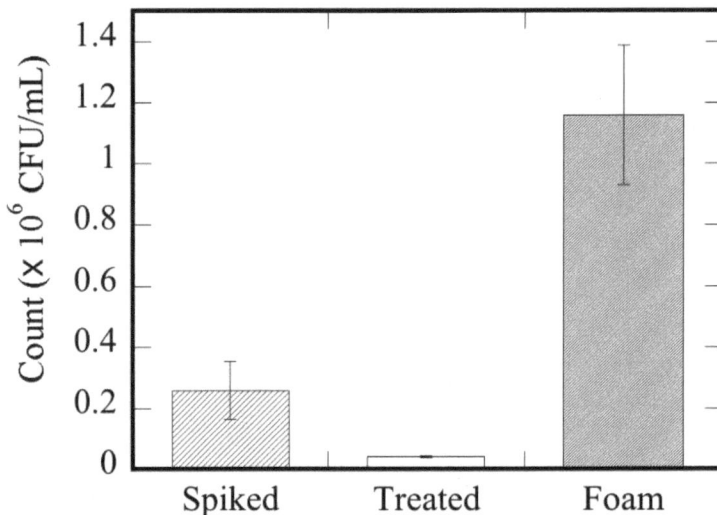

Fig. (3). Counts of *V. nigripulchritudo* in the *Vibrio*-spiked seawater, treated seawater, and foam water after foam separation.

For application to the actual seawater collected from the shrimp farm pond, the removal efficiency of *Vibrio* counts (*Vibrio* genus, not *V. nigripulchritudo*) was 84.6% ± 2.7% (n = 3). According to the mass balance between *Vibrio*-spiked seawater and the foam water, the recovery of *Vibrio* cells in the foam water reached 66.3%. Thus, foam separation is applicable to actual seawater for the removal and concentration of live *Vibrio* cells.

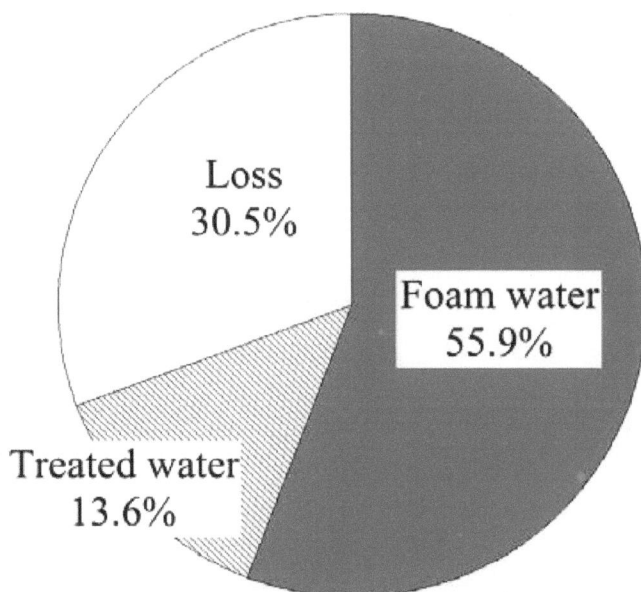

Fig. (4). Mass balance of the treated water and the foam water in terms of *V. nigripulchritudo* counts based on the count in *Vibrio*-spiked seawater.

Removal and Concentration of *Fusarium*

The counts of *F. solani* in the *Fusarium*-spiked seawater, treated seawater, and foam water are presented in Fig. (**5**). The concentration of *F. solani* was 5 CFU/mL in the threated seawater. The efficiency of removal of *F. solani* reached 99.9%. The mean counts of *F. solani* were 5.8×10^4 CFU/mL in the foam water. *Fusarium* was concentrated in the foam water effectively. According to the mass balance between *Fusarium*-spiked seawater and the foam water, the recovery percentage of *Fusarium* cells was more than 100% (135% ± 19%, n = 3) in the foam water. It was assumed that the surplus in the recovery percentage was due to systematic error (dilution and counting imperfections). The foam separation was found to be extremely effective at removing and concentrating live *Fusarium* cells.

Bacterial Removal and Concentration by the Continuous Foam Separation Unit

The counts of viable bacteria and turbidity in harbor seawater, treated seawater, and foam water determined using the continuous foam separation unit are given in Table **1** [11]. The viable bacteria were removed from the harbor seawater and were concentrated in the foam water. The removal efficiency was 49.2%. The bacteria were present at a huge concentration in the foam water, and the concentration factor was 18.5. In addition, the turbidity removal exceeded 80%. Because casein functions as a collector of suspended solids except for bacteria, continuous foam separation by means of casein can be applied to the removal of both microorganisms and suspended particulate matter from seawater.

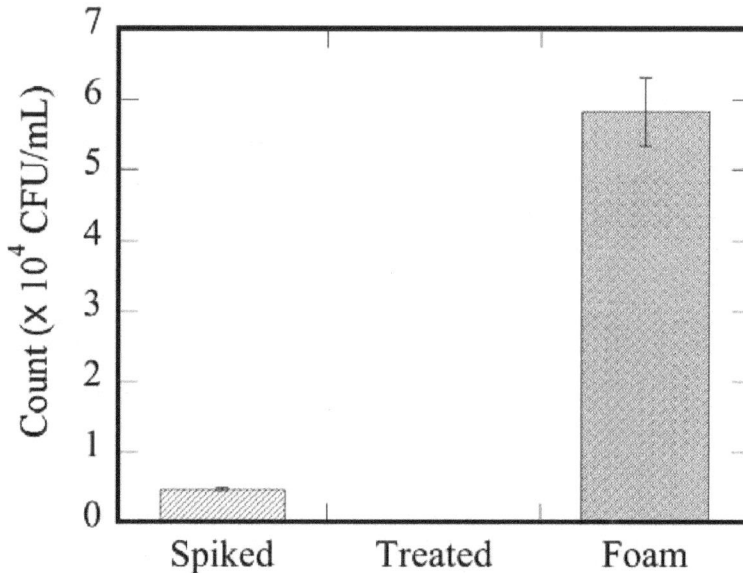

Fig. (5). Counts of *F. solani* in the *Vibrio*-spiked seawater, treated seawater, and foam water after foam separation.

Table 1. Microbial and turbidity analysis for coastal seawater, treated water and foam water in the experiment using a continuous foam separation unit.

Samples	Viable Bacteria	Turbidity
	(CFU/mL)	(Kaolin Unit)
Seawater	335	2.25
(Lower-upper limits)*	(298-372)	
Treated water	170	0.31
(Lower-upper limits)	(140-196)	
Foam water	6200	95.92
(Lower-upper limits)	(4700-7700)	

* Confidence interval (95%) of bacterial count.

CONCLUSION

We developed a membrane-free alternative method for concentration of live microorganisms from seawater by foam separation using casein. The advantages of foam separation are as follows: applicability to the rapid and facile removal and concentration of bacteria and fungi in a live state. The foam separation can be performed for collection of microbes from a large volume of seawater. It should be useful for the studies on microbiological, physiological, and pathogenic properties of seawater because this approach facilitates isolation of microorganisms from seawater.

CONSENT FOR PUBLICATION

Not applicable.

CONFLICT OF INTEREST

The author confirms that this chapter contents have no conflict of interest.

ACKNOWLEDGEMENTS

Declare none.

REFERENCES

[1]   Staley C, Gould TJ, Wang P, Phillips J, Cotner JB, Sadowsky MJ. Evaluation of water sampling methodologies for amplicon-based characterization of bacterial community structure. J Microbiol Methods 2015; 114: 43-50.
       [http://dx.doi.org/10.1016/j.mimet.2015.05.003] [PMID: 25956022]

[2]   Lindquist HD, Harris S, Lucas S, *et al.* Using ultrafiltration to concentrate and detect *Bacillus anthracis, Bacillus atrophaeus* subspecies *globigii*, and *Cryptosporidium parvum* in 100-liter water samples. J Microbiol Methods 2007; 70(3): 484-92.
       [http://dx.doi.org/10.1016/j.mimet.2007.06.007] [PMID: 17669525]

[3]   Zhang Y, Riley LK, Lin M, Hu Z. Determination of low-density *Escherichia coli* and *Helicobacter pylori* suspensions in water. Water Res 2012; 46(7): 2140-8.
       [http://dx.doi.org/10.1016/j.watres.2012.01.030] [PMID: 22342315]

[4]   Gonzales-Gustavson E, Cárdenas-Youngs Y, Calvo M, *et al.* Characterization of the efficiency and uncertainty of skimmed milk flocculation for the simultaneous concentration and quantification of water-borne viruses, bacteria and protozoa. J Microbiol Methods 2017; 134: 46-53.
       [http://dx.doi.org/10.1016/j.mimet.2017.01.006] [PMID: 28093213]

[5]   Southward AJ. Sea foam. Nature 1953; 172: 1059-60.
       [http://dx.doi.org/10.1038/1721059b0]

[6]   Zhou J, Mopper K. The role of surface-active carbohydrates in formation of transparent exopolymer particles by bubble adsorption of seawater. Limnol Oceanogr 1998; 43: 1860-71.
       [http://dx.doi.org/10.4319/lo.1998.43.8.1860]

[7]   Maruyama T, Okuzumi M, Saheki A, Shimamura S. The purification effect of the foam separating system in living fish transportation and preservation. Nippon Suisan Gakkaishi 1991; 57: 219-25.
       [http://dx.doi.org/10.2331/suisan.57.219]

[8]   Chen S, Timmons MB, Bisogni JJ, Aneshansley DJ. Suspended-solids removal by foam fractionation. Prog Fish-Cult 1993; 55: 69-75.
[http://dx.doi.org/10.1577/1548-8640(1993)055<0069:SSRBFF>2.3.CO;2]

[9]   Suzuki Y, Maruyama T, Numata H, Sato H, Asakawa M. Performance of a closed recirculating system with foam separation, nitrification and denitrification units for the intensive culture of eel: Towards zero emission. Aquacult Eng 2003; 29: 165-82.
[http://dx.doi.org/10.1016/j.aquaeng.2003.08.001]

[10]   Suzuki Y, Maruyama T. Removal of suspended solids by coagulation and foam separation using surface-active protein. Water Res 2002; 36(9): 2195-204.
[http://dx.doi.org/10.1016/S0043-1354(01)00439-0] [PMID: 12108712]

[11]   Suzuki Y, Hanagasaki N, Furukawa T, Yoshida T. Removal of bacteria from coastal seawater by foam separation using dispersed bubbles and surface-active substances. J Biosci Bioeng 2008; 105(4): 383-8.
[http://dx.doi.org/10.1263/jbb.105.383] [PMID: 18499055]

A Novel Method Using a Potential-Controlled Electrode for Screening Difficult-to-Cultivate Microorganisms

Sumihiro Koyama[*]

ABLE Co., Ltd., 7-9 Nishigoken-cho, Shinjyuku, Tokyo162-0812, Japan

Abstract: We developed an electrical retrieval method for living environmental microorganisms using a weak electrical potential applied to an optically transparent electrode. Living microorganisms resuspended in non-nutritive solutions such as calcium- and magnesium-free phosphate-buffered saline [PBS(-)] and artificial seawater were attracted by and selectively attached to an indium tin oxide (ITO) electrode surface to which a negative potential between -0.2 and -0.4 V *vs.* Ag/AgCl was applied. The electrically attached microorganisms could be cultured on the agar medium after detachment from the ITO electrode by application of a ± 20 mV *vs.* Ag/AgCl, 9 MHz sine wave potential. When a ± 0.2 V *vs.* Ag/AgCl, 12 MHz sine wave potential was applied to the attached microorganisms on the electrode with an applied negative potential, actinomycetes were selectively cultured and were expanded 60- to 4000-fold in terms of colony-forming units (CFUs) as compared with application of a ± 20 mV *vs.* Ag/AgCl, 9 MHz sine wave potential. Using the electrical retrieval method, we identified and isolated *Nocardiopsis* sp., *Dietzia* sp., *Pseudonocardia* sp., *Brachybacterium* sp., *Nesterenkonia* sp., *Microcella* sp., *Microbacterium* sp., and *Streptomyces* sp. from deep-sea sediment core samples in Suruga Bay, Japan. Phylogenetic analyses based on 16S ribosomal RNA gene sequences indicated that 59% of the obtained strains are novel species or are highly likely to be novel species. The electrical retrieval method for living microorganisms holds promise as a novel screening strategy for hardly cultivable microorganisms.

Keywords: Actinomycetes, Deep-sea microorganisms, Electrical retrieval, Marine sponge, Potential-controlled electrode.

INTRODUCTION

Microorganisms are rich sources of novel therapeutic compounds such as antibiotics [1], anticancer agents [2], and immunosuppressants [3] as well as a wide range of biotechnologically valuable products [4, 5]. However, the majority

[*] **Corresponding author Sumihiro Koyama:** ABLE Co., Ltd., 7-9 Nishigoken-cho, Shinjyuku, Tokyo 162-0812, Japan; Tel: +81-3-3260-0485; Fax; +81-3-3260-0513; E-mail: koyama@able-biott.co.jp

Toshiyuki Takahashi (Ed.)

of microorganisms cannot be cultured under conventional laboratory conditions [6 - 11].

On the one hand, metagenomics has been developed over the past 2 decades to clarify the previously unknown diversity of microorganisms; on the other hand, it has been driven by the increasing biotechnological demand for novel enzymes and biomolecules [6, 12]. Analysis of DNA directly extracted from a soil sediment and biological samples has paved the way to studies on natural microbial communities without the need for cultivation. However, high DNA yields and purity are difficult to achieve due to the coextraction of impurities, such as humic substances, polysaccharides, polyphenols, and lipids that inhibit downstream applications, such as polymerase chain reaction, restriction enzyme digestion, and DNA ligation [13 - 16]. In indirect DNA extraction methods, microorganisms are separated from the samples prior to cell lysis. Higher-molecular-weight and purer environmental DNA is obtained successfully by indirect cell lysis in contrast to direct protocols [13, 17, 18]. However, it has been reported that the DNA obtained is usually derived from only ~25% to 35% of all the microbial species present in the samples [6].

We hypothesized that an electrode with an applied negative potential acts as an energy source for a broad range of microorganisms. Electron donors are often called energy sources because energy is released when they are oxidized. There-fore, an electrode with an applied negative potential *per se* may act as an electron donor that is available for a variety of microorganisms. We tested the working hypothesis and found that a weak negative electric potential attracted living microorganisms to the electrode surface and that they were separated from soil sediments [19] and from cryopreserved marine sponges [20]. Living micro-organisms resuspended in a non-nutritive solution such as calcium- and magn-esium-free phosphate-buffered saline [PBS(-)] and artificial seawater (ASW) are selectively attached to a negative-potential-carrying optically transparent elect-rode against gravitational force [19, 20].

In this chapter, the author reviewed the methods for electrical retrieval of living environmental microorganisms.

Electrical Retrieval of Living Microorganisms using a Potential-Controlled Optically Transparent Electrode

Fig. (**1A**) shows a photograph and schematic illustration of a three-electrode cul-ture system. Working electrodes were constructed by vacuum evaporation of indium tin oxide (ITO; <10 Ω/sq) onto silica glass plates. Electrical potentials were applied to the ITO/glass electrode using an Ag/AgCl reference and a Pt co-unter electrode. To avoid electrochemical reactions, the application of the electrical potential to the working electrode was carried out in the three-electrode system (Fig. **1A**).

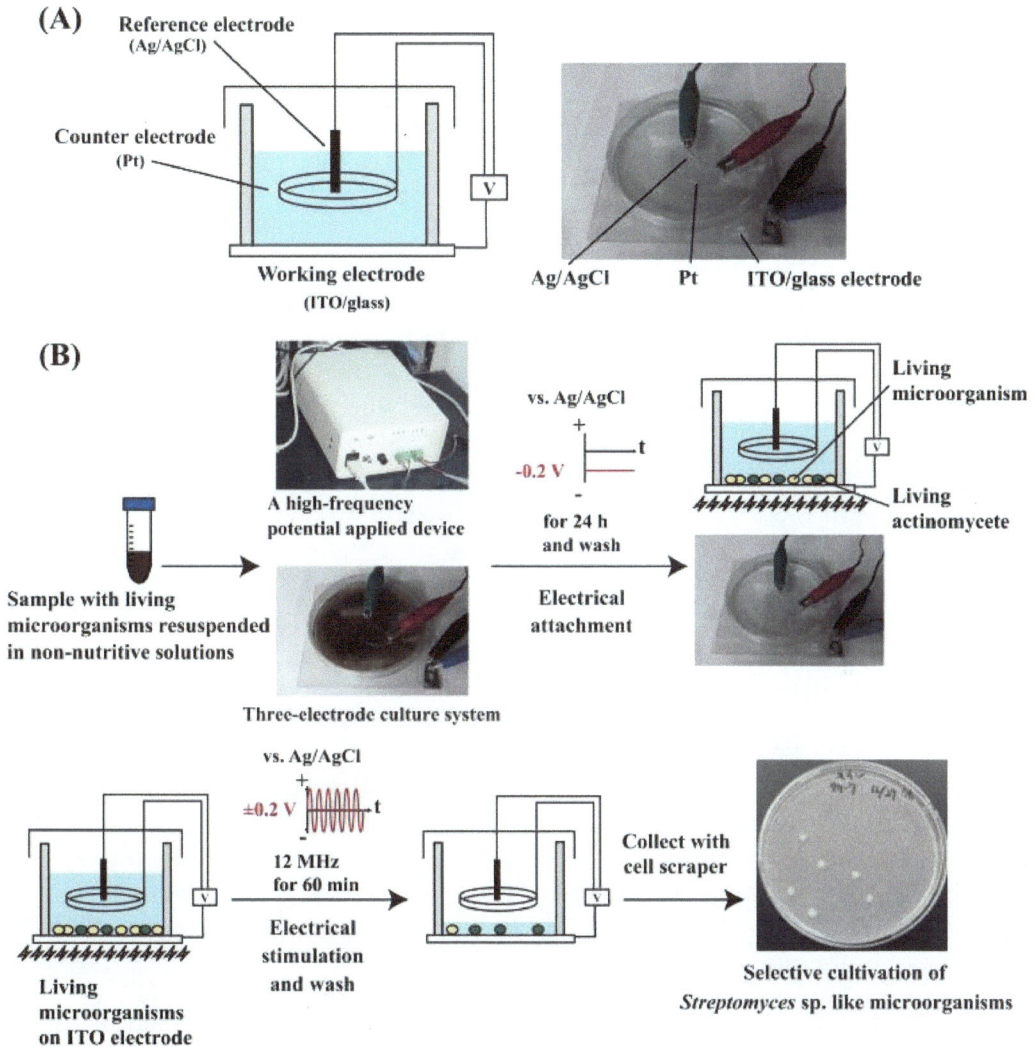

Fig. (1). A screening method for actinomycetes by means of a potential-controlled electrode. (**A**) Schematic illustration and a photograph of the three-electrode culture system. (**B**) Schematic illustration of actinomycetes screening. A high-frequency potential–applying device and the three-electrode culture system were manufactured by ABLE Co., Ltd.

We tested whether microorganisms in soil were attracted by and attached to the ITO electrode with an applied potential against gravitational force [19]. A 10 µg/ml grapery soil suspension in PBS(-) at room temperature (RT) was poured into a three-electrode chamber [19]. The ITO electrode was attached to top of a silicon rubber box on a glass slide. The fabricated silicon rubber box was housed in a sterile square plastic dish. A section (12 mm diameter) of both the Pt ring counterelectrode and Ag/AgCl reference electrode was placed on the plastic lid of the square plastic dish [19]. To analyze respiratory activity of the microorganisms attached to the ITO electrode against gravitational force, we used cyano-ditolyl-tetrazolium chloride (CTC). CTC, a monotetrazolium redox dye that produces red fluorescent formazan when chemically or biologically reduced as in the presence of a dehydrogenase activity, served as an indicator of respiration [21, 22]. A constant potential between +0.6 and −0.6 V *vs.* Ag/AgCl was applied to the ITO electrode in PBS(-) for 24 h at RT [19]. After application of the potential, living soil microorganisms on the electrode were stained with CTC to determine respiratory activity, and red fluorescent formazan production was observed. Living soil microorganisms were attracted by and selectively attached to the ITO electrode surface to which a negative potential between −0.2 and −0.4 V *vs.* Ag/AgCl was applied [19]. Application of a −0.4 V *vs.* Ag/AgCl constant potential induced the maximum attachment of living soil microorganisms to the ITO electrode region [19]. Although most microorganisms have a negative zeta potential at neutral pH in solution, few or no microorganisms selectively attached to the ITO electrode region to which a positive potential was applied [19]. Living microorganisms resuspended in Luria–Bertani medium and glucose solution were not attracted to the electrode [19]. Dead microorganisms were not attracted to the electrode either [19, 20].

After application of a negative potential in the non-nutritive solutions for 24 h, almost none of the soil microorganisms attached to the electrode surface could be detached by scraping several times with a rubber cell scraper [19]. We found that the microorganisms got detached from the electrode after we applied a ±10 to ±20 mV *vs.* Ag/AgCl, 9 MHz high-frequency wave potential in the non-nutritive solution such as PBS(-) or ASW [19, 20]. The mechanisms by which the high-frequency wave potential induced electrical detachment involve both oscillation of the negative zeta-potential–charged cells and insertion of water molecules between the electrode surface and attached cells [19, 20, 23, 24]. The high-frequency wave potential–induced cell detachment is practically noncytotoxic [19, 20, 23, 24]. Namely, 91% of the residual soil microorganisms attached to the ITO electrode remained alive after 60 min application of a ±10 mV *vs.* Ag/AgCl, 9 MHz high-frequency wave potential in PBS(-) [19]. Using the electrical retrieval method, we obtained a wide range of microorganisms including deep-sea sediment bacteria [19], sponge-associated bacteria [20], sponge-associated archaea [20], budding yeast [25], symbiotic bacteria from the deep-sea bivalve *Calyptogena okutanii* [23], and spores of streptomycetes [26].

Electrical Retrieval of Sponge-Associated Living Microorganisms from Marine Sponges Frozen at -80 °C

Marine sponges are a rich source of biologically active compounds with antitumor, antiviral, antibacterial, antifungal, antimalarial, antiprotozoal, and antituberculous activities. There is increasing evidence that some biologically active compounds such as brominated biphenyl ethers, chlorinated peptides, theopalauamid, swinholide A, onnamide A, and psymberin can be produced by sponge-associated microorganisms [27 - 32]. Although cultivation of sponge-associated microorganisms that produce biologically active compounds is the most direct method for large-scale production of these chemicals [11], only a minor fraction of the total sponge-associated microbial community is amenable to cultivation on laboratory media [7 - 11]. Furthermore, some sponge-associated microorganisms simply stop producing the biologically active compounds after a certain period on artificial media [33]. Cultivation of the sponge-associated microbial populations of varied composition was performed immediately after collection to screen the microbes for the presence of diverse biologically active compounds. Therefore, we investigated whether the weak negative electric potential attracted sponge-associated living microorganisms to the electrode surface from marine sponges frozen at −80 °C [20].

The frozen marine sponge *Spirastrella insignis* was transferred into ice-cold sterile ASW and incubated for 2 h. The tissue was dipped into 70% ethanol for 3 s and transferred into sterile ASW containing 1% (*v/v*) antibiotics (penicillin and streptomycin) for 1 min to prevent microbial contamination. The sterilized tissue was minced into 1 mm^3 pieces, transferred to a sterilized quartz mortar, and was brayed for 5 min on ice. To obtain sponge-associated living microorganisms from the homogenized tissue in sterile ASW, a −0.3 V *vs.* Ag/AgCl constant potential was applied to the ITO electrode for 2 h at 9 °C [20]. The −0.3 V *vs.* Ag/AgCl potential induced attachment of living microorganisms derived from *S. insignis* tissue frozen at −80 °C [20]. When the *S. insignis* tissue homogenates were pretreated by autoclaving for 15 min at 121 °C, few or no microorganisms got attached to the ITO electrode [20].

The sponge-associated microorganisms attached to the electrode were detached by application of a ±20 mV *vs.* Ag/AgCl, 9 MHz sine wave potential for 20 min in fresh sterile ASW at 9 °C [20]. The detached microorganisms were collected with a cell scraper and cultured. An originally formulated medium consisting of a mixture of autoclaved microbial media and a filter-sterilized nerve cell culture medium was employed to cultivate the sponge-associated microorganisms under aerobic conditions [20]. We confirmed that the electrically attached microorganisms could be cultured on the originally formulated agar medium after detachment from the ITO electrode by application of the 9 MHz high-frequency wave potential [20]. Using the electrical retrieval methods, we have obtained 32 phyla and 72 classes of living bacteria and three living archaea species, *Crenarchaeota thermoprotei*, *Marine Group I*, and *Thaumarchaeota incertae sedis*, from marine sponges *S. insignis* and *Callyspongia confoederata* [20]. The living sponge-associated microorganisms on the electrode include 40 species of actinomycetes [20].

An Electrical Awakening Method for Dormant Deep-Sea Actinomycete Spores

It is estimated that approximately 40% of biologically active compounds can be isolated from actinobacteria [34]. Actinobacteria, mainly the genus *Streptomyces*, have the ability to produce a variety of biologically active compounds, including antibiotics [34]. However, the rate of novel-compound discovery in terrestrial microorganisms has recently decreased significantly [35]. Most deep-sea floors and deep-sea subsurface sediments remain virtually unexplored. Deep-sea sediments have been found to contain over 1,300 different actinobacterial operational taxonomic units, the majority of which are predicted to represent novel species and genera [36]. Therefore, research has focused on deep-sea environments as a source of novel taxa and, presumably, novel biologically active compounds [37, 38]. The majority of actinomycetes exist for long periods in the form of dormant spores. They germinate in the presence of exogenous nutrients or certain exogenous stimuli, the lack of which prevents the germination of most or all spores [39 - 41]. Because the spore is highly resistant to a variety of physicochemical stressors, exposure to the stressors has been used for preparation and selective cultivation of actinomycetes. To determine whether the *Streptomyces* spores were attached to the ITO/glass electrode region carrying an applied potential, we used the typical streptomycete *Streptomyces albus* and five deep-sea streptomycetes. They were attracted by and selectively attached to the ITO electrode surface to which a −0.2 V *vs.* Ag/AgCl potential was applied [26]. All six streptomycetes produced short fibrous and/or membranous materials and got fixed to the ITO electrode surface [26].

Several researchers have reported that a weak alternating electrical potential that causes no electrolysis of water induces gene expression, protein production, and activation of intracellular signaling pathways thereby contributing to cell growth and induction of differentiation in "animal cells" [42 - 45]. We hypothesized that the weak alternating electrical potential might also activate some kind of microbial functions and may induce a cultivable state in dormant microorganisms. To test this working hypothesis, we determined whether application of a ±20 mV *vs.* Ag/AgCl, sine wave potential increased the number of CFUs of the deep-sea *Streptomyces* sp. ST.28 spores on the ITO electrode surface. Application of a 3,

12, and 15 MHz sine wave potential significantly increased the number of CFUs (*i.e.,* ST.28 spores) on the ITO electrode after collection with the rubber cell scraper. A ±20 mV *vs.* Ag/AgCl, 12 MHz sine wave potential induced the maximum number of CFUs in the ST.28 spores attached to the ITO electrode, with an increase of 3.5-fold as compared with an open circuit [26].

Next, we verified whether the electrical retrieval method yielded *Streptomyces* sp.–like microorganisms from deep-sea sediment core samples [26]. The drilling site C9006A is located in Suruga Bay, Japan (34° 52.4646'N, 138° 34.1639'E) and the water depth at C9006A is 756 m. Thus, 65 m sediment cores were recovered by the drilling vessel *Chikyu* (CK09-01; 19 March 2009) [26].

After application of the −0.2 V *vs.* Ag/AgCl potential for 24 h at RT, a high-frequency sine wave potential was applied for additional 60 min in fresh sterile 1/20 artificial seawater (1/20 ASW) at RT. After the application, the microorganisms attached to the electrode were collected with a rubber cell scraper (Fig. **1B**) [26]. The collected samples were seeded on agar plates and cultured at RT. We found that the ±0.2 V *vs.* Ag/AgCl, 12 MHz sine wave potential induced selective cultivation of *Streptomyces* sp.–like microorganisms on the colloidal chitin agar medium (Fig. **2A**). Little or no *Streptomyces* sp.–like microorganisms' colonies were detected when the attached microorganisms on the electrode were stimulated by a ±20 mV *vs.* Ag/AgCl, 9 MHz sine wave potential for 1 h (Fig. **2A**). The number of CFUs of actinomycetes increased 60- to 4000-fold as compared with application of the 9 MHz sine wave potential. The ±0.2 V *vs.* Ag/AgCl, 12 MHz sine wave potential induced selective cultivation of actinomycetes on agar media (Fig. **2B**). Using the 12 MHz sine wave potential application, we identified and isolated *Nocardiopsis* sp., *Dietzia* sp., *Pseudonocardia* sp., *Brachybacterium* sp., *Nesterenkonia* sp., *Microcella* sp., *Microbacterium* sp., and *Streptomyces* sp. from the deep-sea sediment core samples (Fig. **2C**). Phylogenetic analyses based on 16S ribosomal RNA gene sequences indicated that 59% of the obtained strains were novel species and/or had a high likelihood of being novel species (Fig. **2C**). The results in Fig. (**2**) suggest that the method for electrical retrieval of living microorganisms holds promise as a novel screening strategy for hardly cultivable microorganisms.

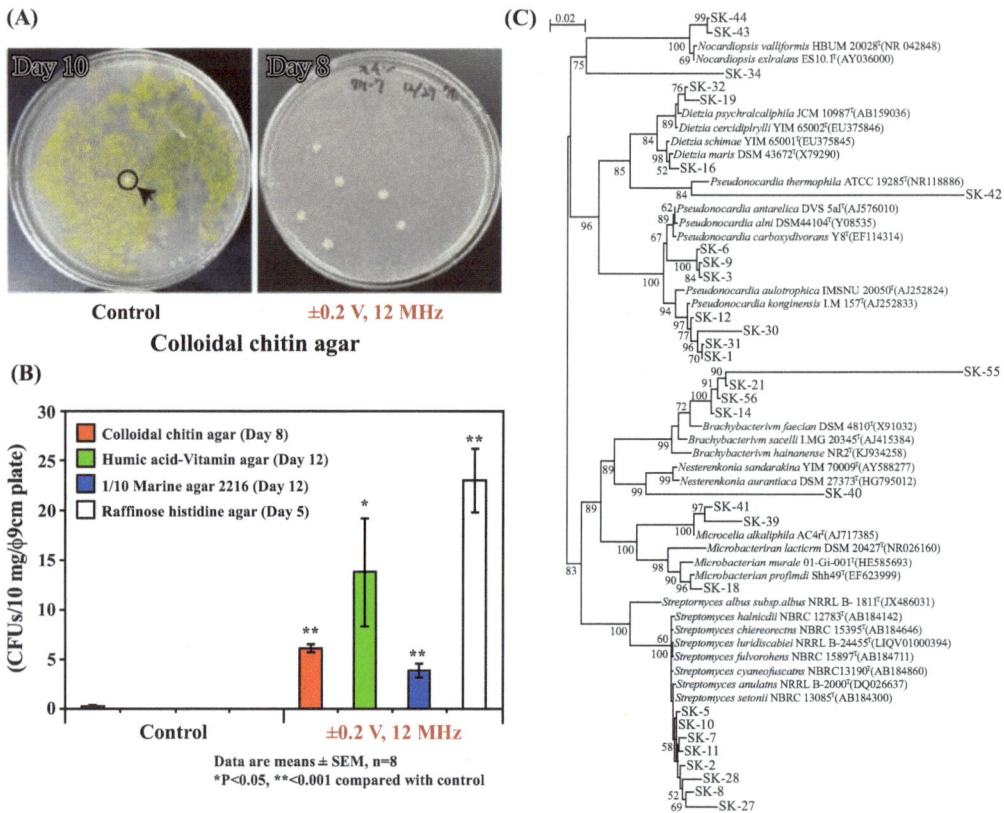

Fig. (2). A novel screening method for hardly cultivable actinomycetes from a deep-sea sediment in Suruga Bay, Japan.
(A) Comparison of photographs of colloidal chitin agar plates. Control: detachment from the ITO electrode by application of a ±20 mV *vs.* Ag/AgCl, 9 MHz sine wave potential. A ±0.2 V *vs.* Ag/AgCl, 12 MHz sine wave potential induced selective cultivation of actinomycetes on an agar medium. (B) CFUs of actinomycetes from agar plates. Colloidal chitin agar is the most selective culture medium for actinomycetes among the four types of agar plates. (C) A neighbor-joining tree based on 16S ribosomal RNA gene sequences showing the phylogenetic position of the isolated strains and related species. Bootstrap values (expressed as percentages of 1000 replications) >50% are shown at branch points.

CONSENT FOR PUBLICATION

Not applicable.

CONFLICT OF INTEREST

The author confirms that this chapter contents have no conflict of interest.

ACKNOWLEDGEMENTS

This review was aided by the generous support provided by Dr. Yuichi Nogi.

REFERENCES

[1]   Raaijmakers JM, Weller DM, Thomashow LS. Frequency of antibiotic-producing *pseudomonas* spp. in natural environments. Appl Environ Microbiol 1997; 63(3): 881-7.
        [PMID: 16535555]

[2]   Shen B, Du L, Sanchez C, Edwards DJ, Chen M, Murrell JM. The biosynthetic gene cluster for the anticancer drug bleomycin from *Streptomyces verticillus* ATCC15003 as a model for hybrid peptide-polyketide natural product biosynthesis. J Ind Microbiol Biotechnol 2001; 27(6): 378-85.
        [http://dx.doi.org/10.1038/sj.jim.7000194] [PMID: 11774003]

[3]   Skoko N, Vujovic J, Savic M, Papic N, Vasiljevic B, Ljubijankic G. Construction of Saccharomyces cerevisiae strain FAV20 useful in detection of immunosuppressants produced by soil actinomycetes. J Microbiol Methods 2005; 61(1): 137-40.
        [http://dx.doi.org/10.1016/j.mimet.2004.11.007] [PMID: 15676204]

[4]   Hatada Y, Mizuno M, Li Z, Ohta Y. Hyper-production and characterization of the ɩ-carrageenase useful for ɩ-carrageenan oligosaccharide production from a deep-sea bacterium, *Microbulbifer thermotolerans* JAMB-A94T, and insight into the unusual catalytic mechanism. Mar Biotechnol (NY) 2011; 13(3): 411-22.
        [http://dx.doi.org/10.1007/s10126-010-9312-0] [PMID: 20686828]

[5]   Ohta Y, Hatada Y. A novel enzyme, lambda-carrageenase, isolated from a deep-sea bacterium. J Biochem 2006; 140(4): 475-81.
        [http://dx.doi.org/10.1093/jb/mvj180] [PMID: 16926183]

[6]   Rajendhran J, Gunasekaran P. Strategies for accessing soil metagenome for desired applications. Biotechnol Adv 2008; 26(6): 576-90.
        [http://dx.doi.org/10.1016/j.biotechadv.2008.08.002] [PMID: 18786627]

[7]   Santavy DL, Willenz P, Colwell RR. Phenotypic study of bacteria associated with the caribbean sclerosponge, Ceratoporella nicholsoni. Appl Environ Microbiol 1990; 56(6): 1750-62.
        [PMID: 2383012]

[8]   Burja AM, Webster NS, Murphy PT, Hill RT. Microbial symbionts of great barrier reef sponges. memoirs. Queensland. Mus 1999; 44: 63-75.

[9]   Webster NS, Hill RT. The culturable microbial community of the Great Barrier Reef sponge *Rhopaloeides odorabile* is dominated by an α-Proteobacterium. Mar Biol 2001; 138: 843-51.
        [http://dx.doi.org/10.1007/s002270000503]

[10]   Friedrich AB, Fischer I, Proksch P, Hacker J, Hentschel U. Temporal variation of the microbial community associated with the Mediterranean sponge *Aplysina aerophoba*. FEMS Microbiol Ecol 2001; 38: 105-13.
        [http://dx.doi.org/10.1111/j.1574-6941.2001.tb00888.x]

[11]   Hill RT. Microbes from marine sponges: A treasure trove of biodiversity for natural products discovery.Microbial diversity and bioprospecting. Washington: ASM Press 2004; pp. 177-90.
        [http://dx.doi.org/10.1128/9781555817770.ch18]

[12]   Simon C, Daniel R. Metagenomic analyses: past and future trends. Appl Environ Microbiol 2011; 77(4): 1153-61.
        [http://dx.doi.org/10.1128/AEM.02345-10] [PMID: 21169428]

[13]   Roh C, Villatte F, Kim BG, Schmid RD. Comparative study of methods for extraction and purification of environmental DNA from soil and sludge samples. Appl Biochem Biotechnol 2006; 134(2): 97-112.
        [http://dx.doi.org/10.1385/ABAB:134:2:97] [PMID: 16943632]

[14]   Demeke T, Adams RP. The effects of plant polysaccharides and buffer additives on PCR. Biotechniques 1992; 12(3): 332-4.
        [PMID: 1571138]

[15]   Wei T, Lu G, Clover G. Novel approaches to mitigate primer interaction and eliminate inhibitors in multiplex PCR, demonstrated using an assay for detection of three strawberry viruses. J Virol Methods 2008; 151(1): 132-9.
[http://dx.doi.org/10.1016/j.jviromet.2008.03.003] [PMID: 18453003]

[16]   Su X, Gibor A. A method for RNA isolation from marine macro-algae. Anal Biochem 1988; 174(2): 650-7.
[http://dx.doi.org/10.1016/0003-2697(88)90068-1] [PMID: 2467581]

[17]   Courtois S, Frostegård A, Göransson P, Depret G, Jeannin P, Simonet P. Quantification of bacterial subgroups in soil: comparison of DNA extracted directly from soil or from cells previously released by density gradient centrifugation. Environ Microbiol 2001; 3(7): 431-9.
[http://dx.doi.org/10.1046/j.1462-2920.2001.00208.x] [PMID: 11553233]

[18]   Gabor EM, de Vries EJ, Janssen DB. Efficient recovery of environmental DNA for expression cloning by indirect extraction methods. FEMS Microbiol Ecol 2003; 44(2): 153-63.
[http://dx.doi.org/10.1016/S0168-6496(02)00462-2] [PMID: 19719633]

[19]   Koyama S, Konishi MA, Ohta Y, *et al.* Attachment and detachment of living microorganisms using a potential-controlled electrode. Mar Biotechnol (NY) 2013; 15(4): 461-75.
[http://dx.doi.org/10.1007/s10126-013-9495-2] [PMID: 23420537]

[20]   Koyama S, Nishi S, Tokuda M, *et al.* Electrical retrieval of living microorganisms from cryopreserved marine sponges using a potential-controlled electrode. Mar Biotechnol (NY) 2015; 17(5): 678-92.
[http://dx.doi.org/10.1007/s10126-015-9651-y] [PMID: 26242755]

[21]   Frederiks WM, van Marle J, van Oven C, Comin-Anduix B, Cascante M. Improved localization of glucose-6-phosphate dehydrogenase activity in cells with 5-cyano-2,3-ditolyl-tetrazolium chloride as fluorescent redox dye reveals its cell cycle-dependent regulation. J Histochem Cytochem 2006; 54(1): 47-52.
[http://dx.doi.org/10.1369/jhc.5A6663.2005] [PMID: 16046670]

[22]   Hiraishi A, Yoshida N. An improved redox dye-staining method using 5-cyano-2,3-ditoryl tetrazolium chloride for detection of metabolically active bacteria in activated sludge. Microbes Environ 2004; 19: 61-70.
[http://dx.doi.org/10.1264/jsme2.19.61]

[23]   Koyama S, Yoshida T. Electrical collection of membrane-intact and dehydrogenase-positive symbiotic bacteria from the deep-sea bivalve *Calyptogena okutanii*. Electrochemistry 2016; 84: 358-60.
[http://dx.doi.org/10.5796/electrochemistry.84.358]

[24]   Koyama S. Electrically modulated attachment and detachment of animal cells cultured on an optically transparent patterning electrode. J Biosci Bioeng 2012; 111: 574-583.

[25]   Koyama S, Tsubouchi T, Usui K, *et al.* Involvement of flocculin in negative potential-applied ITO electrode adhesion of yeast cells. FEMS Yeast Res 2015; 15(6)fov064
[http://dx.doi.org/10.1093/femsyr/fov064] [PMID: 26187908]

[26]   Koyama S, Nishi S, Nagano Y, *et al.* Electrical retrieval of living streptomycete spores using a potential-controlled ITO electrode. Electrochemistry 2017; 85: 297-309.
[http://dx.doi.org/10.5796/electrochemistry.85.297]

[27]   Bewley CA, Faulkner DJ. Lithistid sponges: star performers or hosts to the stars. Angew Chem Int Ed Engl 1998; 37(16): 2162-78.
[http://dx.doi.org/10.1002/(SICI)1521-3773(19980904)37:16<2162::AID-ANIE2162>3.0.CO;2-2] [PMID: 29711453]

[28]   Haygood MG, Schmidt EW, Davidson SK, Faulkner DJ. Microbial symbionts of marine invertebrates: opportunities for microbial biotechnology. J Mol Microbiol Biotechnol 1999; 1(1): 33-43.
[PMID: 10941782]

[29]   Piel J, Hui D, Wen G, *et al.* Antitumor polyketide biosynthesis by an uncultivated bacterial symbiont

of the marine sponge *Theonella swinhoei*. Proc Natl Acad Sci USA 2004; 101(46): 16222-7.
[http://dx.doi.org/10.1073/pnas.0405976101] [PMID: 15520376]

[30]   Schmidt EW, Obraztsova AY, Davidson SK, Faulkner DJ, Haygood MG. Identification of the antifungal peptide-containing symbiont of the marine sponge *Theonella swinhoei* as a novel δ-proteobacterium, *"Candidatus Entotheonella palauensis*. Mar Biol 2000; 136: 969-77.
[http://dx.doi.org/10.1007/s002270000273]

[31]   Hentschel U, Piel J, Degnan SM, Taylor MW. Genomic insights into the marine sponge microbiome. Nat Rev Microbiol 2012; 10(9): 641-54.
[http://dx.doi.org/10.1038/nrmicro2839] [PMID: 22842661]

[32]   Wilson MC, Mori T, Rückert C, *et al.* An environmental bacterial taxon with a large and distinct metabolic repertoire. Nature 2014; 506(7486): 58-62.
[http://dx.doi.org/10.1038/nature12959] [PMID: 24476823]

[33]   Hentschel U, Fieseler L, Wehrl M, *et al.* Microbial diversity of marine sponges. Prog Mol Subcell Biol 2003; 37: 59-88.
[http://dx.doi.org/10.1007/978-3-642-55519-0_3] [PMID: 15825640]

[34]   Bérdy J. Thoughts and facts about antibiotics: where we are now and where we are heading. J Antibiot (Tokyo) 2012; 65(8): 385-95.
[http://dx.doi.org/10.1038/ja.2012.27] [PMID: 22511224]

[35]   Takagi M, Shin-Ya K. Construction of a natural product library containing secondary metabolites produced by actinomycetes. J Antibiot (Tokyo) 2012; 65(9): 443-7.
[http://dx.doi.org/10.1038/ja.2012.52] [PMID: 22739538]

[36]   Stach JEM, Bull AT. Estimating and comparing the diversity of marine actinobacteria. Antonie van Leeuwenhoek 2005; 87(1): 3-9.
[http://dx.doi.org/10.1007/s10482-004-6524-1] [PMID: 15726285]

[37]   Bull AT. Actinobacteria of the extremobiosphere. Extremophiles Handbook. Tokyo: Springer 2011; pp. 1203-40.
[http://dx.doi.org/10.1007/978-4-431-53898-1_58]

[38]   Goodfellow M, Fiedler H-P. A guide to successful bioprospecting: informed by actinobacterial systematics. Antonie van Leeuwenhoek 2010; 98(2): 119-42.
[http://dx.doi.org/10.1007/s10482-010-9460-2] [PMID: 20582471]

[39]   Cross T. Aquatic actinomycetes: a critical survey of the occurrence, growth and role of actinomycetes in aquatic habitats. J Appl Bacteriol 1981; 50(3): 397-423.
[http://dx.doi.org/10.1111/j.1365-2672.1981.tb04245.x] [PMID: 7019182]

[40]   Williams ST, Lanning S, Wellington EMH. Ecology of actinomycetes. The biology of the actinomycetes. London: Academic Press 1983; pp. 481-528.

[41]   Goodfellow M, Haynes JA. Actinomycetes in marine sediments.Biological, biochemical, and biomedical aspects of actinomycetes. London: Academic Press 1984; pp. 453-72.
[http://dx.doi.org/10.1016/B978-0-12-528620-6.50039-2]

[42]   Kimura K, Yanagida Y, Haruyama T, Kobatake E, Aizawa M. Gene expression in the electrically stimulated differentiation of PC12 cells. J Biotechnol 1998; 63(1): 55-65.
[http://dx.doi.org/10.1016/S0168-1656(98)00075-3] [PMID: 9764482]

[43]   Kimura K, Yanagida Y, Haruyama T, Kobatake E, Aizawa M. Electrically induced neurite outgrowth of PC12 cells on the electrode surface. Med Biol Eng Comput 1998; 36(4): 493-8.
[http://dx.doi.org/10.1007/BF02523221] [PMID: 10198536]

[44]   Koyama S, Haruyama T, Kobatake E, Aizawa M. Electrically induced NGF production by astroglial cells. Nat Biotechnol 1997; 15(2): 164-6.
[http://dx.doi.org/10.1038/nbt0297-164] [PMID: 9035143]

[45]   Koyama S, Yanagida Y, Haruyama T, Kobatake E, Aizawa M. Molecular mechanisms of electrically stimulated NGF expression and secretion by astrocytes cultures on the potential controlled electrode surface. Cell Eng 1996; 1: 189-94.

CHAPTER 11

Isolation of Vanadium-Accumulating or -Reducing Bacteria from Ascidians and Their Functional Analysis

Tatsuya Ueki[1,*], **Tri Kustono Adi**[2] and **Romaidi**[3]

[1] *Marine Biological Laboratory, Graduate School of Integrated Sciences for Life, Hiroshima University, Higashi-Hiroshima, Hiroshima, Japan*

[2] *Chemistry Department, Science and Technology Faculty, State Islamic University of Malang, Malang, Indonesia*

[3] *Biology Department, Science and Technology Faculty, State Islamic University of Malang, Malang, Indonesia*

Abstract: Ascidians are known to accumulate extremely high levels of vanadium in their blood cells. The concentration of vanadium in seawater is 35 nM, while the concentration of vanadium reaches 350 mM in blood cells, which corresponds to 10^7 times that of seawater. Ascidians can be regarded as a natural ecosystem that harbors vanadium-related symbiotic bacteria and serves as a useful bacterial resource. Since the 1990s, vanadium-accumulating or -reducing bacteria have been isolated from vanadium-rich ascidians. Recent functional screening and comprehensive molecular studies also identified symbiotic bacteria in the branchial sac and the intestine. These bacteria could contribute to the high accumulation of vanadium by ascidians. In this chapter, the authors overview the vanadium accumulation and reduction in ascidians, review the studies on the isolation and functional analyses of vanadium-accumulating or -reducing bacteria, and provide perspectives on utilization of these bacteria for bioremediation of heavy metals.

Keywords: Accumulation, Marine invertebrate, Metal, Reduction, Symbiotic bacteria, Vanadium.

INTRODUCTION

Ascidians (sea squirts, tunicates) are marine sessile invertebrate animals belonging to the subphylum Urochordata, phylum Chordata. Reflecting their phylogenetic position, they are good models for studying genome evolution, early development, the immune system, and nervous system [1, 2]. A cosmopolitan spe-

* **Corresponding author Tatsuya Ueki:** Marine Biological Laboratory, Graduate School of Integrated Sciences for Life, Hiroshima University, Mukaishima 2445, Onomichi, Hiroshima 722-0073, Japan; Tel: +81-848-44-1434; Fax: +81-848-44-5914; E-mail: ueki@hiroshima-u.ac.jp

cies, *Ciona intestinalis*, is most commonly used for such studies and is the 7th animal species whose genome was determined [3]. [Note: Recent morphological and molecular studies suggest that *C. intestinalis* includes two cryptic species, named "type A" and "type B" [4]. The species *C. intestinalis* in this chapter represent the species widely used in genomic and developmental studies in the last decades and proposed to be *C. intestinalis* type A or *C. robusta*]. On the contrary, ascidians possess unique features that are not found in other chordates: cellulose synthesis [5 - 8], asexual reproduction [9], and metal accumulation.

Ascidians accumulate extremely high levels of vanadium in their blood cells. The vanadium concentration varies among species and can reach 350 mM in *Ascidia gemmata* [10, 11], which is 10^7 times the concentration found in seawater (35 nM) [12, 13]. This is thought to be the highest degree of active accumulation of a metal in any living organism. How and why do ascidians accumulate vanadium at such extremely high levels? To address these questions, our research group has been trying to identify genes and proteins responsible for vanadium accumulation in blood cells as well as the process of vanadium transport from seawater to blood cells through the branchial sac, intestine, and blood plasma [14 - 18]. Recent functional and comprehensive studies on symbiotic bacteria shed light on the mechanism of vanadium uptake from seawater.

In this chapter, the authors overview the history of studies on vanadium accumulation in ascidians, summarize the methods for the isolation and functional analyses of vanadium-accumulating or -reducing bacteria, and provide perspectives on the applications of ascidian genes or the bacteria for bioaccumulation or bioremediation of heavy metals.

VANADIUM

Vanadium is an element with symbol V and atomic number 23. It is the 17th most abundant element in the Earth's crust with a concentration of 0.015% [19]. Metallic vanadium is not found in nature, but its compounds can be obtained as minerals such as vanadinite [$Pb_5(VO_4)_3Cl$; Fig **1**], a lead vanadate ore in which vanadium was first discovered by a Mexican, Andrés Manuel del Río. In 1831, Nils Gabriel Sefström rediscovered this element and he called the element vanadium after Vanadis, another name of the Norse goddess Freyja, who represented beauty and fertility, because of beautifully colored chemical compounds of this element [20].

Vanadium is a multivalent transition metal. Vanadium ions under physiological aqueous conditions are limited to the oxidation states +3, +4, and +5 [21]. Vanadium usually exists in the +5 state as vanadate anions (HVO_4^{2-} or $H_2VO_4^-$; V^V)

in the natural environment [22]. The chemical features of V^V ions resemble those of phosphate anions (HPO_4^{2-} or $H_2PO_4^-$) at low concentration [16, 23].

Fig. (1). Vanadinite [$Pb_5(VO_4)_3Cl$], Morocco.

VANADIUM ACCUMULATION AND REDUCTION BY ASCIDIANS

Approximately a hundred years ago, the German physiological chemist Dr. Martin Henze discovered high levels of vanadium in the blood cells of the ascidian *Phallusia mammillata* collected from the Bay of Naples, Italy [24]. The concentrations of vanadium within the tissues of many ascidian species have been determined by neutron activation analysis, electron spin resonance, or atomic absorption spectrometry. Ascidians belonging to the suborder Phlebobranchia appear to contain higher levels of vanadium than those of the suborder Stolidobranchia [11]. Of the tissues examined, blood cells contain the highest concentrations of vanadium [10, 25]. The highest concentration was found in blood cells of the ascidian *Ascidia gemmata*. The vanadium concentration in this species can reach 350 mM [10].

Ascidians are filter feeders. They take seawater from oral siphon and filter plankton by means of the branchial sac using mucus secreted from the endostyle. Plankton trapped by the mucus is transferred into the stomach together with the mucus. Vanadium absorption from seawater by ascidians was first experimentally demonstrated by Goldberg *et al.* [26]. They administered two ascidian species *Ascidia ceratodes* and *Ciona intestinalis* with radioactive V^{48} as a form of vanadate. Radioautographic analysis indicated that vanadium accumulates primarily in ovaries, the intestinal wall, eggs, and branchial sac. They also observed that high concentration of phosphate ions competed with vanadium uptake. Thereafter, Dingley *et al*., pointed out that the phosphate anion transporter is a candidate for the specific transporter of V^V ions because the influx of vanadate into the cell is a rapid process that can be saturated (K_m = 1.4 mM); meanwhile, phosphate competes with vanadate for transport and is itself taken up by the cell [27]. Nonetheless, for a rebuttal on Goldberg's experiments where phosphate ions cannot completely inhibit the uptake of vanadate ions, Kalk pointed out the possibility of uptake of vanadyl ions complexed with sulfate in a mucosa [28]. Thus, both direct absorption of vanadate and indirect absorption of vanadyl (VO^{2+}; V^{IV}) must be considered.

VANADIUM ACCUMULATION AND REDUCTION BY BACTERIA

Several studies have investigated the importance of vanadium accumulation and reduction by bacteria. Reduction of V^V to V^{IV} was first described by Woolfolk and Whitteley by using the cell extract from *Micrococcus lactilyticus* [29]. Table **1** summarizes the representative articles describing the reduction of V^V by bacteria. Some of the bacterial strains were also examined for accumulation of V. Some of them are reported to be able to remove V from an analytical medium; however, the concentrations in the bacterial cells were not mentioned.

Table 1. Bacterial reduction of V^V to V^{IV} and accumulation of V.

Bacterial Species and Strain	Concentration of V^V in the Analytical Medium	Reduction of V^V	Accumulation of V	References
Geobacter metallireducens	1–5 mM	—	yes	[32]
Micrococcus lactilyticus	42.8 µmoles/mL	—	yes	[29]
Pseudomonas isachenkovii	10 mM	yes	—	[33]
Shewanella oneidensis	5–10 mM	yes	yes	[34]
Shewanella oneidensis MR-1	2 mM	—	yes	[30]
Serratia marcescens PRE01	20–500 mg/L	yes	—	[35]
Vibrio sp. and *Shewanella sp.*	200-500 µM	—	yes	[36]

—: not examined.

There are only a few reports on the mechanism of vanadium reduction and accumulation. Myers *et al*. [30] have demonstrated that menaquinone-, cymA-, and omcB-deficient mutants of *Shewanella oneidensis* MR-1 are defective in the ability to reduce V^V. Carpentier *et al*. performed transposon mutagenesis and screening for a vanadate reduction–deficient phenotype in the same bacterial strain and confirmed that menaquinone and membrane-localized c-type cytochromes play a crucial role in reduction of V^V [31]. Ortiz-Bernad *et al*.

mentioned that a green precipitate accumulates when *Geobacter metallireducens* reduced V^V [32]. They used electron microprobe analysis to suggest that this is a compound of reduced vanadium and phosphorous. They suppose that the compound could be a vanadyl phosphate, such as the green mineral sincosite $[CaV_2(PO_4)_2(OH)_4 \cdot 3H_2O]$.

INTESTINAL BACTERIA ISOLATED FROM ASCIDIANS

Antipov *et al.*, [33] revealed that the facultatively anaerobic bacterium *Pseudomonas isachenkovii* isolated from the intestine of an ascidian *Ascidia phalosia* [Note: The author could not find the relevant species in other publications or public databases] [37] could withstand vanadium toxicity and use vanadate as the electron acceptor during anaerobic respiration. They also identified a vanadium-binding protein associated with V^{IV} at a molar ratio of ~1:20 [38]. The property of the vanadium-binding protein from *P. isachenkovii* is similar to that of vanabins, which the authors' research group isolated from vanadium-rich ascidians, in that both of them are small proteins with a molecular weight of 15–20 kDa and can bind 10–20 vanadium ions per protein molecule [39 - 45]. The molecular nature of the vanadium-binding protein from *P. isachenkovii* is not reported yet.

The intestine harbors many types of bacteria as intestinal microbes. The presence of intestinal microbes in aquatic invertebrates has been investigated in several marine animals such as crustaceans [46, 47], Mollusca [48, 49], and Echinodermata [50, 51]. *Vibrio, Pseudomonas, Flavobacterium, Aeromonas*, and *Shewanella* are the most commonly reported bacteria harbored in the intestine of these marine invertebrates.

The digestion and absorption of food are the dominant functions of the intestine, and several types of close interactions between aquatic invertebrates and their intestinal microbes have been described by Harris [52]. Other types of interactions include an immune response, epithelial development, and pathogenic interactions [46, 47, 53 - 55]. Another important type of host–bacterial interaction is the ability of intestinal bacteria to accumulate heavy metals such as mercury [56], and intestinal bacteria are thought to be the first organisms affected by heavy metal discharge into the environment that results in an increase in the abundance of metal-resistant bacteria in the microenvironment [57].

Recently the authors' research group confirmed that vanadium concentration in the intestinal content of a vanadium-rich ascidian *A. sydneiensis samea* reaches 0.67 mM [36]. This value corresponds to ~20,000-fold higher concentration than that in seawater, indicating that the intestinal content is a vanadium-rich microenvironment. Then, nine strains of vanadium-resistant bacteria were successfully identified (in the intestinal content by functional screening) [36]. Phylogenetic analysis based on the 16S ribosomal RNA (rRNA) gene sequence indicates that five bacterial strains belonged to the genus *Vibrio* and four to the genus *Shewanella*. Two strains, *Vibrio* sp. V-RA-4 and *Shewanella* sp. S-RA-6, accumulated vanadium ions at a higher rate than did the other strains. It was proposed that these bacteria might maintain the vanadium-rich microenvironment and contribute to vanadium uptake by the intestinal cells.

METAGENOMIC STUDIES ON SYMBIOTIC BACTERIA

Metagenomics is a modern and powerful tool to gain a comprehensive understanding of the microbial flora in any sample from which DNA can be extracted. This technique does not require laboratory cultivation or cloning of each bacterial species but starts with robust PCR amplification of a specific region of the 16S rRNA gene. The next-generation sequencing technology enables us to obtain a huge amount of sequence data at high speed and at a relatively low cost, although it should be noted that abundance estimation by 16S rRNA amplicon sequencing is affected by DNA extraction protocols, PCR biases, and copy numbers [58 - 61]. Intestinal microbes are the type of microbial flora that is extensively studied by the metagenomic method.

A comprehensive analysis of the intestinal microbial diversity in the ascidian *C. intestinalis* was achieved using next-generation sequencing [62]. Bacterial communities isolated from the intestine of *C. intestinalis* found in three disparate geographic locations exhibited striking similarity in the abundance of operational taxonomic units (OTUs) consistent with the selection of a core community by the intestinal ecosystem, in which Proteobacteria (80%) were the predominant intestinal bacteria. This is discussed later in relation to intestinal immunity [63].

Though metal ions are one of essential nutrients, little was known about the bacterial contribution to metal absorption in the intestine. Additionally, there are many reports of bacterial metal absorption and utilization, but very few studies have been published on these processes in a symbiotic context. The authors' research group recently published a comprehensive comparative study on the bacterial population in three tissues of ascidians [64]. In this study, metagenomic DNAs were extracted from two organs (branchial sac and intestine), which are the first organs that absorb vanadium from seawater and from the intestinal content (here they are referred to as three tissues). Branchial sac was included because it can harbor bacteria that help the branchial sac cells to absorb vanadium from seawater. The microbial diversity in the three tissues was examined in two vanadium-rich ascidians (*Ascidia ahodori* and *A. sydneiensis samea*) and a vanadium-poor ascidian (*Styela plicata*). The relationship between the bacterial populations and vanadium concentration in these samples was also examined. The authors listed bacteria whose abundance in vanadium-rich ascidians was higher than that in vanadium-poor species in each tissue. Two bacterial genera, *Pseudomonas* and *Ralstonia*, were extremely abundant and highly enriched in the branchial sacs of vanadium-rich ascidians (Table **2**). Two bacterial genera, *Treponema* and *Borrelia*, were abundant and enriched in the intestinal content of vanadium-rich ascidians (Table **3**). It is possible that symbiotic bacteria that are abundant or enriched in vanadium-rich ascidians but not in vanadium-poor ascidians are necessary or beneficial for vanadium accumulation. Further functional studies may reveal the contribution of these abundant or enriched bacteria to the metal accumulation by ascidians.

Table 2. Abundance and abundance ratio to seawater of the two representative bacterial genera in the branchial sac [64].

OTU_ID	Ah Ave	As Ave	Sp Ave	SW Ave	Ah_vs_SW Ratio	As_vs_SW Ratio	Sp_vs_SW Ratio
Pseudomonas	68.6%	65.0%	5.8%	6.1%	1134%	1074%	95%
Ralstonia	17.8%	19.3%	1.8%	1.0%	1807%	1954%	185%

Abbreviations: Aa: *Ascidia ahodori*, As: *Ascidia sydneiensis samea*, Sp: *Styela plicata*. SW: seawater. Ave values are the average abundance in each tissue in each ascidian species. Ratio values are abundance enhancement values for each bacterial taxon in each tissue, as calculated by dividing the abundance in each tissue by that in seawater.

Table 3. Abundance and abundance ratio to seawater of the two representative bacterial genera in the intestinal content [64].

OTU_ID	Ah Ave	As Ave	Sp Ave	SW Ave	Ah_vs_SW Ratio	As_vs_SW Ratio	Sp_vs_SW Ratio
Borrelia	0.5%	0.1%	0.0%	0.0%	1571%	274%	2%
Treponema	0.3%	0.1%	0.0%	0.1%	217%	49%	1%

Abbreviations: Same as in Table **2**.

POSSIBLE USE OF BACTERIA FOR BIOREMEDIATION OF VANADIUM

To recover heavy metal ions from a solution by biosorption or bioremediation, researchers have sought biotechnological application of metal-binding proteins or peptides with the ability to bind heavy metals in various organisms, to improve the metal-binding abilities of microorganisms *via* heterologous expression. Many studies have focused on metallothioneins, which are small cysteine-rich proteins that are widely distributed from prokaryotes to eukaryotes. For example, *Escherichia coli* expressing a metallothionein in the cytoplasm (*e.g.,* [65, 66]) or periplasm (*e.g.,* [66, 67]) can remove heavy metal ions, such as Cd^{II}, Hg^{II}, Pb^{II}, and Cu^{II}, from the culture medium and accumulate them. Several research groups have sought novel small peptides that enhance the bioaccumulation of specific metals (*e.g.,* [68 - 70]).

The authors' research group intended to express ascidian genes in bacteria to construct a bioremediation system for vanadium. The first study was conducted by means of two vanabin genes from *A. sydneiensis samea*. Unfortunately, *E. coli* cells expressing these vanabins could not accumulate V^{IV} significantly, but the absorption of Cu^{II} was enhanced ~20-fold [71]. A subsequent study involving two vanabins (AgVanabin1 and -2) from another ascidian species (*A. gemmata*) was carried out. When AgVanabin2 was expressed in the periplasm of *E. coli*, absorption of both V^{IV} and Cu^{II} was enhanced significantly [40]. Thus, these recombinant bacteria can serve for bioremediation of V or Cu.

The demerit of the heterologous expression system is the limited use of recombinant organisms in open environments. The use of natural organisms could be a straightforward method to overcome this problem. Natural symbiotic bacteria isolated from a vanadium-rich microecosystem could be such a source. For example, the two intestinal bacterial strains V-RA-4 and S-RA-6 mentioned in the previous section can accumulate V^{V} at ~20-fold greater than genetically modified *E. coli* cells expressing AgVanabin2 [36]. In addition, vanadium uptake by the two bacterial strains V-RA-4 and S-RA-6 was significantly higher at pH 3 than under neutral or alkaline conditions (Fig **2**). The optimal pH of 3 for vanadium absorption is consistent with that reported previously for vanadium absorption by *Halomonas sp.* GT-83 [72]. These findings may provide an alternative bioremediation technology for vanadium under acidic conditions.

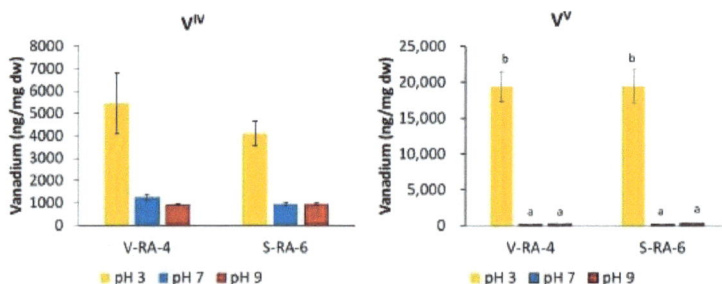

Fig. (2). Effects of pH on vanadium uptake by two vanadium-accumulating bacterial strains [36]. Cells were cultured in 500 μM V^{IV} or V^{V} at pH 3, 7, or 9 and 25°C. The vertical axis indicates the amount of V^{IV} or V^{V} accumulated in the cells per unit dry weight after 6 h of incubation. For each pH, different letters in the same bacterial strain indicate a significant difference between pH levels at $P < 0.05$. Dw: dry weight.

CONCLUDING REMARKS

Vanadium accumulation in ascidians is quite a unique phenomenon. The concentration of vanadium in seawater is 35 nM, while the concentration of this metal in ascidian blood cells reaches 350 mM, which corresponds to 10^7 times that of seawater. There are several possible mechanisms underlying the uptake of vanadium from seawater. One such idea is bacterial accumulation of V or reduction of V^V ions before absorption by the branchial sac or the intestine. Since the 1990s, vanadium-accumulating or -reducing bacteria have been isolated from vanadium-rich ascidians. Ascidians can be regarded as a natural ecosystem that harbor vanadium-related symbiotic bacteria and provide a useful bacterial resource. Recent comprehensive molecular studies identified abundant or enriched symbiotic bacteria in the branchial sac and the intestine. Natural symbiotic bacteria isolated from a vanadium-rich microecosystem could be used for bioremediation of heavy metals.

CONSENT FOR PUBLICATION

Not applicable.

CONFLICT OF INTEREST

The author confirms that this chapter contents have no conflict of interest.

ACKNOWLEDGEMENTS

The authors would like to thank the staff at the International Coastal Research Center, Ocean Research Institute, University of Tokyo, Otsuchi, Iwate, Japan, Kojima Port Fisherman's Association, Kurashiki, Okayama, Japan, and National Bio-Resource Project (NBRP) of the Ministry of Education, Culture, Sports, Science and Technology (MEXT) of Japan for providing adult ascidians.

REFERENCES

[1]   Satoh N, Satou Y, Davidson B, Levine M. *Ciona intestinalis*: an emerging model for whole-genome analyses. Trends Genet 2003; 19(7): 376-81.
       [http://dx.doi.org/10.1016/S0168-9525(03)00144-6] [PMID: 12850442]

[2]   Holland LZ. Genomics, evolution and development of amphioxus and tunicates: the goldilocks principle. J Exp Zool 2014; 9999B: 1-11.
       [PMID: 24665055]

[3]   Dehal P, Satou Y, Campbell RK, *et al.* The draft genome of *Ciona intestinalis*: insights into chordate and vertebrate origins. Science 2002; 298(5601): 2157-67.
       [http://dx.doi.org/10.1126/science.1080049] [PMID: 12481130]

[4]   Brunetti R, Gissi C, Pennati R, Caicci F. Morphological evidence that the molecularly determined *Ciona intestinalis* type a and type b are different species: *Ciona robusta* and *Ciona intestinalis*. J Zool Syst Evol Res 2015; 53: 186-93.

[http://dx.doi.org/10.1111/jzs.12101]

[5] Deck JD, Hay ED, Revel J-P. Fine structure and origin of the tunic of *Perophora viridis*. J Morphol 1966; 120(3): 267-80.
[http://dx.doi.org/10.1002/jmor.1051200304] [PMID: 5970301]

[6] Lübbering-Sommer B, Compère P, Goffinet G. Cytochemical investigations on tunic morphogenesis in the sea peach *Halocynthia papillosa* (Tunicata, Ascidiacea). 1: demonstration of polysaccharides. Tissue Cell 1996; 28(5): 621-30.
[http://dx.doi.org/10.1016/S0040-8166(96)80065-6] [PMID: 18621339]

[7] Koo Y-S, Wang Y-S, You S-H, Kim H-D. Preparation and properties of chemical cellulose from ascidian tunic and their regenerated cellulose fibers. J Appl Polym Sci 2002; 85: 1634-43.
[http://dx.doi.org/10.1002/app.10711]

[8] Sasakura Y, Nakashima K, Awazu S, *et al.* Transposon-mediated insertional mutagenesis revealed the functions of animal cellulose synthase in the ascidian *Ciona intestinalis*. Proc Natl Acad Sci USA 2005; 102(42): 15134-9.
[http://dx.doi.org/10.1073/pnas.0503640102] [PMID: 16214891]

[9] Brown FD, Swalla BJ. Evolution and development of budding by stem cells: ascidian coloniality as a case study. Dev Biol 2012; 369(2): 151-62.
[http://dx.doi.org/10.1016/j.ydbio.2012.05.038] [PMID: 22722095]

[10] Michibata H, Iwata Y, Hirata J. Isolation of highly acidic and vanadium-containing blood cells from among several types of blood cell from ascidiidae species by density-gradient centrifugation. J Exp Zool 1991; 257: 306-13.
[http://dx.doi.org/10.1002/jez.1402570304]

[11] Michibata H, Terada T, Anada N, Yamakawa K, Numakunai T. The accumulation and distribution of vanadium, iron, and manganese in some solitary ascidians. Biol Bull 1986; 171(3): 672-81.
[http://dx.doi.org/10.2307/1541632] [PMID: 29314894]

[12] Cole PC, Eckert J, Williams K. The determination of dissolved and particulate vanadium in sea water by x-ray fluorescence spectrometry. Anal Chim Acta 1983; 153: 61-7.
[http://dx.doi.org/10.1016/S0003-2670(00)85488-4]

[13] Collier RW. Particulate and dissolved vanadium in the North Pacific Ocean. Nature 1984; 309: 441.
[http://dx.doi.org/10.1038/309441a0]

[14] Michibata H, Ueki T. High Levels of Vanadium in Ascidians.Vanadium - Biochemical and molecular biological approaches. Dordrecht: Springer 2012; pp. 51-72.

[15] Michibata H. Vanadium. Dordrecht: Springer 2012.
[http://dx.doi.org/10.1007/978-94-007-0913-3]

[16] Ueki T, Yamaguchi N. Romaidi Isago Y, Tanahashi H. Vanadium Accumulation in Ascidians: A System Overview. Coord Chem Rev 2014; 301-302: 300-8.
[http://dx.doi.org/10.1016/j.ccr.2014.09.007]

[17] Ueki T, Michibata H. Molecular mechanism of the transport and reduction pathway of vanadium in ascidians. Coord Chem Rev 2011; 255: 2249-57.
[http://dx.doi.org/10.1016/j.ccr.2011.01.012]

[18] Ueki T, Michibata H. Vanadium-Protein Interaction Inferred from the Studies on Vanadium-Binding Proteins in Ascidians.Vanadium compounds/vanadate oligomers in biological systems: Chemistry, biochemistry and biological effects. Research SignPost 2008; pp. 135-48.

[19] Emsley J. The Elements (Oxford Chemistry Guides). Oxford: Oxford University Press 1998.

[20] Sefström NG. Ueber das Vanadin, ein Neues Metall, Gefunden im Stangeneisen von Eckersholm, einer Eisenhütte, die Ihr Erz von Taberg in Småland Bezieht. Ann Phys Chem 1831; 97: 43-9.
[http://dx.doi.org/10.1002/andp.18310970103]

[21]   Macara IG. Vanadium — an element in search of a role. Trends Biochem Sci 1980; 5: 92-4.
[http://dx.doi.org/10.1016/0968-0004(80)90256-X]

[22]   McLeod GC, Ladd KV, Kustin K, Toppen DL. Extraction of Vanadium(V) from seawater by tunicates: a revision of concepts. Limnol Oceanogr 1975; 20: 491-3.
[http://dx.doi.org/10.4319/lo.1975.20.3.0491]

[23]   Rehder D. Bioinorganic Vanadium Chemistry. Chichester, UK: John Wiley & Sons, Ltd 2008; pp. 13-51.
[http://dx.doi.org/10.1002/9780470994429.ch2]

[24]   Henze M. Untersuchungen über des Blut der Ascidien. I. Mitteilung. Die Vanadiumverbindung der Blutkörperochen. Hoppe Seylers Z Physiol Chem 1911; 72: 494-501.
[http://dx.doi.org/10.1515/bchm2.1911.72.5-6.494]

[25]   Ueki T, Takemoto K, Fayard B, *et al.* Scanning x-ray microscopy of living and freeze-dried blood cells in two vanadium-rich ascidian species, *Phallusia mammillata* and *Ascidia sydneiensis samea.* Zool Sci 2002; 19(1): 27-35.
[http://dx.doi.org/10.2108/zsj.19.27] [PMID: 12025401]

[26]   Goldberg ED, McBlair W, Taylor BJ. The uptake of vanadium by tunicates. Biol Bull 1951; 101: 84-94.
[http://dx.doi.org/10.2307/1538503]

[27]   Dingley AL, Kustin K, Macara IG, McLeod GC. Accumulation of vanadium by tunicate blood cells occurs *via* a specific anion transport system. Biochim Biophys Acta 1981; 649: 493-502.
[http://dx.doi.org/10.1016/0005-2736(81)90152-8]

[28]   Kalk M. Absorption of vanadium by tunicates. Nature 1963; 198: 1010-1.
[http://dx.doi.org/10.1038/1981010a0]

[29]   Woolfolk CA, Whiteley HR. Reduction of inorganic compounds with molecular hydrogen by *Micrococcus lactilyticus*. I. Stoichiometry with compounds of arsenic, selenium, tellurium, transition and other elements. J Bacteriol 1962; 84: 647-58.
[PMID: 14001842]

[30]   Myers JM, Antholine WE, Myers CR. Vanadium(V) reduction by *Shewanella oneidensis* MR-1 requires menaquinone and cytochromes from the cytoplasmic and outer membranes. Appl Environ Microbiol 2004; 70(3): 1405-12.
[http://dx.doi.org/10.1128/AEM.70.3.1405-1412.2004] [PMID: 15006760]

[31]   Carpentier W, De Smet L, Van Beeumen J, Brigé A. Respiration and growth of *Shewanella oneidensis* MR-1 using vanadate as the sole electron acceptor. J Bacteriol 2005; 187(10): 3293-301.
[http://dx.doi.org/10.1128/JB.187.10.3293-3301.2005] [PMID: 15866913]

[32]   Ortiz-Bernad I, Anderson RT, Vrionis HA, Lovley DR. Vanadium respiration by *Geobacter metallireducens*: novel strategy for *in situ* removal of vanadium from groundwater. Appl Environ Microbiol 2004; 70(5): 3091-5.
[http://dx.doi.org/10.1128/AEM.70.5.3091-3095.2004] [PMID: 15128571]

[33]   Antipov AN, Lyalikova NN, Khijniak TV, L'vov NP. Vanadate reduction by molybdenum-free dissimilatory nitrate reductases from vanadate-reducing bacteria. IUBMB Life 2000; 50(1): 39-42.
[http://dx.doi.org/10.1080/15216540050176575] [PMID: 11087119]

[34]   Carpentier W, Sandra K, De Smet I, Brigé A, De Smet L, Van Beeumen J. Microbial reduction and precipitation of vanadium by *Shewanella oneidensis*. Appl Environ Microbiol 2003; 69(6): 3636-9.
[http://dx.doi.org/10.1128/AEM.69.6.3636-3639.2003] [PMID: 12788772]

[35]   Wang L, Lin H, Dong Y, He Y, Liu C. Isolation of vanadium-resistance endophytic bacterium PRE01 from *Pteris vittata* in stone coal smelting district and characterization for potential use in phytoremediation. J Hazard Mater 2018; 341: 1-9.
[http://dx.doi.org/10.1016/j.jhazmat.2017.07.036] [PMID: 28759788]

[36] Romaidi U, Ueki T. Bioaccumulation of vanadium by vanadium-resistant bacteria isolated from the intestine of ascidia sydneiensis samea. Mar Biotechnol (NY) 2016; 18(3): 359-71.
[http://dx.doi.org/10.1007/s10126-016-9697-5] [PMID: 27177911]

[37] Yurkova N, Lyalikova NN. New facultative chemolithotrophic bacteria reducing vanadate. Mikrobiologiâ 1990; 59: 968-75.

[38] Antipov AN, Lyalikova NN, L'vov NP. Vanadium-binding protein excreted by vanadate-reducing bacteria. IUBMB Life 2000; 49(2): 137-41.
[http://dx.doi.org/10.1080/15216540050022467] [PMID: 10776597]

[39] Kanda T, Nose Y, Wuchiyama J, Uyama T, Moriyama Y, Michibata H. Identification of a vanadium-associated protein from the vanadium-rich ascidian, *Ascidia sydneiensis samea.* Zool Sci 1997; 14(1): 37-42.
[http://dx.doi.org/10.2108/zsj.14.37] [PMID: 9200977]

[40] Samino S, Michibata H, Ueki T. Identification of a novel vanadium-binding protein by EST analysis on the most vanadium-rich ascidian, *Ascidia gemmata.* Mar Biotechnol (NY) 2012; 14(2): 143-54.
[http://dx.doi.org/10.1007/s10126-011-9396-1] [PMID: 21748343]

[41] Trivedi S, Ueki T, Yamaguchi N, Michibata H. Novel vanadium-binding proteins (vanabins) identified in cDNA libraries and the genome of the ascidian *Ciona intestinalis.* Biochim Biophys Acta 2003; 1630(2-3): 64-70.
[http://dx.doi.org/10.1016/j.bbaexp.2003.09.007] [PMID: 14654236]

[42] Ueki T, Satake M, Kamino K, Michibata H. Sequence variation of Vanabin2-like vanadium-binding proteins in blood cells of the vanadium-accumulating ascidian *Ascidia sydneiensis samea.* Biochim Biophys Acta 2008; 1780(7-8): 1010-5.
[http://dx.doi.org/10.1016/j.bbagen.2008.04.001] [PMID: 18466774]

[43] Ueki T, Adachi T, Kawano S, *et al.* Vanadium-binding proteins (vanabins) from a vanadium-rich ascidian *Ascidia sydneiensis samea.* Biochim Biophys Acta 2003; 1626(1-3): 43-50.
[http://dx.doi.org/10.1016/S0167-4781(03)00036-8] [PMID: 12697328]

[44] Yamaguchi N, Kamino K, Ueki T, Michibata H. Expressed sequence tag analysis of vanadocytes in a vanadium-rich ascidian, *Ascidia sydneiensis samea.* Mar Biotechnol (NY) 2004; 6(2): 165-74.
[http://dx.doi.org/10.1007/s10126-003-0024-6] [PMID: 14595550]

[45] Yoshihara M, Ueki T, Watanabe T, Yamaguchi N, Kamino K, Michibata H. VanabinP, a novel vanadium-binding protein in the blood plasma of an ascidian, *Ascidia sydneiensis samea.* Biochim Biophys Acta 2005; 1730(3): 206-14.
[http://dx.doi.org/10.1016/j.bbaexp.2005.07.002] [PMID: 16084607]

[46] Rungrassamee W, Klanchui A, Maibunkaew S, Chaiyapechara S, Jiravanichpaisal P, Karoonuthaisiri N. Characterization of intestinal bacteria in wild and domesticated adult black tiger shrimp (*Penaeus monodon*). PLoS One 2014; 9(3)e91853
[http://dx.doi.org/10.1371/journal.pone.0091853] [PMID: 24618668]

[47] Li P, Burr GS, Gatlin DM III, *et al.* Dietary supplementation of short-chain fructooligosaccharides influences gastrointestinal microbiota composition and immunity characteristics of Pacific white shrimp, *Litopenaeus vannamei,* cultured in a recirculating system. J Nutr 2007; 137(12): 2763-8.
[http://dx.doi.org/10.1093/jn/137.12.2763] [PMID: 18029496]

[48] Tanaka R, Ootsubo M, Sawabe T, Ezura Y, Tajima K. Biodiversity and *in situ* abundance of gut microflora of abalone (*Haliotis discus hannai*) determined by culture-independent techniques. Aquaculture 2004; 241: 453-63.
[http://dx.doi.org/10.1016/j.aquaculture.2004.08.032]

[49] Simon CA, McQuaid C. Extracellular digestion in two co-occurring intertidal mussels (*Perna perna* (L.) and *Choromytilus meridionalis* (Kr)) and the role of enteric bacteria in their digestive ecology. J Exp Mar Biol Ecol 1999; 234: 59-81.

[http://dx.doi.org/10.1016/S0022-0981(98)00141-5]

[50] da Silva SG, Gillan DC, Dubilier N, De Ridder C. Characterization by 16S rRNA Gene Analysis and *in situ* Hybridization of Bacteria Living in the Hindgut of a Deposit-Feeding Echinoid (Echinodermata). J Mar Biol Assoc UK 2006; 86: 1209-13.
[http://dx.doi.org/10.1017/S0025315406014202]

[51] Thorsen MS. Abundance and biomass of the gut-living microorganisms (Bacteria, Protozoa and Fungi) in the irregular sea urchin *Echinocardium cordatum* (Spatangoida: Echinodermata). Mar Biol 1999; 133: 353-60.
[http://dx.doi.org/10.1007/s002270050474]

[52] Harris JM. The presence, nature, and role of gut microflora in aquatic invertebrates: A synthesis. Microb Ecol 1993; 25(3): 195-231.
[http://dx.doi.org/10.1007/BF00171889] [PMID: 24189919]

[53] Hooper LV, Wong MH, Thelin A, Hansson L, Falk PG, Gordon JI. Molecular analysis of commensal host-microbial relationships in the intestine. Science 2001; 291(5505): 881-4.
[http://dx.doi.org/10.1126/science.291.5505.881] [PMID: 11157169]

[54] Brune A, Friedrich M. Microecology of the termite gut: structure and function on a microscale. Curr Opin Microbiol 2000; 3(3): 263-9.
[http://dx.doi.org/10.1016/S1369-5274(00)00087-4] [PMID: 10851155]

[55] Jayasree L, Janakiram P, Madhavi R. Characterization of *vibrio* spp. associated with diseased shrimp from culture ponds of Andhra Pradesh (India). J World Aquacult Soc 2006; 37: 523-32.
[http://dx.doi.org/10.1111/j.1749-7345.2006.00066.x]

[56] Kaschak E, Knopf B, Petersen JH, Bings NH, König H. Biotic methylation of mercury by intestinal and sulfate-reducing bacteria and their potential role in mercury accumulation in the tissue of the soil-living *Eisenia foetida*. Soil Biol Biochem 2014; 69: 202-11.
[http://dx.doi.org/10.1016/j.soilbio.2013.11.004]

[57] Silver S. Bacterial resistances to toxic metal ions-a review. Gene 1996; 179(1): 9-19.
[http://dx.doi.org/10.1016/S0378-1119(96)00323-X] [PMID: 8991852]

[58] Rubin BER, Sanders JG, Hampton-Marcell J, Owens SM, Gilbert JA, Moreau CS. DNA extraction protocols cause differences in 16S rRNA amplicon sequencing efficiency but not in community profile composition or structure. Microbiology Open 2014; 3(6): 910-21.
[http://dx.doi.org/10.1002/mbo3.216] [PMID: 25257543]

[59] Poretsky R, Rodriguez-R LM, Luo C, Tsementzi D, Konstantinidis KT. Strengths and limitations of 16S rRNA gene amplicon sequencing in revealing temporal microbial community dynamics. PLoS One 2014; 9(4)e93827
[http://dx.doi.org/10.1371/journal.pone.0093827] [PMID: 24714158]

[60] Větrovský T, Baldrian P. The variability of the 16S rRNA gene in bacterial genomes and its consequences for bacterial community analyses. PLoS One 2013; 8(2)e57923
[http://dx.doi.org/10.1371/journal.pone.0057923] [PMID: 23460914]

[61] Sun C, Zhao Y, Li H, Dong Y, MacIssac HJ,. ZHAN A. Unreliable quantitation of species abundance based on high-throughput sequencing data of zooplankton communities. Aquat Biol 2015; 24: 9-15.
[http://dx.doi.org/10.3354/ab00629]

[62] Dishaw LJ, Flores-Torres J, Lax S, *et al.* The gut of geographically disparate *Ciona intestinalis* harbors a core microbiota. PLoS One 2014; 9(4)e93386
[http://dx.doi.org/10.1371/journal.pone.0093386] [PMID: 24695540]

[63] Dishaw LJ, Cannon JP, Litman GW, Parker W. Immune-directed support of rich microbial communities in the gut has ancient roots. Dev Comp Immunol 2014; 47(1): 36-51.
[http://dx.doi.org/10.1016/j.dci.2014.06.011] [PMID: 24984114]

[64] Ueki T, Fujie M, Romaidi, Satoh N. Symbiotic bacteria associated with ascidian vanadium

accumulation identified by 16S rRNA amplicon sequencing. Mar Genomi 2019; 43: 33-42.
[http://dx.doi.org/10.1016/j.margen.2018.10.006] [PMID: 30420273]

[65]   Yoshida N, Kato T, Yoshida T, Ogawa K, Yamashita M, Murooka Y. Bacterium-based heavy metal biosorbents: enhanced uptake of cadmium by *E. coli* expressing a metallothionein fused to beta-galactosidase. Biotechniques 2002; 32(3): 551-552, 554, 556 passim.
[http://dx.doi.org/10.2144/02323st08] [PMID: 11911659]

[66]   Pazirandeh M, Chrisey LA, Mauro JM, Campbell JR, Gaber BP. Expression of the *Neurospora crassa* metallothionein gene in *Escherichia coli* and its effect on heavy-metal uptake. Appl Microbiol Biotechnol 1995; 43(6): 1112-7.
[http://dx.doi.org/10.1007/BF00166934] [PMID: 8590662]

[67]   Mauro JM, Pazirandeh M. Construction and expression of functional multi-domain polypeptides in *Escherichia coli*: expression of the *Neurospora crassa* metallothionein gene. Lett Appl Microbiol 2000; 30(2): 161-6.
[http://dx.doi.org/10.1046/j.1472-765x.2000.00697.x] [PMID: 10736021]

[68]   Samuelson P, Wernérus H, Svedberg M, Ståhl S. Staphylococcal surface display of metal-binding polyhistidyl peptides. Appl Environ Microbiol 2000; 66(3): 1243-8.
[http://dx.doi.org/10.1128/AEM.66.3.1243-1248.2000] [PMID: 10698802]

[69]   Kotrba P, Dolecková L, de Lorenzo V, Ruml T. Enhanced bioaccumulation of heavy metal ions by bacterial cells due to surface display of short metal binding peptides. Appl Environ Microbiol 1999; 65(3): 1092-8.
[PMID: 10049868]

[70]   Mejàre M, Ljung S, Bülow L. Selection of cadmium specific hexapeptides and their expression as OmpA fusion proteins in *Escherichia coli*. Protein Eng 1998; 11(6): 489-94.
[http://dx.doi.org/10.1093/protein/11.6.489] [PMID: 9725628]

[71]   Ueki T, Sakamoto Y, Yamaguchi N, Michibata H. Bioaccumulation of copper ions by *Escherichia coli* expressing vanabin genes from the vanadium-rich ascidian *Ascidia sydneiensis samea*. Appl Environ Microbiol 2003; 69(11): 6442-6.
[http://dx.doi.org/10.1128/AEM.69.11.6442-6446.2003] [PMID: 14602598]

[72]   Tajer Mohammad Ghazvini P, Ghorbanzadeh Mashkani S. Effect of salinity on vanadate biosorption by *Halomonas* sp. GT-83: preliminary investigation on biosorption by micro-PIXE technique. Bioresour Technol 2009; 100(8): 2361-8.
[http://dx.doi.org/10.1016/j.biortech.2008.11.025] [PMID: 19117752]

CHAPTER 12

Electrochemical Detection of Microorganisms Using an Electrochemical Quartz Crystal Microbalance

Takeshi Kougo[*]

Department of Materials Science and Engineering, National Institute of Technology (KOSEN), Suzuka College, Suzuka, Japan

Abstract: We attempted evaluation of biofilm formation by the electrochemical quartz crystal microbalance (EQCM) method. The EQCM method is applicable to *in situ* settings and may be able to determine corrosion and a chemical reaction on materials under the influence of biofilm formation. We used Au having high chemical stability to measure frequency and changes of the potential with biofilm formation alone. Additionally, sample surfaces were examined by a conventional biofilm formation assay such as fluorescent X-ray analysis and Raman spectroscopy. We conducted three types of experiments, 1st: rest-potential measurement, 2nd: spontaneous-potential measurement, and 3rd: constant-potential measurement. From the results of these measurements, it was concluded that frequency changes depend on changes in the potential. We investigated biofilm formation by conventional methods of evaluation and found that the above phenomenon was caused by concentration of ions such as Ca^{2+} and Cl^- from tap water in biofilm. In the positive voltage range, the magnitude of biofilm formation was inhibited by a halogen ion such as Cl^-, whereas in the negative range, cations such as Ca^{2+} were concentrated at an accelerated rate. Therefore, electrochemical evaluation of biofilm formation was carried out successfully by means of Au, with high chemical stability. In conclusion, the EQCM method appears to be able to measure biofilm formation.

Keywords: Biofilm, Electrochemical, EQCM, Evaluation, *In-situ* measurement.

INTRODUCTION

A biofilm is a sticky and gel-like substance produced by bacteria and has been an attractive subject in several fields of research [1 - 4]. For example, biofilms have caused serious problems in various industrial and medical fields [5]. In industrial fields, mineralization of a cooling pipe is caused by a biofilm, because mineral io-

[*] **Corresponding author Takeshi Kougo:** Department of Materials Science and Engineering, National Institute of Technology (KOSEN), Suzuka College, Suzuka, Japan; Tel; +81-59-368-1849; Fax; +81-59-368-1849; E-mail: kougo@mse.suzuka-ct.ac.jp

Toshiyuki Takahashi (Ed.)

ns, such as Ca^{2+}, Si^{4+}, and Na^+ are concentrated in the biofilm and are deposited as hydroxide and carbonate [6]. This mechanism is mediated by biofilm stickiness. In addition, biofilms function as a protective layer and become a hotbed of microbial activity. However, biofilms have been gainfully utilized for purification of food and the environment [7] (Fig. **1**) shows the process of biofilm formation. At first, some organic matter and inorganic matter adhere to a substance, then conditioning films are formed (a). Next, floating bacteria adhere to the substrate (b) and produce extracellular polymeric substances (EPSs) with quorum sensing (c). This colony grows up further and finally becomes a biofilm (d), and this biofilm collapses and scatters to various places (e).

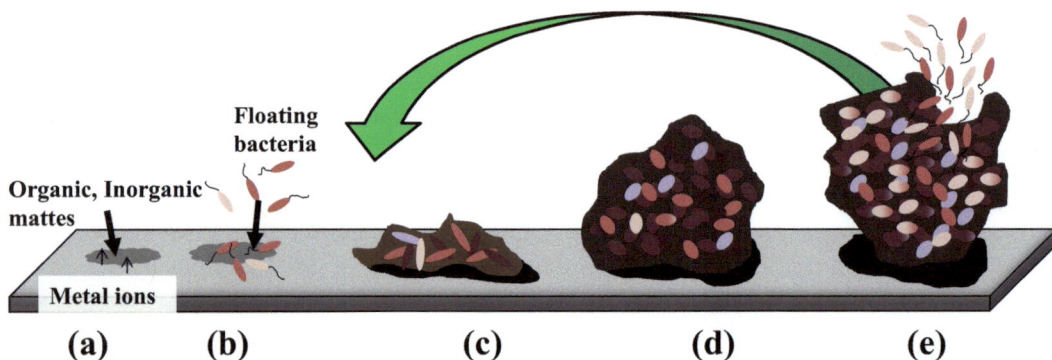

Fig. (1). The process of biofilm formation.

The evaluation of biofilm formation has been conducted for a long time by many researchers, just as we have. The most popular and simple evaluation method is staining with a dye such as Crystal Violet and Alcian Blue [8 - 11]. Nonetheless, this method poses a problem because the dye adsorbs on a substance without a biofilm. There are several reasons such as chemical bonding to the substance, over-adsorption, and aggregation. In particular, these problems have been occurring on porous materials and materials that contain metal oxides. These mechanisms have been researched as a possible dye-sensitized solar cell [12, 13]. As another method, we have utilized Raman spectroscopy, Fourier transform infrared spectroscopy with attenuated total reflection (FT-IR ATR), and X-ray fluorescence analysis (XRF) [14 - 18]. These methods are very effective at identifying a biofilm and attributing phenomena to a biofilm. However, biofilm formation has to be stopped, and continuous evaluation methods are not always suitable. In addition, these methods can analyze a local area or spot; therefore, researchers have hoped to evaluate these phenomena more widely, comprehensively, and quantitatively with *in situ* measurement. As a possible new assay that may solve these problems, we focused on the electrochemical quartz crystal microbalance (EQCM) method [19]. The EQCM method is an assay of behavior of organic matter on an electrode with electrochemical measurement. If a quartz oscillator is used as a working electrode, it is possible to evaluate the phenomena occurring on the quartz oscillator *via* a frequency change and potential change. For these reasons, it is possible to study an electrochemical reaction on a substance with *in situ* measurement of biofilm formation. As a matter of fact, some authors have reported evaluation of biofilm formation by QCM and EQCM [20, 21]. However, most of these studies involve a system with a single bacterial species, and those authors did not evaluate a biofilm formed by ubiquitous bacteria as we did. In this study, we evaluated a biofilm formation process by the EQCM method and discussed the results.

EXPERIMENT

Fig. (**2**) shows a scheme of evaluation of biofilm formation using EQCM. We measured biofilm formation with an EQCM instrument (HQ-101B, HOKUTO DENKO Co., Ltd.). The working electrode (QCM math sensor), the counterelectrode (Pt), and the reference electrode (Ag-AgCl) were installed in the biofilm reactor. The conditions of biofilm formation and the scheme are described elsewhere [22]. The amount of matter on an electrode can be estimated by Sauerbrey's equation as follows [19]:

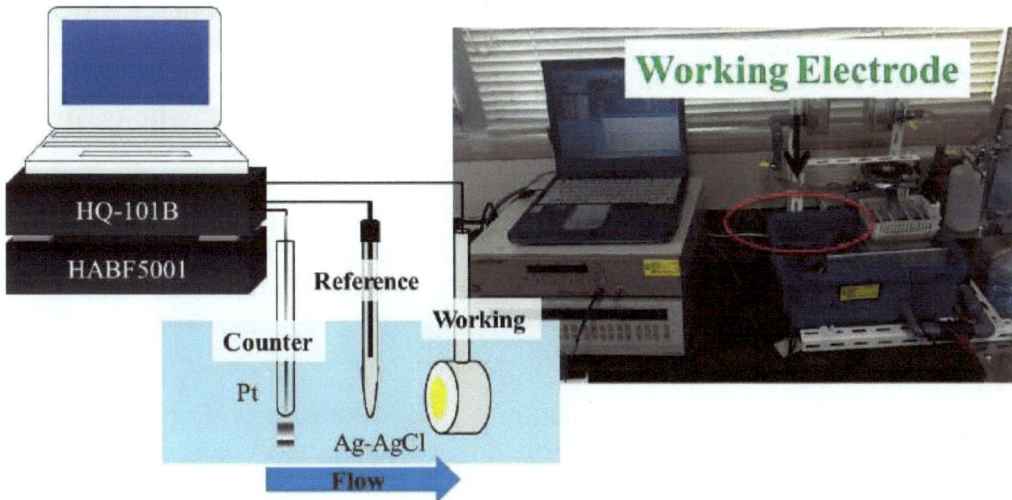

Fig. (2). The scheme of a biofilm formation evaluation system based on EQCM.

$$\Delta F = \frac{2F_0^2 \Delta m}{A\sqrt{\rho_q \mu_q}}$$

(1)

Fig. (**3**) shows the mechanism of detection of biofilm formation by QCM, where ΔF is the frequency shift (Hz), Δm is the mass change, $F0$ is the primary frequency of the quartz crystal (6.0 MHz), A is the Au electrode area (1.33 cm^2), ρ_q is the density of quartz (2.648 g/cm^3), and μ_q is the shear modulus of quartz (2.947 × 10^{11} dyn/cm^2). At the first step, we confirmed that the effect of biofilm formation can be evaluated with Au, which was stable chemically, and the potential was not accelerated (0 mV, resting potential). At the next step, we evaluated a natural immersion potential. The solution for biofilm formation was tap water, which was kept at 27±1°C, and water was flowed gently.

Fig. (3). The mechanism of detection of biofilm formation by QCM.

As other assays of biofilm formation, we used the following methods. A dye staining method was employed with a 0.05 wt% Crystal Violet (WAKO Co., Ltd.) solution. After biofilm formation samples were immersed in the Crystal Violet solution for 30 min, they were rinsed with distilled water. These samples were examined in detail under a video microscope (VW-9100, Keyence Co., Ltd.). Detailed morphological and elemental analyses of the biofilm were carried out under a low-vacuum scanning electron microscope (SEM) (TM-1000, Hitachi High-Technologies) with X-ray fluorescence analysis. To identify functional groups, a laser Raman spectrometer (NRS-3000, Nippon Bunko) was used.

RESULTS AND DISCUSSION

First, we applied EQCM at a resting potential to a biofilm reactor. Fig. (**4**) shows the result of EQCM at 0 mV. It was confirmed that frequency decreased

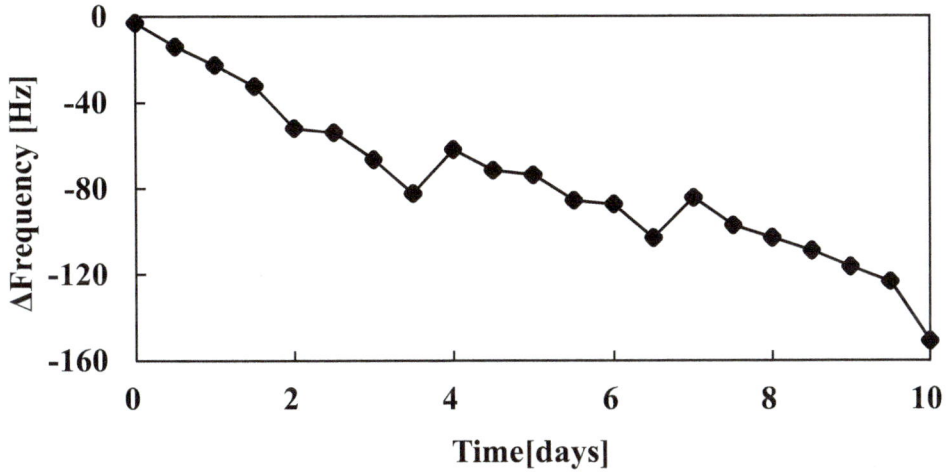

Fig. (4). The results of QCM on frequency change at 0 mV.

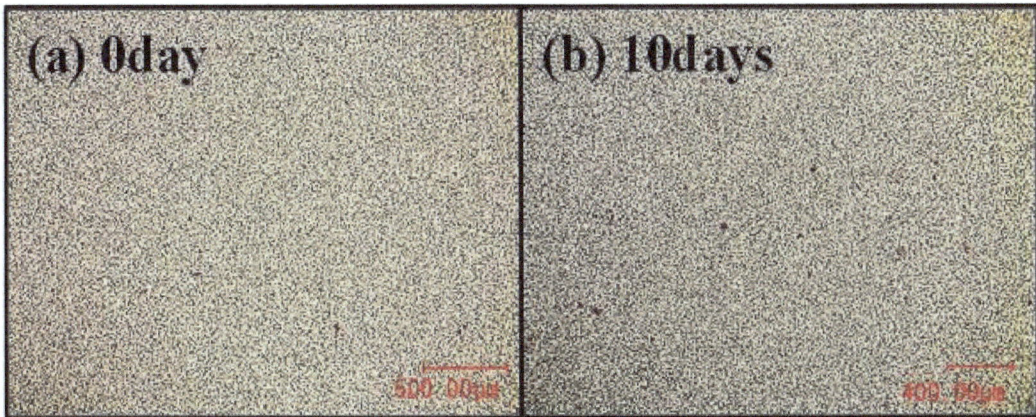

Fig. (5). The images from a video microscope. **(a)** Before biofilm formation, **(b)** 10 days after biofilm formation.

Table 1. The concentrations of elements in biofilm.

	Au	Na	Si	Cl	Ca
0 days	100.0	0.0	0.0	0.0	0.0
1 days	98.7	0.1	0.0	0.5	0.7
3 days	98.3	0.0	0.5	0.7	0.4
10 days	88.9	0.0	1.2	2.3	7.6

Next, we investigated the biofilm formation with electrical behavior. Fig. (**8**) shows the results of measuring the natural immersion potential and frequency change by the EQCM method. The primary potential was found to be +530 mV. This value was generated by the Au-Pt electrode together with the Ag-AgCl reference electrode. The potential underwent a negative shift, and frequency decreased with time. We focused on these behaviors, and it was confirmed that the behavior of the frequency changed at +250 mV. The frequency and the potential were observed to rapidly decrease above +250 mV. However, the potential shifted to negative slowly, and the frequency increased and decreased at +250 mV. These results suggested that the frequency behavior correlates with the potential. Therefore, to investigate the biofilm formation with the potential, the frequency change was measured at a constant potential. Fig. (**9**) shows the result of EQCM measurement at -250, 0, and 250 mV. Compared with 0 mV, it was confirmed that the frequency decrease was suppressed at +250 mV. However, the frequency decreased greatly at -250 mV. It was concluded from these results that the biofilm formation process was affected by ionic and electrochemical actions on the electrode. Therefore, the surface data from QCM were compared for each voltage by Raman spectroscopy and XRF measurement.

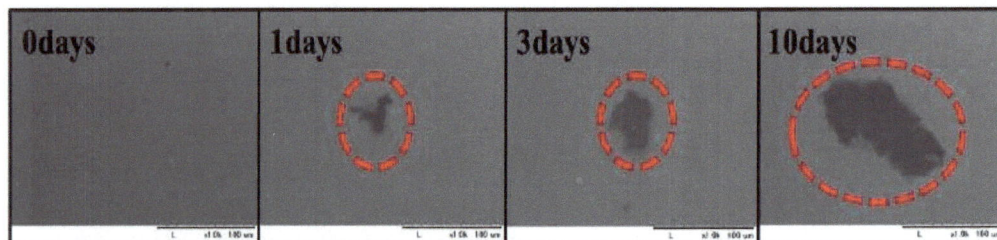

Fig. (6). Examination of a biofilm on an electrode by SEM.

Table 2. The relation between the potential and biofilm composition.

	Au	Na	Si	Cl	Ca	others
0 mV	27.5	5.2	7.4	21.0	12.5	19.2
+250 mV	85.4	1.9	2.4	4.3	3.4	0.8
-250 mV	0.0	0.1	0.4	0.0	99.5	0.0

Table **2** and Fig. (**10**) show the morphology of the biofilm after SEM examination and elemental analysis. It was confirmed that the biofilm size at 0 mV was over 100 μm wide, but at +250 mV, the biofilm size was smaller than that at 0 mV. However, it was observed that an acicular crystal was deposited on the electrode at -250 mV. The data in Table **2** confirm that the biofilm included components of tap water such as Ca and Cl. Compared with 0 mV and +250 mV, the biofilm became smaller; thus, it was demonstrated that the proportion of Au as a base electrode increased and that the acicular crystal mostly consists of Ca. (Fig. **11**) shows the results of Raman spectroscopy. It was confirmed that no peaks were present on day 0, because there was only Au as the electrode. After 10 days passed, some peaks were present because of the deposit on the electrode. At 0 mV, some peaks were detected as follows: 1) 1550 to 1450 cm^{-1}, 2) 1300 to 1260 cm^{-1}, 3) 1180 to 1110 cm^{-1}, and 4) 1050 to 900 cm^{-1}. At +250 mV, these peaks were at the same positions as at 0 mV but were smaller and broader. However, at -250 mV, a sharp and big peak was confirmed: 5) at 1080 to 1070 cm^{-1}. According to the report of Takagi, this peak was attributed to $CaCO_3$ [23]. What is the reason for this result? We propose the following mechanism (Fig. **12**): Ca^{2+} in water was attracted to the electrode at a negative potential and was drawn into the biofilm.

There were many bacterial species in the biofilm, CO_2 was produced mainly by aerobic bacteria, and $CaCO_3$ was deposited in the biofilm. Therefore, the big frequency decrease at a negative potential was dependent on deposited $CaCO_3$. However, Cl^- in water was attracted to the electrode at a positive potential. These halogenic ions are toxic to creatures, especially at high concentrations. Therefore, it was assumed that the growth of bacteria was suppressed by Cl^- and biofilm formation was suppressed. Of course, it is possible that quorum sensing and direct redox action on bacteria could be affected, and this topic warrants further research in the future. However, we concluded that EQCM was very effective at evaluating biofilm formation.

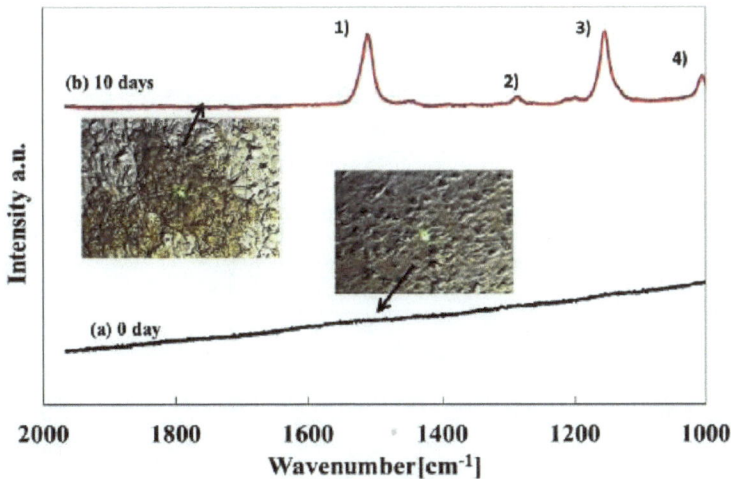

Fig. (7). The results of Raman spectroscopy.
(**a**) before biofilm formation, (**b**) 10 days after biofilm formation.

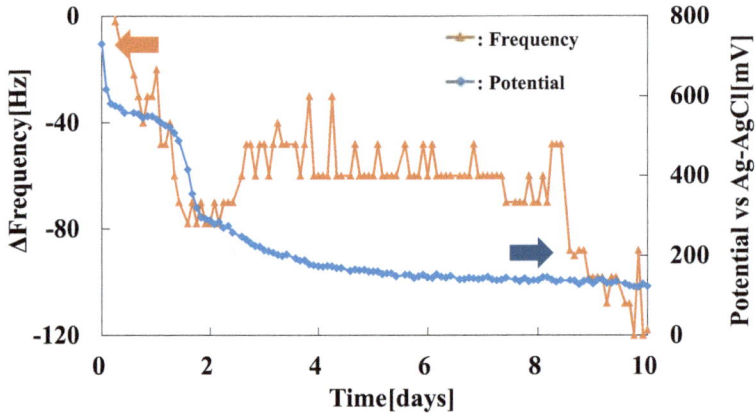

Fig. (8). The results of EQCM with a frequency change and potential change.

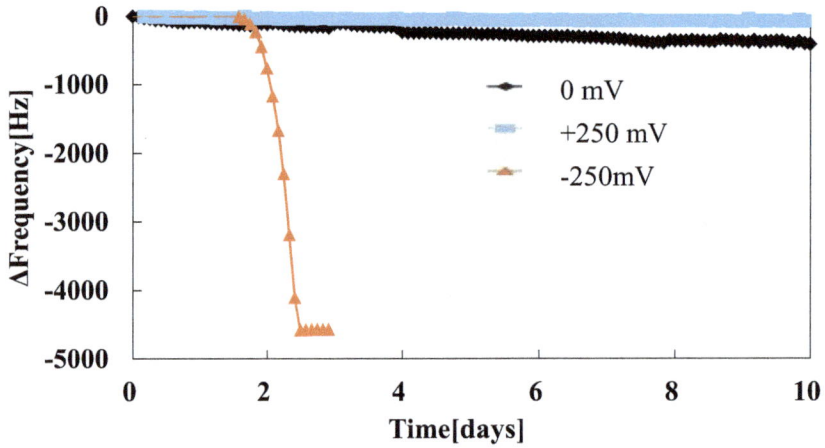

Fig. (9). The result of EQCM measurement depending on the potential.

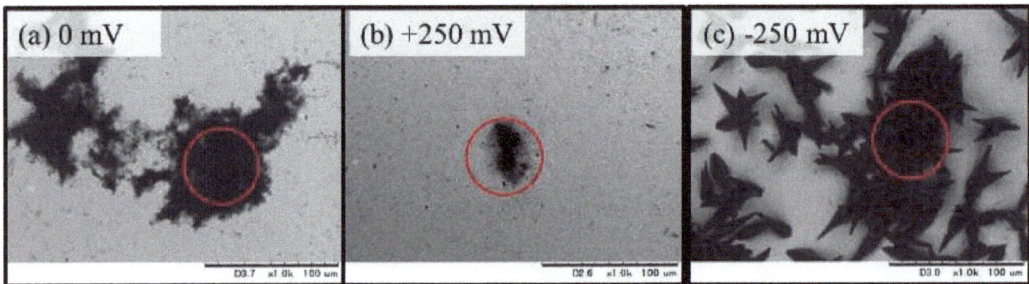

Fig. (10). Examination of a biofilm on the electrode by SEM.
(**a**) 0 mV, (**b**) +250 mV, (**c**) -250 mV.

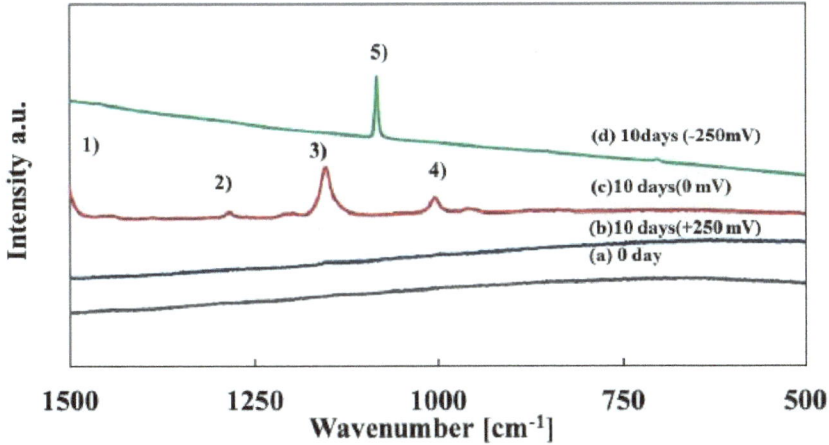

Fig. (11). Examination of biofilm on the electrode by Raman spectroscopy.
(**a**) before biofilm formation, and biofilm formation after 10 days at (**b**) +250 mV, (**c**) 0 mV,(**d**) -250 mV.

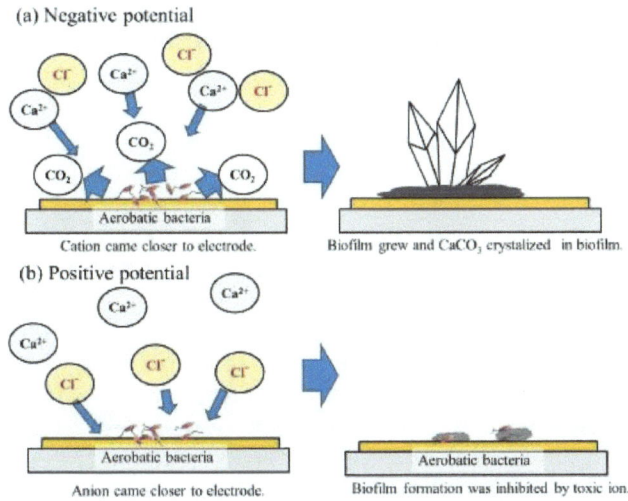

Fig. (12). The mechanism of biofilm formation depending on a potential.

CONCLUSION

In this study, we tested a biofilm formation assay based on the EQCM method. Chemically stable Au was used to confirm the effects on biofilm formation *via* three patterns, 1st: resting-potential measurement, 2nd: spontaneous-potential measurement, 3rd: constant-potential measurement. These results confirmed (because of the frequency change) that the formation behavior of the biofilm is related to the potential. Therefore, it can be concluded that evaluation of biofilm formation by the EQCM method is possible.

CONSENT FOR PUBLICATION

Not applicable.

CONFLICT OF INTEREST

The author confirms that this chapter contents have no conflict of interest.

ACKNOWLEDGEMENTS

Declare none.

REFERENCES

[1]   Costerton JW, Lewandowski Z, Caldwell DE, Korber DR, Lappin-Scott HM. Microbial biofilms. Annu Rev Microbiol 1995; 49: 711-45.
[http://dx.doi.org/10.1146/annurev.mi.49.100195.003431] [PMID: 8561477]

[2]   Costerton JW, Stewart PS, Greenberg EP. Bacterial biofilms: A common cause of persistent infections. Science 1999; 284(5418): 1318-22.
[http://dx.doi.org/10.1126/science.284.5418.1318] [PMID: 10334980]

[3]   Characklis WG, Marschall KC. Biofilms. New York, U.S.: John Wiley Sons, Inc. 1999.

[4]   Mittleman MW. Biological fouling of purified-water system: Part 1, Bacterial Growth and Replication. Microcontamination 1985; 3(10): 51-5.

[5]   Flemming HC. Biofouling in water systems-cases, causes and countermeasures. Appl Microbiol Biotechnol 2002; 59(6): 629-40.
[http://dx.doi.org/10.1007/s00253-002-1066-9] [PMID: 12226718]

[6]   Kougo T, Yamamoto Y, Naiki A, *et al.* Proposal for evaluation to biofilm formation on metal oxide in atmospheric exposure type circulation water system Kosen Kiyou NIT SUZUKA college 2013; 44: 81-3.

[7]   Shukla SK, Mangwani N, Rao TS, Das S. Biofilm-Mediated Bioremediation of Polycyclic Aromatic Hydrocarbons, Microbial Biodegradation and Bioremediation. 1st ed. Oxford: Elsevier 2014; pp. 203-32.

[8]   Christensen GD, Simpson WA, Younger JJ, *et al.* Adherence of coagulase-negative staphylococci to plastic tissue culture plates: A quantitative model for the adherence of staphylococci to medical devices. J Clin Microbiol 1985; 22(6): 996-1006.
[PMID: 3905855]

[9]   Erukhimovitch V, Tsror L, Hazanovsky M, *et al.* Identification of fungal phyto-pathogens by Fourier-transform infrared (FTIR) microscopy. J Agric Sci Technol 2005; 1: 145-52.

[10]   Rayner J, Veeh R, Flood J. Prevalence of microbial biofilms on selected fresh produce and household surfaces. Int J Food Microbiol 2004; 95(1): 29-39.
[http://dx.doi.org/10.1016/j.ijfoodmicro.2004.01.019] [PMID: 15240072]

[11]   Hiraki A, Tsuchiya Y, Fukuda Y, Yamamoto T, Kurniawan A, Morisaki H. Analysis of How a Biofilm Forms on the Surface of the Aquatic Macrophyte Phragmites australis. Microbes Environ 2009; 24(3): 265-72.
[http://dx.doi.org/10.1264/jsme2.ME09122] [PMID: 21566383]

[12]   Zheng H, Tachibana Y, Kalantar-Zadeh K. Dye-sensitized solar cells based on WO$_3$. Langmuir 2010; 26(24): 19148-52.
[http://dx.doi.org/10.1021/la103692y] [PMID: 21077615]

[13]   El-Agez TM, Taya SA, Elrefi KS, Abdel-Latif MS. Dye-sensitized solar cells using some organic dyes as photosensitizers. Opt Appl 2014; 44(2): 345-51.

[14]   Schmitt J, Flemming HC. FTIR-spectroscopy in microbial and material analysis. Int Biodeterior Biodegradation 1998; 41(1): 1-11.
[http://dx.doi.org/10.1016/S0964-8305(98)80002-4]

[15]   Sano K, Kanematsu H, Kogo T, Hirai N, Tanaka T. Corrosion and biofilm for a composite coated iron observed by FTIR-ATR and Raman spectroscopy Transactions of the IMF. 139-45.
[http://dx.doi.org/10.1080/00202967.2016.1167315]

[16]   Chao Y, Zhang T. Surface-enhanced Raman scattering (SERS) revealing chemical variation during biofilm formation: from initial attachment to mature biofilm. Anal Bioanal Chem 2012; 404(5): 1465-75.
[http://dx.doi.org/10.1007/s00216-012-6225-y] [PMID: 22820905]

[17]   Kanematsu H, Kudara H, Kanesaki S, *et al.* Application of a Loop-Type Laboratory Biofilm Reactor to the Evaluation of Biofilm for Some Metallic Materials and Polymers such as Urinary Stents and Catheters. Material 2016; 9(10): 824-34.
[http://dx.doi.org/10.3390/ma9100824]

[18]   Mages M. Tumpling jung, Veen, A. V. D., Baborowski, M., Element determination in natural biofilms of mine drainage water by total reflection X-ray fluorescence spectrometry. Spectrochim Acta B At Spectrosc 2006; 61(10-11): 1146-52.
[http://dx.doi.org/10.1016/j.sab.2006.05.007]

[19]   Sauerbrey G. Verwendung von Schwingquarzen zur Wägung dünner Schichten und zur Mikrowägung. Z Phys 1959; 55(2): 206-22.
[http://dx.doi.org/10.1007/BF01337937]

[20]   Nivens DE, Chambers JQ, Anderson TR, White DC. Long-term, on-line monitoring of microbial biofilms using a quartz crystal microbalance. Anal Chem 1993; 65: 65-9.
[http://dx.doi.org/10.1021/ac00049a013]

[21]   Tam K, Kinsinger N, Ayala P, Qi F, Shi W, Myung NV. Real-time monitoring of Streptococcus mutans biofilm formation using a quartz crystal microbalance. Caries Res 2007; 41(6): 474-83.
[http://dx.doi.org/10.1159/000108321] [PMID: 17851235]

[22]   Kanematsu H, Kougo T, Kuroda D, Itho H, Ogino Y, Yamamoto Y. Biofilm formation derived from ambient air and the characteristics of apparatus Journal of Physics: Conference Series 2013; 433(6): 012031.1-.

[23]   Takagi H. Differentiation of inorganic salt by Raman spectroscopy - Focusing on calcium salt distributed in cultural assets. Bunkazai Jouhougaku Kenkyuu 2012; 9: 7-11.

CHAPTER 13

In-situ Observation of Biofilms in Physiological Salt Water by Scanning Ion Conductance Microscopy

Nobumitsu Hirai[1,*], Futoshi Iwata[2] and Hideyuki Kanematsu[3]

[1] *Department of Chemistry and Biochemistry, National Institute of Technology (KOSEN), Suzuka College, Suzuka, Japan*

[2] *Graduate School of Integrated Science and Technology, Shizuoka University, Hamamatsu, Japan*

[3] *Department of Materials Science and Engineering, National Institute of Technology (KOSEN), Suzuka College, Suzuka, Japan*

Abstract: Biofilms formed on glass plates or polystyrene petri dishes were investigated *in-situ* in physiological salt water by scanning ion conductance microscopy (SICM). Environmental biofilms formed in a laboratory biofilm reactor (LBR) as well as biofilms formed by a single bacterium, *Aliivibrio fischeri*, could be observed by SICM. Maintaining consistency of salt concentration in the water used for SICM observation and that used in the LBR and the addition of glutaraldehyde to water before observation might be effective for clear and stable observation of biofilms by SICM.

Keywords: Biofilm, Extracellular polymeric substances (EPSs), *In situ*, Microorganisms, Scanning ion conductance microscopy (SICM), Scanning probe microscopy (SPM).

INTRODUCTION

Biofilms consist of components, such as water, microorganisms, and extracellular polymeric substances (EPSs) produced by the microorganisms, and they exist at the interface of material surfaces and water. It is well known that EPSs work as a barrier to biocides and/or environmental change. Visualization of the initial stage of biofilm formation can provide key information for preventing biofilm formation. Several techniques, such as optical microscopy (OM) [1], environmental scanning electron microscopy (SEM) [2], and atomic force microscopy (AFM) [3], can be used to visualize biofilms formed on various subs-

---

* Corresponding author Nobumitsu Hirai: Department of Chemistry and Biochemistry, National Institute of Technology (KOSEN), Suzuka College, Suzuka, Japan; Tel: +81-59-368-1823; Fax: +81-59-368-1820; E-mail: hirai@chem.suzuka-ct.ac.jp

Toshiyuki Takahashi (Ed.)

trates. In particular, confocal laser scanning microscopy is one of the most powerful tools for biofilm observation in water at the micron scale, and a lot of interesting information regarding the early stage of biofilm formation has been revealed using this technique [4, 5]. However, confocal laser scanning microscopy presents some difficulties in observation at a very early stage of biofilm formation when the biofilm thickness is less than a few microns, because reflection of the laser from the substrate affects the observation.

AFM, which is a kind of scanning probe microscopy (SPM), is one of the most powerful tools for observing material surfaces in vacuum, air, and water. In particular, its resolution is superior to that of other techniques, due to which, this technique is suitable for observing very early stages of biofilm formation [6 - 9]. Microorganisms in the biofilm can be clearly detected by AFM; however, AFM observation of EPSs in the biofilm can be affected by the AFM probe because EPSs may be much softer than microorganisms. Scanning ion conductance microscopy (SICM) is also a type of SPM that can yield a three-dimensional topography of the material surface placed in water at the submicron scale by detecting ion currents [10, 11]. The effect of the SICM probe is much smaller than that of the AFM probe [11], which is very important for biofilm observation by SICM. To our knowledge, there have been no experimental trials for the observation of biofilms by SICM. In this paper, biofilms formed on glass plates or polystyrene petri dishes were investigated *in-situ* in physiological salt water by means of SICM.

EXPERIMENTAL

Two kinds of substrates, a glass plate (10 mm × 10 mm × 1 mm) and a polystyrene petri dish, were set into a laboratory biofilm reactor (LBR, Fig. (**1**)) to form environmental biofilms on the substrates. The LBR was structured as follows [12]. A transparent acrylic column of 50 mm diameter and 80 mm length was prepared. A long support was inserted over the entire length of the column, so that its several artificial projections could allow the fixing of the substrates with screws and holes. Both ends of the column had flanges of 60 mm diameter, which allowed pipes from/to the circulation system to be closely connected to the column. The bottom tank was filled with tap water or artificial seawater, whose temperature was controlled at 30°C by a heater, and a pump was used to circulate water through the system at a specified flow rate (5 L min^{-1}). At a specific point in the circulation system, after water flowed out of the column and just above the bottom tank, the system was designed to pour water over the patterned target surface of a plate. On the plate, the organisms in the ambient air were mixed with water. Air was also blown onto the plate by a fan set just above it, resulting in an increased number of bacteria in the circulating water. After circulation for 1 week, the plate was moved from the LBR to the SICM system. The 16S rRNA gene analysis of the microorganisms in the biofilm formed on the glass plate produced almost the same results as reported previously [13].

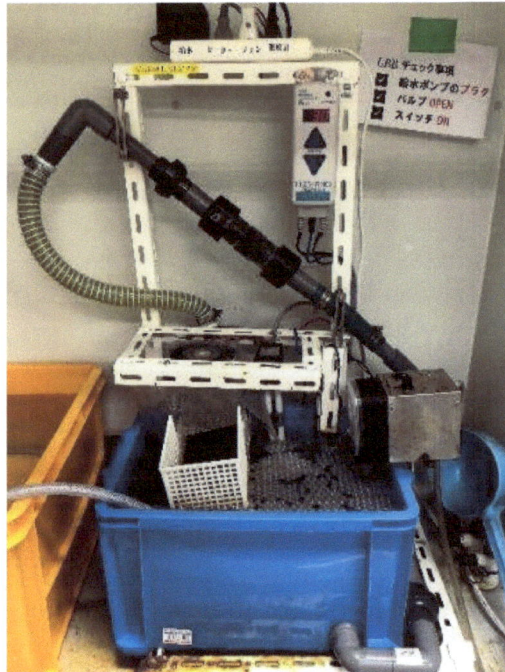

Fig. (1). The laboratory biofilm reactor (LBR) used in this study.

The procedure to form the biofilm with a single kind of bacterium, *Aliivibrio fischeri* (JCM18803), was as follows; *Aliivibrio fischeri* was cultivated in Marine Broth at 22°C for 2 days. Then, a glass plate (10 mm × 10 mm × 1 mm) was placed into 16 times-diluted Marine Broth at 22°C for 2 days. The plate was then moved to the SICM system.

In-situ SICM observation was performed in physiological salt water. The SICM probe consists of a glass micropipette filled with an electrolyte with an Ag/AgCl electrode. Obtaining one image of 32 × 32, 64 × 64, 128 × 128, or 256 × 256 pixels required about 7, 30, 120, or 480 minutes, respectively.

RESULTS AND DISCUSSION

SICM Observation of an Environmental Biofilm Formed in Tap Water in an LBR

Fig. (**2**) shows the SICM images (21 μm × 19 μm) of environmental biofilms formed on a glass surface placed in physiological salt water (**A**) immediately, (**B**) at 6 min, or (**C**) at 28 min after formation of a biofilm in tap water in an LBR for 1 week. The pixels of the images shown in Figs. (**2A, B and C**) are 32 × 32, 32 × 32, and 64 × 64, respectively. These figures indicate that the environmental biofilms formed on a glass plate in an LBR can be observed using SICM.

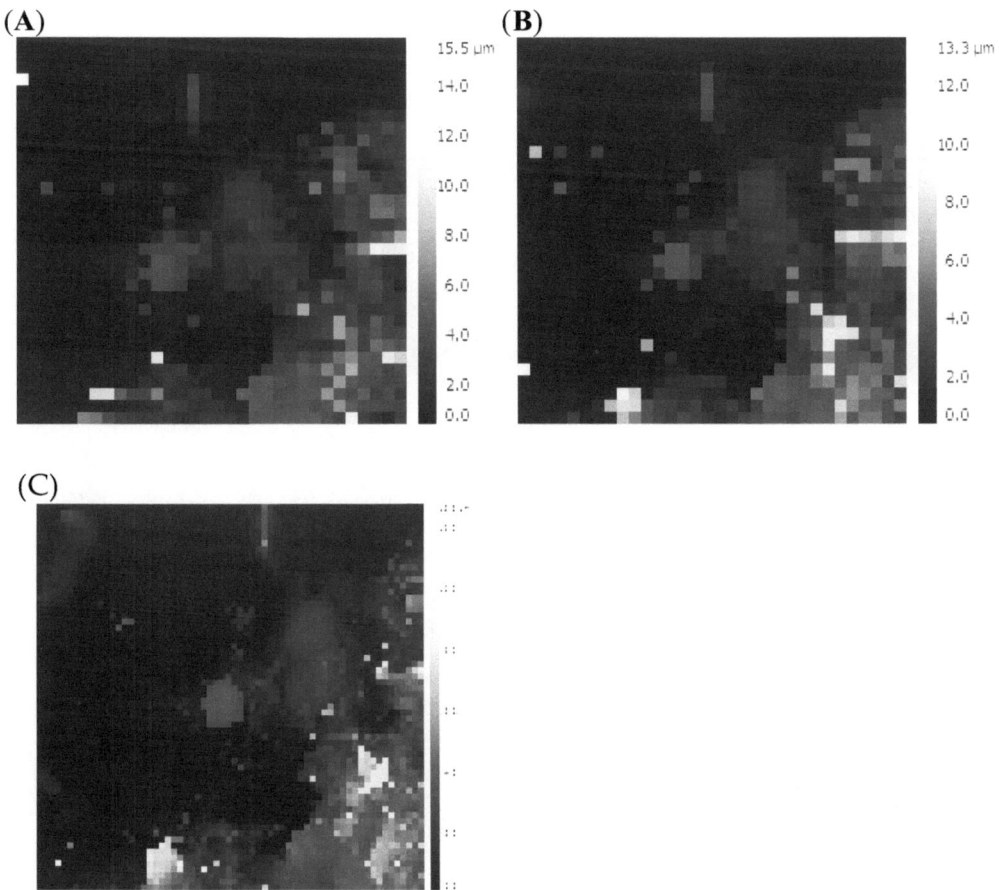

Fig. (2). SICM images (21 μm × 19 μm) of environmental biofilm formed on the glass surface in physiological salt water. (**A**) Immediately, (**B**) at 6 min, and (**C**) at 28 min after biofilm formation in tap water in an LBR for 1 week. The pixels of the images obtained in Figs. (**2A, B and C**) are 32 × 32, 32 × 32, and 64 × 64, respectively.

Fig. (**3**) shows the SICM images (21 μm × 19 μm) of environmental biofilms formed on the glass surface in physiological salt water at (**A**) 3 h, (**B**) 3 h 8 min, or (**C**) 3 h 36 min after biofilm formation in tap water in an LBR for 1 week. The pixels of the images obtained in Figs. (**3A, B and C**) are 32 × 32, 32 × 32, and 64

× 64, respectively. Fig. (**3**) indicates that the biofilm collapsed in a few minutes. The instability of the biofilm may have been caused by the difference in the water used for biofilm formation (tap water) and that used for SICM observation (physiological salt water).

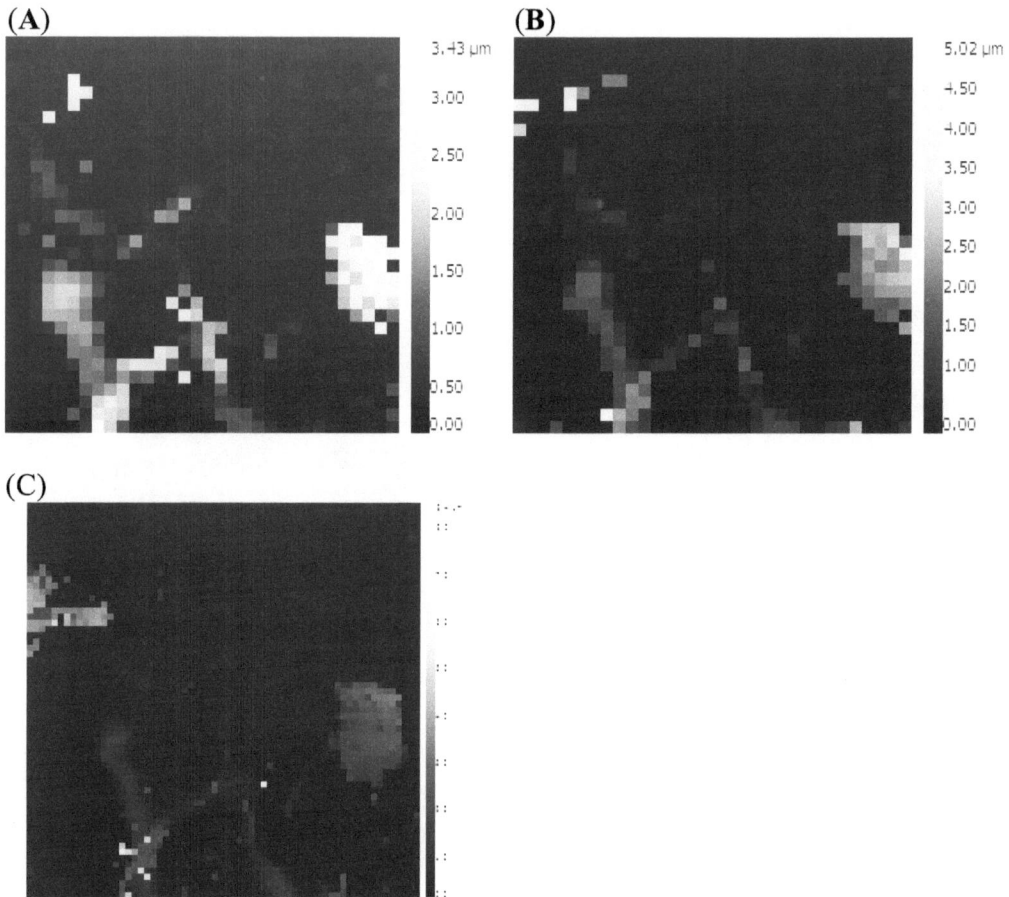

Fig. (3). SICM images (21 μm × 19 μm) of the environmental biofilm formed on a glass surface in physiological salt water. (A) 3 h, (B) 3 h 8 min, and (C) 3 h 36 min after biofilm formation in tap water in an LBR for 1 week. The pixels of the images obtained in Figs. (**3A, B and C**) are 32 × 32, 32 × 32, and 64 × 64, respectively.

SICM Observation of the Environmental Biofilm Formed in Artificial Seawater in an LBR

Fig. (**4**) shows the SICM images (40 μm × 40 μm) of the environmental biofilm formed on a glass surface in physiological salt water at (**A**) 3-4 h, (**B**) 4-7 h, or (**C**) 16-17 h after biofilm formation in diluted artificial seawater (salinity concentration; 1%) in an LBR for 1 week. The pixels of the obtained images in Figs. (**4A, B and C**) are 128 × 128, 256 × 256, and 128 × 128, respectively. As shown in Fig. (**4**), twisted and tangled "wires" of about 1-2 μm thickness was observed on the glass surface. The observed "wires" might be Actinomycetes. Notably, the detailed structure of the biofilm is clearly observed in the images obtained by SICM in comparison with the images in Figs. (**2** and **3**). This indicates that the consistency of salt concentration of the water used for SICM observation with that of the water used in the LBR might be important for the clear and stable observation of biofilms by SICM.

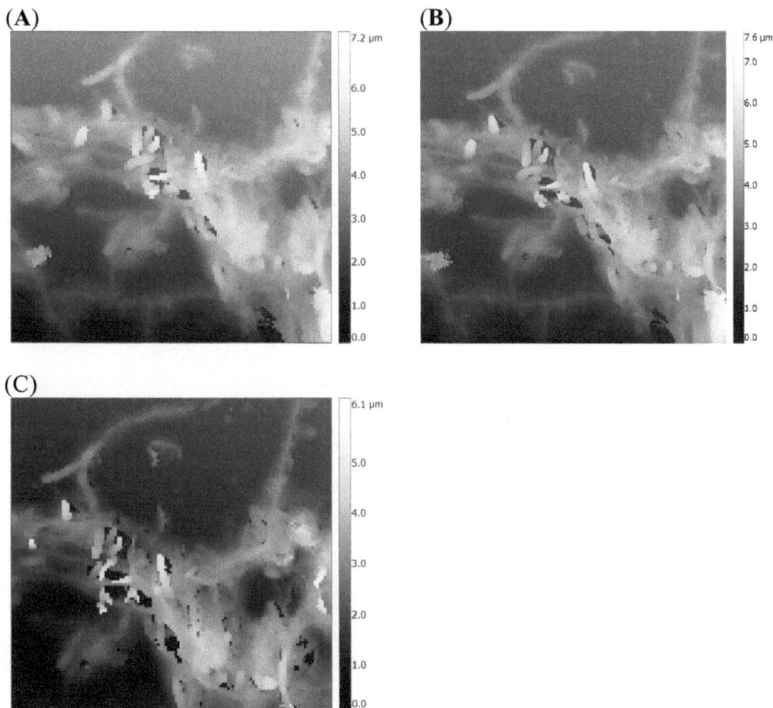

Fig. (4). SICM images (40 μm × 40 μm) of the environmental biofilm formed on a glass surface in physiological salt water. (**A**) 3-4 h, (**B**) 4-6 h, and (**C**) 16-17 h after biofilm formation in diluted artificial seawater (salinity concentration; 1%) in LBR for 1 week. The pixels of the images obtained in Figs. (**4A, B and C**) are 128 × 128, 256 × 256, and 128 × 128, respectively.

Effect of Glutaraldehyde Addition to Water before Environmental Biofilm Observation

Fig. (**5**) shows the SICM images (48 μm× 53 μm) (**A**) of the environmental biofilm formed on glass or (**B**) a polystyrene petri dish in physiological salt water at 1 day after biofilm formation in diluted artificial seawater (salinity concentration: 1%) in an LBR for 1 week. Immediately after biofilm formation in the LBR, the substrates were dipped into physiological salt water with 4% glutaraldehyde for 1 day.

(A)   **(B)**

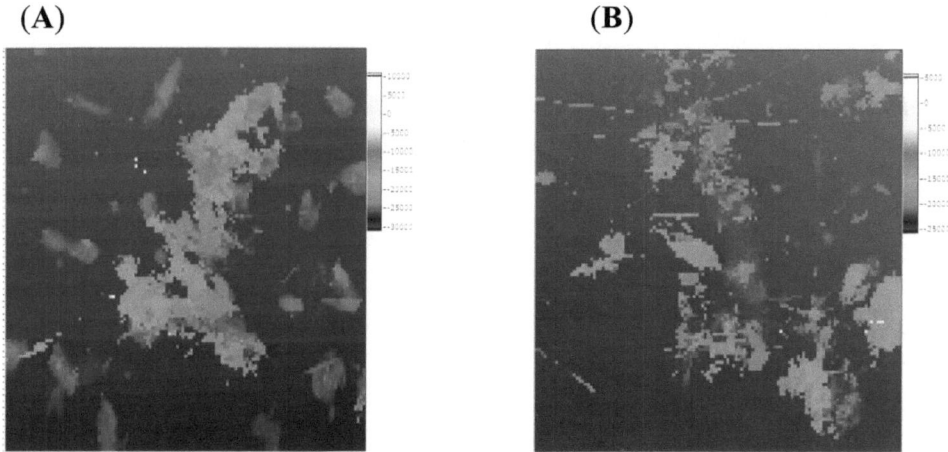

Fig. (5). SICM images (48 μm× 53 μm) of the environmental biofilm formed in physiological salt water. (**A**) Image of the biofilm on the glass or (**B**) on the polystyrene petri dish in physiological salt water after biofilm formation in diluted artificial seawater (salinity concentration; 1%) in an LBR for 1 week. Immediately after biofilm formation in LBR, the substrates were dipped in physiological salt water with 4% glutaraldehyde for 1 day. The pixels of the images shown in Figs. (**5A and B**) are 128 × 128.

The pixels of the obtained images shown in Figs. (**5A and B**) are 128 × 128. As shown in Fig. (**5A**), the biofilm formed on the glass dipped in physiological salt water with 4% glutaraldehyde could be clearly observed by SICM. Addition of glutaraldehyde to the water before observation was found to be effective for clear and stable observation of the biofilm by SICM.

As shown in Fig. (**5B**), the biofilm formed on polystyrene was clearly observed by SICM. It was also found that environmental biofilms formed on polystyrene can be observed using SICM.

SICM Observation of *Aliivibrio fischeri* Biofilms

Figs. (**6A and B**) show an optical microscopy image and an SICM image (48 μm × 53 μm), respectively, of an *Aliivibrio fischeri* biofilm observed in physiological salt water, after biofilm formation on glass in 16 times-diluted Marine Broth for 2 days. Immediately after biofilm formation, the substrates were dipped into physiological salt water with 4% glutaraldehyde for 1 day. The pixels of the obtained image in Fig. (**6B**) are 256 × 256. At left hand side of Fig. (**6A**), the SICM probe can also be observed. The red square in Fig. (**6A**) roughly indicates the area observed in Fig. (**6B**). Fig. (**6B**) shows that the observed *Aliivibrio fischeri* biofilm consists of many small protrusions with a height of about 1-3 μm.

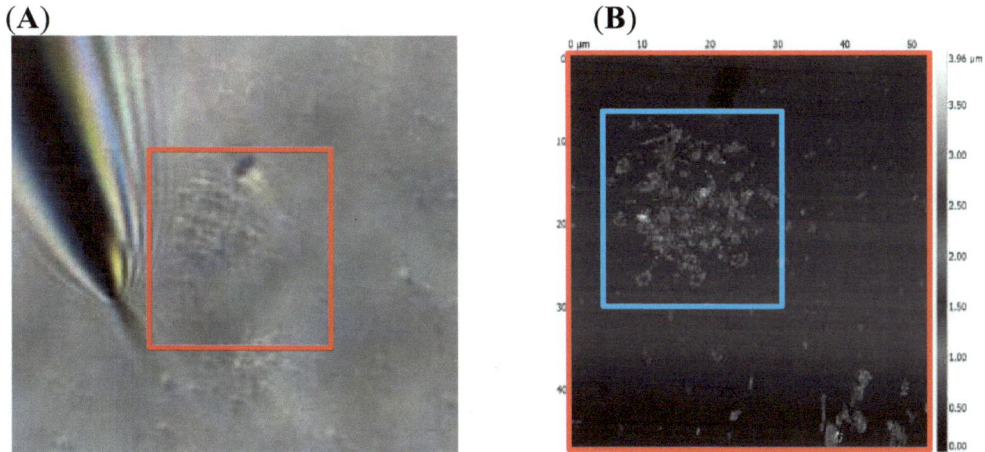

Fig. (6). Optical microscopy image and SICM image (48 μm × 53 μm) of the *Aliivibrio fischeri* biofilm observed in physiological salt water, after biofilm formation on glass in 16 times-diluted Marine Broth for 2 days. (**A**) Optical microscope image and (**B**) SICM image (48 μm × 53 μm). Immediately after biofilm formation, the substrates were dipped into physiological salt water with 4% glutaraldehyde for 1 day. The pixels of the obtained image in Fig. (**6B**) are 256 × 256.

In order to observe the detailed structure of the formed biofilm, a higher resolution image was also obtained, as shown in Fig. (**7**). Fig. (**7B**) shows an SICM image (10.6 μm× 10.6 μm) with a higher resolution. The pixels of the obtained image in Fig. (**7B**) are 128 × 128. In order to show the area observed, a digitally zoomed image is also shown in Fig. (**7A**). The blue square in Fig. (**6B**) roughly indicates the area observed in Fig. (**7A**), and the green square in Fig. (**7A**) roughly indicates the area observed in Fig. (**7B**). Notably, Fig. (**7B**) indicates that the size of the smaller protrusions is roughly the same as the size of the microorganisms and that the size of the larger protrusions is much larger than that of the microorganisms, indicating that the larger protrusions might be microorganisms covered with EPSs.

(A)

(B)

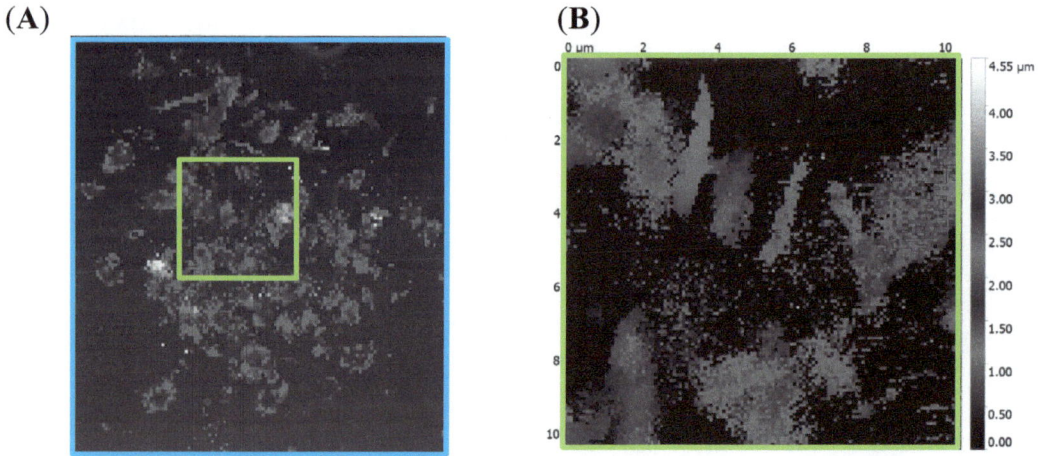

Fig. (7). SICM images of Aliivibrio fischeri biofilm formed on glass in physiological salt water, after biofilm formation in 16 times-diluted Marine Broth for 2 days. (**A**) Digitally zoomed image of Fig. (**6B**). (**B**) SICM image (10.6 μm × 10.6 μm). The pixel of the obtained image in Fig. (**7B**) is 128 × 128.

CONCLUSIONS

Environmental biofilms formed on a glass plate or polystyrene petri dish in a LBR, as well as biofilms formed by a single bacterium, *Aliivibrio fischeri*, were investigated *in-situ* in physiological salt water by SICM. The main results obtained are indicated below:

1. Environmental biofilms formed on glass plates or polystyrene petri dishes in an LBR can be observed by SICM.
2. The consistency of salt concentration in the water used for SICM observation and the water used in the LBR and the addition of glutaraldehyde to the water before observation might be effective for the clear and stable observation of the biofilm by SICM.
3. Biofilms formed by a single kind of bacterium, such as *Aliivibrio fischeri*, can be also observed by SICM. The observed *Aliivibrio fischeri* biofilm consisted of many small protrusions with a height of about 1-3 μm.

CONSENT FOR PUBLICATION

Not applicable.

CONFLICT OF INTEREST

The author confirms that this chapter contents have no conflict of interest.

ACKNOWLEDGEMENTS

This study was partly supported by Grants-in-Aid for Scientific Research (C) (JSPS KAKENHI; Grant Numbers 16K06819 and 25420795) and Collaborative Research Projects of the Research Center for Biomedical Engineering. I would like to thank the students at Iwata Lab. in Shizuoka University, Yusuke Eguchi, Masayoshi Yoshioka, Itsuki Shirasawa, and Katsura Yuichiro, and the students at Hirai Lab. at NIT, Suzuka College for supporting the research presented in this paper.

REFERENCES

[1] Bakke R, Olsson PQ. Biofilm thickness measurements by light microscopy. J Microbiol Methods 1986; 5: 93-8.
[http://dx.doi.org/10.1016/0167-7012(86)90005-9]

[2] Little B, Wagner P, Ray R, Pope R, Scheetz R. Biofilms: An ESEM evaluation of artifacts introduced during SEM preparation. J Ind Microbiol 1991; 8: 213-21.
[http://dx.doi.org/10.1007/BF01576058]

[3] Bremer PJ, Geese GG, Drake B. Atomic force microscopy examination of the topography of a hydrated bacterial biofilm on a copper surface. Curr Microbiol 1992; 24: 223-30.
[http://dx.doi.org/10.1007/BF01579285]

[4] Lawrence JR, Korber DR, Hoyle BD, Costerton JW, Caldwell DE. Optical sectioning of microbial biofilms. J Bacteriol 1991; 173(20): 6558-67.
[http://dx.doi.org/10.1128/jb.173.20.6558-6567.1991] [PMID: 1917879]

[5] Yawata Y, Nomura N, Uchiyama H. Development of a novel biofilm continuous culture method for simultaneous assessment of architecture and gaseous metabolite production. Appl Environ Microbiol 2008; 74(17): 5429-35.
[http://dx.doi.org/10.1128/AEM.00801-08] [PMID: 18606794]

[6] Beech IB, Smith JR, Steele AA, Penegar I, Campbell SA. The use of atomic force microscopy for studying interactions of bacterial biofilms with surfaces. Colloids Surf B Biointerfaces 2002; 23: 231-47.
[http://dx.doi.org/10.1016/S0927-7765(01)00233-8]

[7] Hansma HG, Pietrasanta LI, Auerbach ID, Sorenson C, Golan R, Holden PA. Probing biopolymers with the atomic force microscope: a review. J Biomater Sci Polym Ed 2000; 11(7): 675-83.
[http://dx.doi.org/10.1163/156856200743940] [PMID: 11011766]

[8] Webb HK, Crawford RJ, Sawabe T, Ivanova EP. Poly(ethylene terephthalate) polymer surfaces as a substrate for bacterial attachment and biofilm formation. Microbes Environ 2009; 24(1): 39-42.
[http://dx.doi.org/10.1264/jsme2.ME08538] [PMID: 21566352]

[9] Wright CJ, Shah MK, Powell LC, Armstrong I. Application of AFM from microbial cell to biofilm. Scanning 2010; 32(3): 134-49.
[http://dx.doi.org/10.1002/sca.20193] [PMID: 20648545]

[10] Hansma PK, Drake B, Marti O, Gould SA, Prater CB. The scanning ion-conductance microscope. Science 1989; 243(4891): 641-3.
[http://dx.doi.org/10.1126/science.2464851] [PMID: 2464851]

[11] Ushiki T, Nakajima M, Choi M, Cho SJ, Iwata F. Scanning ion conductance microscopy for imaging biological samples in liquid: a comparative study with atomic force microscopy and scanning electron microscopy. Micron 2012; 43(12): 1390-8.
[http://dx.doi.org/10.1016/j.micron.2012.01.012] [PMID: 22425359]

[12]   Kanematsu H, Kougo T, Kuroda D, Itoh H, Ogino H, Yamamoto Y. Biofilm formation derived from ambient air and the characteristics of apparatus. J Phys Conf Ser 2013; 433012031
[http://dx.doi.org/10.1088/1742-6596/433/1/012031]

[13]   Ogawa A, Noda M, Kanematsu H, Sano K. Application of bacterial 16S rRNA gene analysis to a comparison of the degree of biofilm formation on the surface of metal coated glasses. Mater Technol 2015; 30: B61-5.
[http://dx.doi.org/10.1179/1753555714Y.0000000230]

CHAPTER 14

Fluorescence-Activated Cell Sorting (FACS)-Based Characterization of Microalgae

Bogdan I. Gerashchenko[*]

R.E. Kavetsky Institute of Experimental Pathology, Oncology and Radiobiology, National Academy of Sciences of Ukraine, Ukraine

Abstract: Fluorescence-activated cell sorting (FACS) that is often interchangeably termed as flow cytometry allows for rapid multiparametric analysis of the physical characteristics of individual particles or cells as they flow in a fluid stream one by one through a fixed laser beam. FACS also implies physical sorting of particles with selected optical properties out of the flow stream and collecting them for subsequent studies. Unicellular microalgae of freshwater and seawater habitats represent a very convenient object for FACS analysis. The current chapter summarizes major achievements of FACS in the study of morpho-functional peculiarities and taxonomic affiliation of microalgae representing eukaryotic organisms. Life cycle analysis of green microalgae using a combination of key life cycle parameters such as cell size/volume, DNA and chlorophyll content distributions is also considered here.

Keywords: FACS, Fluorescence, Forward light scatter, Life cycle, Side light scatter, Unicellular microalgae.

INTRODUCTION

Microalgae inhabiting freshwater and marine systems are efficient absorbers of radiant energy and are major consumers of CO_2 that, *via* photosynthesis, is converted into carbohydrates. Moreover, due to photosynthesis, these organisms significantly contribute to atmospheric O_2 production. There are great number of microalgal species that widely vary in size, typically from a couple micrometers (μm) to a few hundred micrometers, except for a couple of species such as marine green alga *Ostreococcus tauri* [1] and freshwater green alga *Medakamo hakoo* [2], both only ~0.8 μm across (notably, the former was discovered by fluorescence-activated cell sorting [FACS] [1]). Further, all these organisms are very different in shape. Depending on the species, microalgae may contain various types of biomolecules (*e.g.*, carbohydrates, peptides, lipids, pigments, and vitamins) and

---
[*] **Corresponding author Bogdan I. Gerashchenko:** R.E. Kavetsky Institute of Experimental Pathology, Oncology and Radiobiology, National Academy of Sciences of Ukraine, Ukraine; Tel: +380442571177; Fax: +380442581656; E-mail: biger63@yahoo.com

minerals [3]. The microalgal chemical composition can be drastically changed depending on cultivation conditions (illumination, temperature, pH, CO_2 supply, salinity, and nutrients), thus providing a unique opportunity to substantially accumulate (in algae) the desired products having potential commercial applications [3]. However, a hostile and toxic environment can readily impair algal biochemistry and growth, and, in this regard, green microalgae are often used as a convenient and sensitive model to monitor the quality of aquatic ecosystems [4]. Microalgae may also be subjected to viral infection impairing the vital functions of these organisms [5]. It is worth noting that interactions of algae with viruses, bacteria, and protozoa as well as interactions among algae themselves are widespread in nature, playing significant roles in their development and structuring algal communities [6 - 9]. Some microalgae can form symbiotic associations with host organisms: a phenomenon that is not fully understood yet [10]. Apparently, the morpho-functional state of microalgae is not constant and changes dynamically with life cycle progression. Moreover, the life cycle pattern is unique in every single algal species. Thus, microalgae represent a series of challenges to varying degrees that can be solved by means of FACS based on the study of the optical properties of algal cells such as light scatter and fluorescence (with or without fluorescence staining), thereby characterizing their life cycle and other morpho-functional features. Finally, the number of discovered species or created strains of microalgae are continuously growing and their characterization accordingly requires quick and reliable methodologies. Due to the unicellularity of most microalgae known to date, these organisms are well suited for FACS analysis. In the current era, the era of biotechnological progress, FACS with flow sorting capabilities becomes increasingly popular especially in aquatic microbiology.

HISTORICAL BACKGROUND AND MAJOR ACHIEVEMENTS

The first FACS instruments were built at the turn of the 1960s and 1970s [11 - 13], the event that immediately attracted much interest especially of cancer researchers. A few years later, FACS became an important technique for aquatic microbiologists. The first aquatic microorganisms analyzed by FACS were microalgae such as *Euglena* spp. and *Chlorella* spp., including several more species of marine and freshwater habitats [14 - 19]. These microalgae, which were maintained under standard cultivation conditions, were largely analyzed for DNA [14 - 16, 18, 19], protein [16], and chlorophyll (CHL) contents [15, 16, 19], in addition to cell size and shape. Cell cycle phase-dependent redistribution of DNA contents in algae was monitored by measuring fluorescence intensity of DNA-binding dyes such as ethidium bromide (EB), propidium iodide (PI), Hoechst 33342, DAPI (4',6-diamidino-2-phenylindole), and mithramycin, whose binding and fluorescence features were well described [20]. As for proteins, their intracellular amounts were analyzed by measuring the fluorescence intensity of protein-coupled fluorescein isothiocyanate (FITC) [21]. Contrary to DNA and protein content analyses, CHL content analysis did not require any staining procedures, because CHL, namely CHL *a*, is an endogenous pigment that brightly fluoresces (>620 nm) after excitation with blue light (488 nm, a wavelength characteristic for argon ion lasers) [15]. In addition to the aforementioned parameters, determination of the content of neurotoxic saxitoxin in the dinoflagellate *Gonyaulax tamarensis* was also proposed, a method based on the detection of an oxidized form of saxitoxin that fluoresces at 380–410 nm when excited at 330 nm [17]. Notably, in some *Gonyaulax* spp., the circadian clock directs substantial changes in luciferin content (maximum excitation and emission wavelengths are 390 and 470 nm, respectively) that can be measured by FACS [22]. Sometime later, in the second half of the 1980s, the commercially available lipophilic stain Nile Red was applied to monitor amounts of neutral lipids in microalgae by FACS [23]. At the time, more specific properties of microalgae were examined using carbohydrate-binding probes such as FITC-labeled lectins and monoclonal antibodies designed for detecting defects in cell surface glycoproteins, a type of study that was initially performed on *Chlamydomonas* cells [24]. As for assessing cell viability and the degree of metabolic activity, fluorescein diacetate (FDA) staining was carried out to detect the presence of active esterases that cleave FDA into highly fluorescent products [25]. Esterase activity together with CHL fluorescence, cell size, and cell division rate are valuable ecotoxic criteria [26]. With the development of multicolor FACS, simultaneous analysis of several cell fluorescence parameters became possible.

Microalgal communities may consist of a mixture of different types of algae that can be distinguished to some extent from one another and from the possible coexisting bacteria (*e.g.,* cyanobacteria) based on differences in size/volume, DNA ploidy, and peculiarities in fluorescence spectra of endogenous pigments such as CHL, phycoerythrin (PE), phycocyanin (PC), and allophycocyanin (APC) [27 - 31]. In addition, all these microorganisms and even viruses can be quantified using an internal reference consisting of very small (0.95 μm) fluorescent beads of a known concentration [30]. To determine the taxonomic affiliation of microalgae, specific fluorescence staining is likely to be needed. In the 1990s, a series of studies were launched aimed at generation of polyclonal and monoclonal antibodies to specifically label certain types of microalgae for FACS analysis. Vrieling *et al.* [32], regardless of the physiological state of toxic dinoflagellates *Gyrodinium aureolum* and *Gymnodinium nagasakiense*, were able to detect them by FACS using FITC-conjugated monoclonal antibodies against cell surface antigens. However, the achievements in the field of immunofluorescent FACS detection of other strains of microalgae are scarce so far due to apparent difficulties with preparing strain-specific antibodies with satisfactory identification. Nevertheless, there are a few more relatively new "success stories" that describe FACS-based immunodetection of another species of marine microalgae such as dinoflagellate *Alexandrium minutum* [33], red tide alga *Heterosigma akashiwo* [34], and brown tide algae *Aureococcus anophagefferens* [35] and *Aureoumbra lagunensis* [36]. There is also an alternative approach based on whole-cell *in situ* hybridization with taxon-specific fluorescent rRNA-targeted oligonucleotide probes that was successfully tested by FACS on several microalgal species [37 - 39]. However, this approach does not seem to be superior to immunodetection in terms of stability and strength of the fluorescence signal [39].

Since the middle of the 1990s, there has been an increasing number of reports regarding FACS-based detection of reactive oxygen species (ROS) (in some freshwater and seawater microalgae) generated in response to various environmental stressors. For this purpose, special indicators for ROS such as dihydroethidium, dihydrorhodamine 123, 2',7'-dichlorofluorescein diacetate (DCFDA), 2',7'-dichlorodihydrofluorescein diacetate (H_2DCFDA), chloromethyl derivative of H_2DCFDA (CM-H_2DCFDA), and fluorinated carboxylated variant of H_2DCFDA (carboxy-H_2DFFDA), were used. [40 - 43]. Notably, intracellular content of another stress-related product—nitric oxide (NO), playing a definite role in mortality of some marine microalgae (diatoms and dinoflagellates)—was only recently determined with NO indicator 4-amino-5-methylamino-2',7'-difluorofluorescein diacetate (DAF-FM diacetate) [8, 44, 45]. There is also an interesting report by Rioboo *et al.* [46] proposing to use membrane-penetrating 5,6-carboxyfluorescein diacetate succinimidyl ester (CFDA-SE) to monitor cell proliferation alterations in *Chlorella* cells under stress conditions. Moreover, after staining with CFDA-SE, the release of daughter cells can be quantitatively estimated [46].

As for the flow sorting designed to sort cells with selected optical characteristics out of the flow stream for subsequent studies, this technique is being increasingly employed in microalgal research. Particularly, it has become helpful for taxonomic identification of microalgae [47], obtaining axenic clonal cultures, and isolation of overproducing strains [48, 49].

BASIC PRINCIPLES OF FACS

FACS is an optical method that quantitatively characterizes individual particles or cells in a rapid and multiparametric manner as they flow in a fluid stream one by one (thousands of events per second) through a sensing point. A conventional FACS system consists of three integral parts, namely fluidics, optics, and electronics, which have been described in detail [20].

Usually, a benchtop FACS instrument represents a combination of such components as:

1. a source of light,
2. a lens system to focus the light into the flow cell and to collect the light exiting the flow cell,
3. a set of the optical components to separate and direct the light of the different wavelengths onto the detectors,
4. electronics to amplify and process the resulting signals,
5. and a computer to collect and analyze data.

The instrument performs cell analysis as cells travel in a moving stream through a fixed laser beam and can measure the following parameters: forward light scatter (FSC), 90° side light scatter (SSC), and fluorescence (FL), including the pulse height (H), area (A), and width (W). As a cell passes through the laser beam, several measurements can be made at once based on the physical characteristics of the cell, thus providing information on how it scatters the light and emits fluorescence. FSC, which is composed of the light diffracted around a cell, is largely predetermined by cell's size/volume and shape, whereas SSC, which is composed of the light refracted and reflected by a cell, is largely predetermined by intra- and extracellular structures and the refractive index [50]. The magnitude of FSC signals is roughly proportional to the cell's size/volume. Contrary to FSC, the interpretation of SSC signal intensities appears to be more complex. Nevertheless, it is well recognized that the magnitude of SSC signals is proportional to intracellular granularity or internal complexity. As for the fluorescence emitted by a cell labeled either intrinsically (*e.g.*, assuming autofluorescence) or extrinsically, this parameter characterizes the relative fluorescence intensity. There is an optical configuration as in a "four-color" BD FACSCalibur instrument (BD Biosciences, San Jose, CA, USA) enabling separate collection of fluorescence signals from four types of natural photosynthetic pigments: PE (585/42 nm band-pass filter, FL2 detector), CHL (670 nm long-pass filter, FL3 detector), and PC or APC (661/16 nm band-pass filter, FL4 detector) [31]. PE and CHL are excited by the blue laser (488 nm), whereas PC or APC are excited by the red laser (635 nm).

At present, many modern FACS instruments are equipped with a special cell-sorting module enabling identification and sorting of cell groups of interest for subsequent analyses. There are two main sorting principles: (1) fluidic sorting with a closed flow cell, and (2) stream-in-air droplet sorting, each of which has definite advantages and disadvantages (more details on this issue can be found in other literature sources [20, 47]). As for BD FACSCalibur, which has been used in our studies, it can be supplemented with a closed fluidic sorting system (*i.e.*, the former sorting principle). Thus, after the sample is acquired and cells pass through the laser beam, a unique catcher tube mechanism moves in and out of the sample core stream at a rate of ~300 times/s to capture cells of interest and direct them to a collection tube or to an optional cell concentrator module for further processing.

LIFE CYCLE ANALYSIS

The fact that cell size of vegetating green microalgae can be monitored in association with their DNA and endogenous CHL contents prompted us to propose a FACS protocol for algal life cycle analysis [51]. This protocol is largely based on our previous studies with unicellular green algae such as free-living *Chlorella* or *Chlorella*-like algae ex-symbiotic from ciliated protozoan cells of *Paramecium bursaria* [52, 53]. *Chlorella* spp. naturally occurring in freshwater and marine habitats, and even in soil, is currently of great scientific and commercial value. In addition to ready availability, *Chlorella* cells are easy to culture. Typically, they are spherical with the diameter ranging from 2 to 10 μm; accordingly, they are perfectly suitable for FACS analysis. *Chlorella*, an asexually reproducing haploid organism, is characterized by three major vegetative life cycle stages: (1) growth, (2) ripening, and (3) division or autospore liberation [54]. Except for the last stage, the first two stages are light dependent, during which an algal cell increases its volume and mass along with accumulation of nuclear substances for a subsequent division stage. The *Chlorella* cell usually cleaves more than once predominantly resulting in four to eight daughter cells (autospores).

Using a "three-color" BD FACSCalibur equipped with a single laser (488 nm), we in addition to light scattering (FSC-H and SSC-H) were able to detect fluorescence signals that arise from DNA-bound PI (FL2-H) and endogenous CHL (FL3-H). As an example, Fig. (1) depicts the multiparameter analysis of asynchronous stationary-phase ex-symbiotic algae (strain SA-1 isolated from *P. bursaria*) morphologically resembling *Chlorella* cultured in a low-nutrient citric-acetate medium under constant light of ~2,000 lux at 24°C as described elsewhere [55]. The plots FSC-H *vs.* SSC-H (Fig. **1A**) and FSC-H *vs.* FL3-H (Fig. **1C**) show that the algae, depending on the life cycle stage, widely vary in size, internal complexity, and CHL content per cell. In addition, the plot FSC-H *vs.* FL2-H (Fig. **1B**) indicates that predominantly larger cells are hyperhaploid (>1C) corresponding to 2C and 4C peaks in DNA content histogram (FL2-H; Fig. **1D**). These hyperhaploid cells, after being sorted out (sort regions R_1 and R_2; Fig. **1B**), were stained with DAPI followed by microscopic examination of their general/nuclear morphology. Sorted cells mainly represent ripening or fully ripened algae with two or four nuclei (Fig. **1E**). The algae having four nuclei are likely to be fully ripened sporangia with four autospores. Thus, based on the aforementioned findings, the life cycle of algae can be concisely interpreted as follows (schematic illustrations depicted in Fig. **1A** and Fig. **1C**). Small daughter cells with low light scattering, after they have been released from a mother cell (sporangium), begin to grow in size/volume accompanied by DNA and CHL syntheses. In this stage, an increase in SSC may also be associated with expansion of large subcellular structures such as nuclei and chloroplasts, an assumption that is likely to be supported by the finding that the logarithmically growing algae include a distinct group of cells with a stronger SSC signal [53]. Upon completion of nuclear DNA doublings, mitoses, and chloroplast divisions, algae enter the stage of ripening, which leads to formation of autospores within mother cells. This stage is manifested by the emergence of discrete hyperhaploid DNA peaks (particularly characteristic for stationary-phase cultures, as shown in Fig. **1D**) and further increase of FSC with no apparent changes in CHL content (Fig. **1C**). It is worth noting that the life cycle optical pattern obtained by FACS is species/strain-specific that dynamically changes with the phase of algal growth [52, 53]. An increase in the number of cells concordantly synthesizing DNA and CHL takes place in actively proliferating algal cultures [52], which is a well-documented observation about other species as well [19]. However, to compartmentalize asynchronously growing algae into discrete cell groups representing algae of a particular life cycle stage, and then quantify algae in each of these groups, they probably should not be fixed to avoid poor compartmentation of light scatter signals. In the aforementioned study, the light scattering of algae was likely to be altered by ethanol fixation, which is commonly used to remove CHL before staining of DNA with PI. To address this issue, we deemed it necessary to conduct experiments with unfixed *Chlorella* cells. Indeed, they showed much better compartmentation of light scatter signals, so the percentages of cells that constitute each optically compartmentalized subgroup can be quantified [53].

Using exactly the same instrument settings and culture conditions, we next studied the life cycle patterns and growth rates of *C. vulgaris* (c-27) and *Chlorella*-like algae (SA-3) maintained in the presence of colloidal silica (0.01%) [56]. In that study, in addition to a significant expansion of populations of both algal strains, silica remarkably affected the life cycle patterns particularly in logarithmically growing cultures. Of note, in contrast to silica-treated SA-3 algae, silica-treated *C. vulgaris* (c-27) released many more very small daughter cells (<2 μm), which were much smaller than regularly produced daughter cells. Perhaps the accelerated growth rate led to a premature release of daughter cells.

Finally, the life cycle of algae residing in host organisms as symbionts can be affected by these organisms. Obvious alterations in cell size and DNA and CHL content distributions in symbiotic algae residing in *P. bursaria* have also been reported [57, 58]. Host-regulated expansion of algae is likely to predetermine these alterations [59]. In conclusion, FACS-related one methodological aspect, in our view, should be pointed out for further experiments focused on the research into the algal life cycle. To simultaneously analyze the changes in DNA and CHL (or any other photosynthetic pigment) contents within one sample without addition of any fixatives, algae could be stained with membrane-penetrating UV-excitable DNA dyes such as Hoechst or DAPI emitting blue light (UV laser is required). Perhaps without permeabilization of cell membranes, algae, namely *Chlorella* cells, can be stained with SYBR Green I [60], another dye that is DNA-specific but excitable with a blue laser (488 nm).

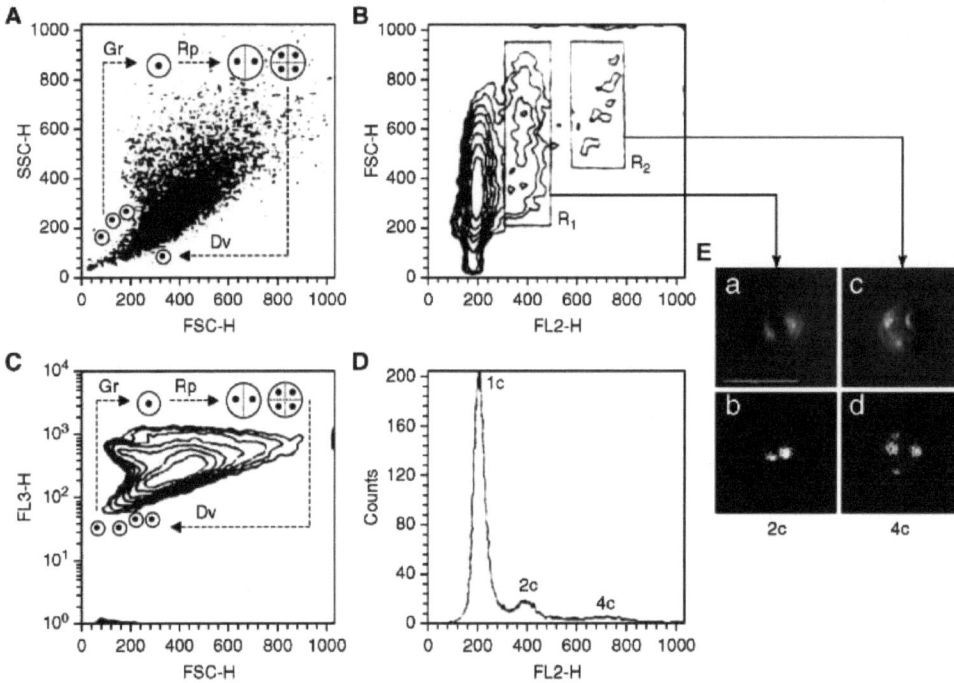

Fig. (1). FACS analysis of asynchronous culture of SA-1 algae in the stationary growth phase by means of the magnitude of their light scattering (FSC-H and SSC-H) and fluorescence intensity signals from CHL (FL3-H) and DNA-bound PI (FL2-H). FL2-H signals were collected through a 585/42 nm band-pass filter, whereas FL3-H signals were collected through a 650 nm long-pass filter. The algae analyzed with the plots presented in panels **A**, **B**, and **D**, were fixed with ethanol, whereas the algae analyzed *via* the plot presented in panel **C** were fixed with formaldehyde. Abbreviations: Dv: stage of "division" or "autospore liberation"; Gr: stage of "growth"; Rp: stage of "ripening". Photomicrographs in panel **E** represent Nomarski differential interference contrast (**a**, **c**) and fluorescence microscopy (**b**, **d**) images of sorted algae with two (**a**, **b**) and four (**c**, **d**) nuclei. Scale bar = 8 μm. Adapted from Gerashchenko *et al*., 2001 with permission [52].

CONSENT FOR PUBLICATION

Not applicable.

CONFLICT OF INTEREST

The author confirms that this chapter contents have no conflict of interest.

ACKNOWLEDGEMENTS

Declare none.

REFERENCES

[1]   Courties C, Vaquer A, Troussellier M, *et al*. Smallest eukaryotic organism. Nature 1994; 370: 255.
        [http://dx.doi.org/10.1038/370255a0]

[2]   Kuroiwa T, Ohnuma M, Imoto Y, *et al*. Genome size of the ultrasmall unicellular freshwater green
        alga, *Medakamo hakoo* 311, as determined by staining with 4′,6-diamidino-2-phenylindole after
        microwave oven treatments: II. Comparison with *Cyanidioschyzon merolae, Saccharomyces
        cerevisiae (n, 2n),* and *Chlorella variabilis.* Cytologia (Tokyo) 2016; 81: 69-76.
        [http://dx.doi.org/10.1508/cytologia.81.69]

[3]   Kim S-K, Chojnacka K, Eds. Marine algae extracts: Processes, products, and applications. Weinheim,
        Germany: Wiley-VCH Verlag GmbH & Co. KGaA 2015; Vol. 1,2.

[4]   Blaise C, Férard J-F, Eds. Small-scale freshwater toxicity investigations. Toxicity test methods.
        Dordrecht, Netherlands: Springer 2005; Vol. 1.

[5]   Van Etten JL, Lane LC, Meints RH. Viruses and viruslike particles of eukaryotic algae. Microbiol Rev
        1991; 55(4): 586-620.
        [PMID: 1779928]

[6]   Cole JJ. Interactions between bacteria and algae in aquatic ecosystems. Annu Rev Ecol Syst 1982; 13:
        291-314.
        [http://dx.doi.org/10.1146/annurev.es.13.110182.001451]

[7]   Tillmann U. Interactions between planktonic microalgae and protozoan grazers. J Eukaryot Microbiol
        2004; 51(2): 156-68.
        [http://dx.doi.org/10.1111/j.1550-7408.2004.tb00540.x] [PMID: 15134250]

[8]   Bidle KD. The molecular ecophysiology of programmed cell death in marine phytoplankton. Annu
        Rev Mar Sci 2015; 7: 24.1-24.35.
        [http://dx.doi.org/10.1146/annurev-marine-010213-135014]

[9]   Borowitzka MA. Chemically-mediated interactions in microalgae.The physiology of microalgae.
        Berlin, Heidelberg: Springer 2016; pp. 321-57.
        [http://dx.doi.org/10.1007/978-3-319-24945-2_15]

[10]   Grube M, Seckbach J, Muggia L, Eds. Algal and cyanobacteria symbioses. Singapore: World
        Scientific 2017.
        [http://dx.doi.org/10.1142/q0017]

[11]   Van Dilla MA, Trujillo TT, Mullaney PF, Coulter JR. Cell microfluorometry: a method for rapid
        fluorescence measurement. Science 1969; 163(3872): 1213-4.
        [http://dx.doi.org/10.1126/science.163.3872.1213] [PMID: 5812751]

[12]   Dittrich W, Göhde W. Impulsfluorometrie bei Einzelezellen in Suspensionen. Z Naturforsch B 1969;
        24(3): 360-1.

[http://dx.doi.org/10.1515/znb-1969-0326] [PMID: 4389205]

[13] Hulett HR, Bonner WA, Barrett J, Herzenberg LA. Cell sorting: automated separation of mammalian cells as a function of intracellular fluorescence. Science 1969; 166(3906): 747-9.
[http://dx.doi.org/10.1126/science.166.3906.747] [PMID: 4898615]

[14] Falchuk KH, Krishan A, Vallee BL. DNA distribution in the cell cycle of *Euglena gracilis*. Cytofluorometry of zinc deficient cells. Biochemistry 1975; 14(15): 3439-44.
[http://dx.doi.org/10.1021/bi00686a023] [PMID: 807244]

[15] Trask BJ, van den Engh GJ, Elgershuizen JHBW. Analysis of phytoplankton by flow cytometry. Cytometry 1982; 2(4): 258-64.
[http://dx.doi.org/10.1002/cyto.990020410] [PMID: 6799265]

[16] Hutter K-J, Eipel HE. Flow cytometric determinations of cellular substances in algae, bacteria, moulds and yeasts. Antonie van Leeuwenhoek 1978; 44(3-4): 269-82.
[http://dx.doi.org/10.1007/BF00394305] [PMID: 378116]

[17] Yentsch CM. Flow cytometric analysis of cellular saxitoxin in the dinoflagellate *Gonyaulax tamarensis* var. *excavata*. Toxicon 1981; 19(5): 611-21.
[http://dx.doi.org/10.1016/0041-0101(81)90099-4] [PMID: 7197816]

[18] Yentsch CM, Mague FC, Horan PK, Muirhead K. Flow cytometric DNA determinations on individual cells of the dinoflagellate *Gonyaulax tamarensis* var. *excavata*. J Exp Mar Biol Ecol 1983; 67: 175-83.
[http://dx.doi.org/10.1016/0022-0981(83)90088-6]

[19] Olson RJ, Frankel SL, Chisholm SW, Shapiro HM. An inexpensive flow cytometer for the analysis of fluorescence signals in phytoplankton: Chlorophyll and DNA distributions. J Exp Mar Biol Ecol 1983; 68: 129-44.
[http://dx.doi.org/10.1016/0022-0981(83)90155-7]

[20] Shapiro HM. Practical Flow Cytometry. 4th ed., Hoboken, New Jersey: John Wiley & Sons 2003.
[http://dx.doi.org/10.1002/0471722731]

[21] Crissman HA, Oka MS, Steinkamp JA. Rapid staining methods for analysis of deoxyribonucleic acid and protein in mammalian cells. J Histochem Cytochem 1976; 24(1): 64-71.
[http://dx.doi.org/10.1177/24.1.56392] [PMID: 56392]

[22] Johnson CH, Inoué S, Flint A, Hastings JW. Compartmentalization of algal bioluminescence: autofluorescence of bioluminescent particles in the dinoflagellate *Gonyaulax* as studied with image-intensified video microscopy and flow cytometry. J Cell Biol 1985; 100(5): 1435-46.
[http://dx.doi.org/10.1083/jcb.100.5.1435] [PMID: 4039325]

[23] Cooksey KE, Guckert JB, Williams SA, Callis PR. Fluorometric determination of the neutral lipid content of microalgal cells using Nile Red. J Microbiol Methods 1987; 6: 333-45.
[http://dx.doi.org/10.1016/0167-7012(87)90019-4]

[24] Bloodgood RA, Salomonsky NL, Reinhart FD. Use of carbohydrate probes in conjunction with fluorescence-activated cell sorting to select mutant cell lines of *Chlamydomonas* with defects in cell surface glycoproteins. Exp Cell Res 1987; 173(2): 572-85.
[http://dx.doi.org/10.1016/0014-4827(87)90296-5] [PMID: 3691676]

[25] Dorsey J, Yentsch CM, Mayo S, McKenna C. Rapid analytical technique for the assessment of cell metabolic activity in marine microalgae. Cytometry 1989; 10(5): 622-8.
[http://dx.doi.org/10.1002/cyto.990100518] [PMID: 2776579]

[26] Stauber JL, Franklin NM, Adams MS. Applications of flow cytometry to ecotoxicity testing using microalgae. Trends Biotechnol 2002; 20(4): 141-3.
[http://dx.doi.org/10.1016/S0167-7799(01)01924-2] [PMID: 11906740]

[27] Cunningham A, Leftley JW. Application of flow cytometry to algal physiology and phytoplankton ecology. FEMS Microbiol Rev 1986; 32: 159-64.
[http://dx.doi.org/10.1111/j.1574-6968.1986.tb01190.x]

[28]   Troussellier M, Courties C, Vaquer A. Recent applications of flow cytometry in aquatic microbial ecology. Biol Cell 1993; 78(1-2): 111-21.
[http://dx.doi.org/10.1016/0248-4900(93)90121-T] [PMID: 8220221]

[29]   Veldhuis MJ, Cucci TL, Sieracki ME. Cellular DNA content of marine phytoplankton using two new fluorochromes: Taxonomic and ecological implications. J Phycol 1997; 33: 527-41.
[http://dx.doi.org/10.1111/j.0022-3646.1997.00527.x]

[30]   Marie D, Partensky F, Vaulot D. Enumeration of phytoplankton, bacteria, and viruses in marine samples. Curr Protoc Cytom 1999; 10: 11.11.1-11.11.15.

[31]   Collier JL. Flow cytometry and single cell in phycology. J Phycol 2000; 36(4): 628-44.
[http://dx.doi.org/10.1046/j.1529-8817.2000.99215.x] [PMID: 29542146]

[32]   Vrieling EG, van de Poll WH, Vriezekolk G, Gieskes WWC. Immuno-flow cytometric detection of the ichthyotoxic dinoflagellates *Gyrodinium aureolum* and *Gymnodinium nagasakiense*: Independence of physiological state. J Sea Res 1997; 37: 91-100.
[http://dx.doi.org/10.1016/S1385-1101(96)00006-8]

[33]   Carrera M, Garet E, Barreiro A, *et al.* Generation of monoclonal antibodies for the specific immunodetection of the toxic dinoflagellate *Alexandrium minutum* Halim from Spanish waters. Harmful Algae 2010; 9: 272-80.
[http://dx.doi.org/10.1016/j.hal.2009.11.004]

[34]   Huang J, Wen R, Bao Z, *et al.* Application of immune-magnetic bead and immunofluorescent flow cytometric technique for the quantitative detection of HAB microalgae. Chin J Oceanology Limnol 2012; 30: 433-9.
[http://dx.doi.org/10.1007/s00343-012-1208-6]

[35]   Stauffer BA, Schaffner RA, Wazniak C, Caron DA. Immunofluorescence flow cytometry technique for enumeration of the brown-tide alga, *Aureococcus anophagefferens*. Appl Environ Microbiol 2008; 74(22): 6931-40.
[http://dx.doi.org/10.1128/AEM.00996-08] [PMID: 18820052]

[36]   Koch F, Kang Y, Villareal TA, Anderson DM, Gobler CJ. A novel immunofluorescence flow cytometry technique detects the expansion of brown tides caused by *Aureoumbra lagunensis* to the Caribbean Sea. Appl Environ Microbiol 2014; 80(16): 4947-57.
[http://dx.doi.org/10.1128/AEM.00888-14] [PMID: 24907319]

[37]   Simon N, LeBot N, Marie D, Partensky F, Vaulot D. Fluorescent *in situ* hybridization with rRNA-targeted oligonucleotide probes to identify small phytoplankton by flow cytometry. Appl Environ Microbiol 1995; 61(7): 2506-13.
[PMID: 7618862]

[38]   Lange M, Guillou L, Vaulot D, *et al.* Identification of the class Prymnesiophyceae and the genus *Phaeocystis* with ribosomal RNA-targeted nucleic acid probes detected by flow cytometry. J Phycol 1996; 32: 858-68.
[http://dx.doi.org/10.1111/j.0022-3646.1996.00858.x]

[39]   Anderson DM, Kulis DM, Keafer BA, Berdalet E. Detection of the toxic dinoflagellate *Alexandrium fundyense* (Dinophycea) with oligonucleotide and antibody probes: Variability in labeling intensity with physiologic condition. J Phycol 1999; 35: 870-83.
[http://dx.doi.org/10.1046/j.1529-8817.1999.3540870.x]

[40]   Lesser MP. Acclimation of phytoplankton to UV-B radiation: Oxidative stress and photoinhibition of photosynthesis are not prevented by UV-absorbing compound in the dinoflagellate *Prorocentrum micans*. Mar Ecol Prog Ser 1996; 132: 287-97.
[http://dx.doi.org/10.3354/meps132287]

[41]   Knauert S, Knauer K. The role of reactive oxygen species in copper toxicity to two freshwater green alage. J Phycol 2008; 44(2): 311-9.

[http://dx.doi.org/10.1111/j.1529-8817.2008.00471.x] [PMID: 27041187]

[42] Sheyn U, Rosenwasser S, Ben-Dor S, Porat Z, Vardi A. Modulation of host ROS metabolism is essential for viral infection of a bloom-forming coccolithophore in the ocean. ISME J 2016; 10(7): 1742-54.
[http://dx.doi.org/10.1038/ismej.2015.228] [PMID: 26784355]

[43] Szivák I, Behra R, Sigg L. Metal-induced reactive oxygen species production in *Chlamydomonas reinhardtii* (Chlorophyceae). J Phycol 2009; 45(2): 427-35.
[http://dx.doi.org/10.1111/j.1529-8817.2009.00663.x] [PMID: 27033821]

[44] Thompson SEM, Taylor AR, Brownlee C, Callow ME, Callow JA. The role of nitric oxide in diatom adhesion in relation to substratum properties. J Phycol 2008; 44(4): 967-76.
[http://dx.doi.org/10.1111/j.1529-8817.2008.00531.x] [PMID: 27041615]

[45] Hawkins TD, Davy SK. Nitric oxide production and tolerance differ among *Symbiodinium* types exposed to heat stress. Plant Cell Physiol 2012; 53(11): 1889-98.
[http://dx.doi.org/10.1093/pcp/pcs127] [PMID: 22992385]

[46] Rioboo C, O'Connor JE, Prado R, Herrero C, Cid A. Cell proliferation alterations in *Chlorella* cells under stress conditions. Aquat Toxicol 2009; 94(3): 229-37.
[http://dx.doi.org/10.1016/j.aquatox.2009.07.009] [PMID: 19679360]

[47] Reckermann M. Flow sorting in aquatic ecology. Sci Mar 2000; 64: 235-46.
[http://dx.doi.org/10.3989/scimar.2000.64n2235]

[48] Cho D-H, Ramanan R, Kim B-H, *et al.* Novel approach for the development of axenic microalgal cultures from environmental samples. J Phycol 2013; 49(4): 802-10.
[http://dx.doi.org/10.1111/jpy.12091] [PMID: 27007211]

[49] Pereira H, Schulze PSC, Schüler LM. Fluorescence activated cell-sorting principles and applications in microalgal biotechnology. Algal Res 2018; 30: 113-20.
[http://dx.doi.org/10.1016/j.algal.2017.12.013]

[50] Kerker M. Elastic and inelastic light scattering in flow cytometry. Cytometry 1983; 4(1): 1-10.
[http://dx.doi.org/10.1002/cyto.990040102] [PMID: 6617390]

[51] Gerashchenko BI, Takahashi T, Kosaka T, Hosoya H. Life cycle analysis of unicellular algae. Curr Protoc Cytom 2010; 11: 11.19.1-11.19.6.
[http://dx.doi.org/10.1002/0471142956.cy1119s52]

[52] Gerashchenko BI, Kosaka T, Hosoya H. Growth kinetics of algal populations exsymbiotic from *Paramecium bursaria* by flow cytometry measurements. Cytometry 2001; 44(3): 257-63.
[http://dx.doi.org/10.1002/1097-0320(20010701)44:3<257::AID-CYTO1118>3.0.CO;2-V] [PMID: 11429776]

[53] Gerashchenko BI, Kosaka T, Hosoya H. Optical compartmentation of vegetating algae species as a basis for their growth-specific characterization. Cytometry 2002; 48(3): 153-8.
[http://dx.doi.org/10.1002/cyto.10120] [PMID: 12116361]

[54] Morimura Y. Synchronous culture of *Chlorella*: I. Kinetic analysis of the life cycle of *Chlorella ellipsoidea* as affected by changes of temperature and light intensity. Plant Cell Physiol 1959; 1: 49-62.

[55] Nishihara N, Horiike S, Takahashi T, *et al.* Cloning and characterization of endosymbiotic algae isolated from *Paramecium bursaria.* Protoplasma 1998; 203: 91-9.
[http://dx.doi.org/10.1007/BF01280591]

[56] Gerashchenko BI, Gerashchenko II, Kosaka T, Hosoya H. Stimulatory effect of aerosil on algal growth. Can J Microbiol 2002; 48(2): 170-5.
[http://dx.doi.org/10.1139/w01-143] [PMID: 11958570]

[57] Gerashchenko BI, Nishihara N, Ohara T, Tosuji H, Kosaka T, Hosoya H. Flow cytometry as a strategy

to study the endosymbiosis of algae in *Paramecium bursaria.* Cytometry 2000; 41(3): 209-15.
[http://dx.doi.org/10.1002/1097-0320(20001101)41:3<209::AID-CYTO8>3.0.CO;2-U]   [PMID: 11042618]

[58]   Kadono T, Kawano T, Hosoya H, Kosaka T. Flow cytometric studies of the host-regulated cell cycle in algae symbiotic with green paramecium. Protoplasma 2004; 223(2-4): 133-41.
[http://dx.doi.org/10.1007/s00709-004-0046-6] [PMID: 15221518]

[59]   Takahashi T, Shirai Y, Kosaka T, Hosoya H. Arrest of cytoplasmic streaming induces algal proliferation in green paramecia. PLoS ONE 2007; 2(12): e1352.
[http://dx.doi.org/10.1371/journal.pone.0001352] [PMID: 18159235]

[60]   Yamamoto M, Fujishita M, Hirata A, Kawano S. Regeneration and maturation of daughter cell walls in the autospore-forming green alga *Chlorella vulgaris* (Chlorophyta, Trebouxiophyceae). J Plant Res 2004; 117(4): 257-64.
[http://dx.doi.org/10.1007/s10265-004-0154-6] [PMID: 15108033]

CHAPTER 15

Single-Cell Imaging and Sequencing-Based Detection of Microorganisms Using Highly Sensitive Fluorescence *in situ* Hybridization (FISH)

Tsuyoshi Yamaguchi[1] and Shuji Kawakami[2,*]

[1] *Department of Civil and Environmental Engineering, National Institute of Technology (KOSEN), Matsue College, Matsue, Japan*

[2] *Department of Construction Systems Engineering, National Institute of Technology (KOSEN), Anan College, Anan, Japan*

Abstract: Fluorescence *in situ* hybridization (FISH) has become a standard technique in the visual detection and phylogenetic identification of environmental microorganisms in marine microbiology. Thirty years have passed since the appearance of FISH, and now FISH can detect not only ribosomal RNA but also mRNA and functional genes with high sensitivity. This chapter describes the principles, drawbacks, and applications of FISH and highly sensitive FISH. In particular, we introduce two high-sensitivity FISH methods developed by the authors. The first is *in situ* DNA hybridization chain reaction (HCR), a highly sensitive and highly penetrating detection technique for ribosomal-RNA–targeting FISH, and the second is two-pass tyramide signal amplification (TSA)-FISH that can detect single-copy genes. Finally, we discuss how the FISH technique has contributed to the field of marine microbiology.

Keywords: CARD-FISH, Environmental microbiology, FISH, Functional genes, GeneFISH, HCR-FISH, Highly sensitive FISH, *in situ* DNA-HCR, ISH, Marine bacteria, Microscopic observation, Oligonucleotide probe, Phylogenetic identification, Polynucleotide probe, Probe permeabilization, Single cell imaging, Two-pass TSA-FISH, Visual detection.

INTRODUCTION

Fluorescence *in situ* hybridization (FISH) is a widely used technique in the field of marine microbiology for visual detection of bacteria without cultivation [1, 2]. FISH can detect target bacteria in their environment by hybridizing a fluorescently labeled DNA probe to bacterium ribosomal RNA (rRNA) in a base sequence–specific manner. As with all techniques, there have been technical Dra-

---

* **Corresponding author Shuji Kawakami:** Department of Construction Systems Engineering, National Institute of Technology (KOSEN), Anan College, Anan, Japan; Tel/Fax: +81-884-23-7189; E-mail: shuji@anan-nct.ac.jp

Toshiyuki Takahashi (Ed.)

wbacks associated with FISH, but over the past 30 years since FISH was first reported, these issues have been overcome by various researchers. Many methods have been reported for increasing FISH signal intensity, *e.g.*, the use of polynucleotide probes and enzymatic catalytic reactions, and thus FISH can now detect not only rRNA but also mRNA and functional genes. In addition, assay sensitivity and resolution of microscopes are increasing, which expand the application range of FISH. This chapter will describe the principles, limitations, and applications of FISH and will briefly present several contributions of the FISH technique to marine microbiology.

FISH

FISH allows for the identification of target bacteria by hybridizing fluorescently labeled target-specific oligonucleotide probes to rRNA, followed by microscopy analysis. There are four main steps in the detection of target bacteria by FISH: probe selection, sample fixation, probe and target hybridization, and microscopy Fig. (**1**).

Oligonucleotide probes, which are usually 18–24 bp long and are labeled at the 5′ end with a fluorescent label (*e.g.*, Atto, Cyanine, and Alexa dyes), are widely used in rRNA-targeting FISH.

To achieve specific detection, a probe that hybridizes only to the rRNA of a target microorganism must be selected. Representative probes to detect marine microorganisms have been previously published (Table **1**), and probes targeting other groups can be found in the "probeBase" database [12, 13]. Furthermore, the ARB software is usually used for designing new probes [14].

Sample fixation is performed for increasing probe penetration into the target bacterial cells. Bacterial fixation usually employs 1–4% (wt/vol) paraformaldehyde in PBS [2]. The aldehyde creates a mesh-type structure within the cells by crosslinking cellular proteins [15]. In addition, various fixation methods employing glutaraldehyde, ethanol, and methanol have been described due to differences in microbial cell structure [16].

After fixation of microbial cells, fluorescently labeled probes hybridize to the target rRNA of microorganisms in hybridization buffer. Reaction parameters, including temperature, salt concentration, and pH, are important for specific hybridization. Various buffer recipes have been reported, but most buffers are mainly composed of Tris buffer and NaCl as well as formamide as a denaturant [2]. Formamide can alter hybridization reaction temperatures if researchers change formamide concentration in the buffer (1% formamide in a buffer raises the reaction temperature by 0.69 °C). Thus, the mismatch discrimination potential of a FISH probe is defined as the difference between the formamide-dependent melting temperature of perfectly complementary duplexes and that of mismatched duplexes with base mismatches in the target sequence. Specific hybridization of the probe to target rRNA is achieved by optimizing formamide concentrations. The optimal concentrations have been determined in experimental studies using purely cultured target microorganisms. However, it is difficult to determine the optimal conditions when targeting uncultured microorganisms. To overcome this problem, Clone-FISH, which assesses parameters *via* clones incorporating the target gene [17], and mathFISH, which predicts optimal conditions using a thermodynamic approach [18], have been developed.

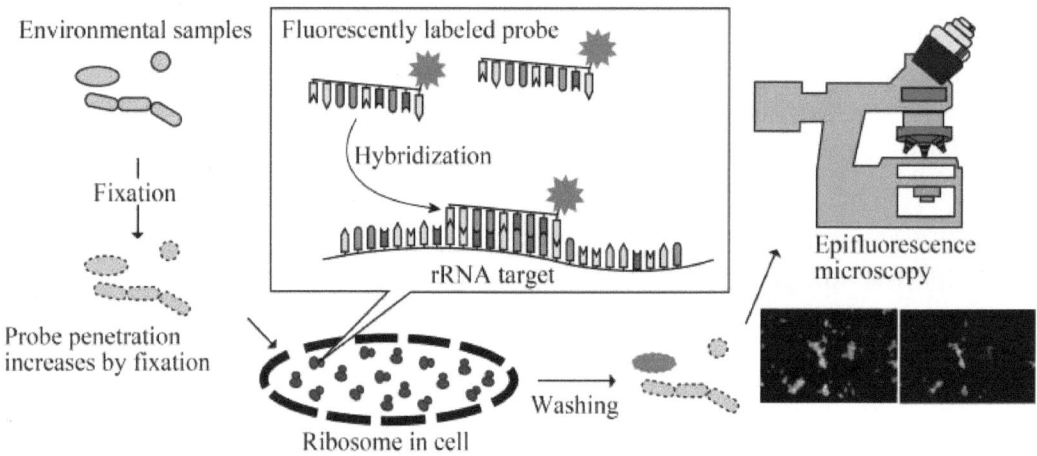

Fig. (1). Principles of FISH.

Table 1. Major FISH probe sequences for detecting marine microbes in their environments.

Samples	Probe Name	Target Microorganisms	`Probe Sequence 5'-3'	Reference
Marine surface	EUB338	Most *Bacteria*	GCTGCCTCCCGTAGGAGT	[3]
	Gam42a	*Gammaproteobacteria*	GCCTTCCCACATCGTTT	[4]
	BET42a	*Betaproteobacteria*	GCCTTCCCACTTCGTTT	[4]
	CF319a	*Cytophaga-flavobacterium*	TGGTCCGTGTCTCAGTAC	[5]
	SAR11-152R	SAR11 clade	ATTAGCACAAGTTTCCYCGTGT	[6]
	SAR11-441R	SAR11 clade	TACAGTCATTTTCTTCCCCGAC	[6]
	SAR11-542R	SAR11 clade	TCCGAACTACGCTAGGTC	[6]
	SAR11-732R	SAR11 clade	GTCAGTAATGATCCAGAAAGYTG	[6]
	ALF968	*Alpaprotebacteria*	GGTAAGGTTCTGCGCGTT	[7]
	PLA886	*Planctomycetales*	GCCTTGCGACCATACTCCC	[8]

(Table 1) cont.....

Samples	Probe Name	Target Microorganisms	`Probe Sequence 5'-3'	Reference
Marine sediment	DSR651	*Desulforhopalus* spp.	CCCCCTCCAGTACTCAAG	[9]
	DSS658	*Desulfosarcina* spp, *Desulfobaba* spp, *Desulfococccus* spp, *Desulfofrigus* spp.	TCCACTTCCCTCTCCCAT	[9]
	ARC915	*Archaea*	GTGCTCCCCCGCCAATTCCT	[10]
	NON338	EUB338 antisense	ACTCCTACGGGAGGCAGC	[11]

Finally, the specific detection of target bacteria by FISH is confirmed by microscopy. For general samples, a phase contrast microscope or a differential interference microscope to which a fluorescent device is added can be used. For thick biofilms or microbial lumps, a confocal laser microscope is more helpful for visualization. Microscope fluorescence filters make it possible to distinguish different fluorescent labels and to observe a plurality of microorganisms through multiple staining [19]. Moreover, fluorescent cells can also be sorted and recovered by a cell sorter or micromanipulator. The sorted cells can then be subjected to single-cell genome analysis [20].

Major Drawbacks and Solutions

At times, FISH cannot obtain sufficient fluorescence intensities for the detection due to a lack of probe permeabilization, low cellular ribosome content, or inaccessibility to the probe binding site based on the higher-order structure of the rRNA [1].

If probe penetration is low, sufficient fluorescence intensity cannot be obtained-even if rRNA is present in the cell in ample amounts-because the probe cannot be retained in the cells. Paraformaldehyde fixation for most gram-negative bacteria is enough for successful FISH outcomes; however, some gram-negative and many gram-positive bacteria as well as archaea require other permeabilization agents, such as enzymes, solvents, detergents, or even HCl [16]. Unfortunately, there is no standard permeabilization protocol for all microbial cells. Development of a cell wall treatment that is appropriate for all cell structures has been attempted, but its utility is still being studied. However, based on the fact that many environmental microorganisms cannot be isolated and cultured, it is necessary to keep in mind that many analysis results contain false negatives.

The signal intensity of cells hybridized with oligonucleotide probes directly correlates with cellular rRNA content. Therefore, if there is not enough rRNA in the cell, positive fluorescence signals cannot be obtained; this drawback has been recognized since the inception of the FISH technique [21]. The average number of ribosomes in *Escherichia coli* cells is 37,000 ± 4,500 copies/cell during the stationary growth phase [22], which means FISH can sufficiently detect rRNA through microscopy. However, it is often reported that marine bacteria cannot be detected by FISH using oligonucleotide probes [23] because most marine bacteria in aquatic habitats are small, slow growing, or starved. Small bacteria are unable to retain ample amounts of rRNA in the cell owing to space constrains and only several hundred rRNAs can exist in bacteria [1]. In addition, the amount of intracellular rRNA decreases during poor nutrition and slow growth conditions. Thus, the obtained signal intensities are frequently below the detection limit or drowned out by strong background fluorescence. To overcome this problem, various high-sensitivity FISH methods have been developed, including probes fluorescently labeled at both ends (double labeling of oligonucleotide probes [DOPE]-FISH [24]), helper oligonucleotide probes [25], multiply labeled polyribonucleotide probes [26], and signal amplification with reporter enzymes (catalyzed reporter deposition [CARD]-FISH [27]), which will be described later.

After probe permeation into the cells, positive fluorescence may not be obtained in some cases when the affinity between the probe and the target molecule is low. Fuchs *et al.* examined the differences in fluorescence intensities by designing 170 sequential probes for *E. coli* rRNA and found that only approximately 20% of the probes provided sufficient fluorescence signals for detection [25]. This problem derives from the higher-order structure of rRNA and the proteins binding to rRNA, meaning that it is difficult for probes to bind rRNA at some points along its structure. Yilmaz *et al.* proposed to evaluate the binding affinity between a probe and the target molecule before FISH by means of an affinity index obtained *via* the thermodynamic approach and then simulating probe design [28]. We can now design new probes using specially adapted software tools, such as the ARB software package [14], and test for specificity and binding conditions with programs such as probeCheck [29] or mathFISH [18]. For increasing binding affinity, artificially synthesized DNA, such as peptide nucleic acid (PNA [30];) and locked nucleic acid (LNA [31];), is also utilized. By replacement of DNA with artificially synthesized DNA without changing the base sequence, the obtained fluorescence intensity can be dramatically increased. These techniques have made probe design much more flexible than ever before.

HIGHLY SENSITIVE FISH FOR TARGETING RRNA

Highly sensitive FISH has been developed to overcome the limitations of low rRNA content in cells. Several techniques have been reported including CARD-FISH [32], which is widely used in environmental microbiology. However, there are limitations to CARD-FISH, and to overcome them, we devised *in situ* DNA-hybridization chain reaction (HCR) in 2015 as a new highly sensitive FISH technique [33, 34]. *in situ* DNA-HCR achieves high sensitivity *via* a network of nucleic acid probes instead of the enzymatic catalysis utilized by CARD-FISH. The following section will discuss the principles, applications, and limitations of CARD-FISH as well as introduce the concept and applications of *in situ* DNA-HCR.

CARD-FISH

CARD-FISH has been utilized for the detection of microorganisms with low rRNA content [27]. It is a type of sensitive FISH where signal amplification is mediated by the interaction of horseradish peroxidase (HRP)-labeled probes and fluorescently labeled tyramide; this principle is illustrated in Fig (**2A**). First, HRP-labeled probes hybridize to the target molecules, after which HRP converts tyramide into a radical intermediate in the presence of hydrogen peroxide. As this radical reaction occurs near HRP molecules, a large number of tyramides will be deposited around HRP, leading to tyramide signal amplification (TSA). Thus, the signal intensity of CARD-FISH is more than 26- to 41-fold higher than that of standard FISH [22]. As a result, this method has been applied to the detection of microorganisms in oligotrophic environments, such as marine samples [35 - 37].

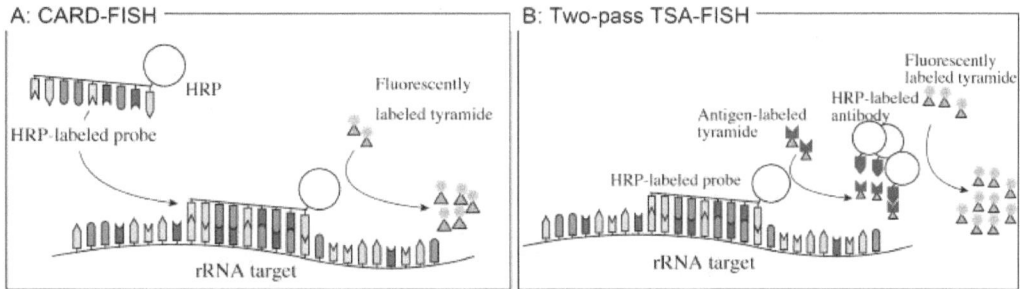

Fig. (2). Schematic depiction of **A**: CARD-FISH and **B**: two-pass TSA-FISH.
The illustration depicts tyramide immobilization on tyrosine. HRP, horseradish-peroxidase

Despite high sensitivity, a number of technical issues have been reported for CARD-FISH. For example, it is difficult for the HRP-labeled probe to penetrate fixed cells because the molecular weight of HRP (approximately 40 kDa) is higher than that of fluorochromes (mostly 500–1000 Da). Therefore, special permeabilization steps for CARD-FISH—usually enzymatic reactions—are required before application of HRP-labeled probes. Lysozyme treatment is most commonly performed for bacterial permeabilization [38], whereas proteinase K is used for Archaea [27]. Moreover, several gram-positive bacteria cannot be permeabilized efficiently with lysozyme treatment alone and instead, simultaneous treatment with achromopeptidase and lysozyme is more effective

[39 - 41]. Contrastingly, methanogens with an S-layer and gram-negative bacteria as well as *Planctomycetes* in marine sediments do not require the permeabilization step [36]. These findings indicate that permeabilization methods must be optimized for each target microorganism; this task is very difficult because the optimization range is very narrow.

Furthermore, endogenous peroxidase poses a major problem for CARD-FISH [38], especially because it is present in anaerobic ammonium oxidation (Anammox) bacteria and microorganisms in deep-sea sediments [38, 42]. As TSA proceeds *via* hydrogen peroxide oxidation into hydroxyl radicals by peroxidase activity, endogenous peroxidase should be inactivated, otherwise CARD-FISH will produce false-positive signals. To inactive endogenous peroxidase, a number of treatments have been reported, which include HCl, H_2O_2, and diethylpyrocarbonate [43, 44]. This inactivation step is necessary for the optimization and successful application of CARD-FISH.

In situ DNA-HCR

In situ DNA-HCR combines FISH and HCR to detect environmental microorganisms in oligotrophic samples [33]. The principle relies on HCR for signal amplification (Fig. **3**). The HCR method involves two hairpin probes. Once a single-stranded initiator is added to the mixture, it opens one hairpin, which exposes a single-stranded sequence. This sequence then interacts with the other hairpin species, opening it up as well [45]. As a result, an HCR product is obtained by the extension of the two hairpin probes from the initiator sequence [46]. For *in situ* DNA-HCR, an initiator probe that is contained in the HCR initiator sequence and probe sequence, hybridizes to target 16S rRNA molecules (Fig. **3**, step 1). Next, the two hairpin probes begin extending from this initiator sequence (Fig. **3**, step 2). After the excess initiator probes are washed away, signal amplification is mediated by the chain reaction of fluorescently labeled hairpin probes extending from the initiator.

The signal intensity of *in situ* DNA-HCR is 8- to 16-fold higher than that of standard FISH [33] and after optimization of hybridization and amplification buffers, the signal intensity of quickHCR-FISH is 2-fold higher than that of *in situ* DNA-HCR [34]. Moreover, the advantage of *in situ* DNA-HCR over CARD-FISH is the use of fluorescently labeled hairpin probes for signal amplification, in contrast to enzymatic reactions of high-molecular-weight HRP-labeled probes. Due to these advantages, *in situ* DNA-HCR has been carried out for detecting not only pure cultured microorganisms but also microorganisms on marine surfaces and sediments (Fig. **4**). Moreover, the initiator probe sequence can be synthesized from the list of major FISH probes and thus simultaneous detection of bacteria,

Archaea, and methanogens in anaerobic samples can be successfully performed with *in situ* DNA-HCR [33].

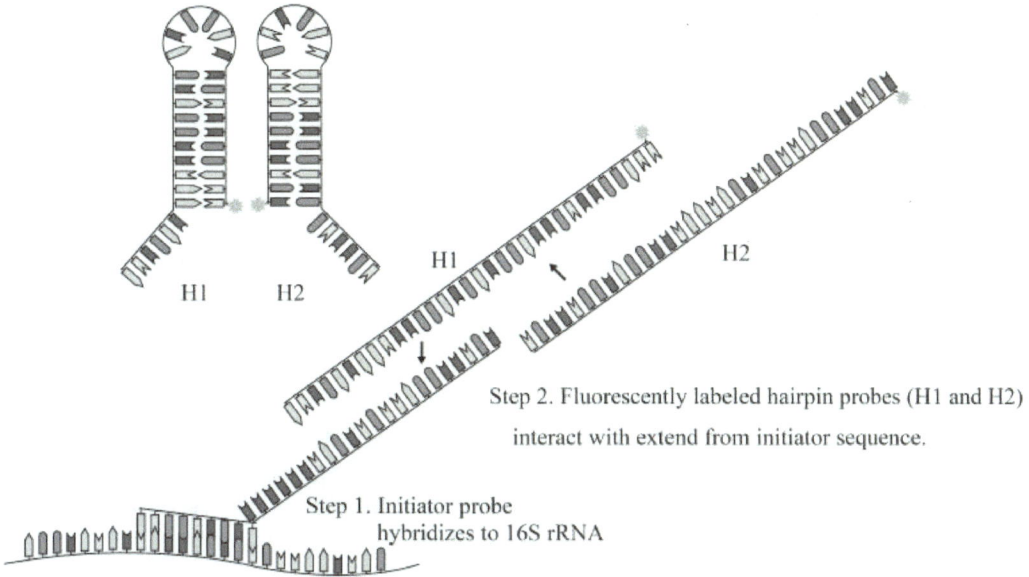

Step 2. Fluorescently labeled hairpin probes (H1 and H2) interact with extend from initiator sequence.

Step 1. Initiator probe hybridizes to 16S rRNA

Fig. (3). The principle of *in situ* DNA-HCR.

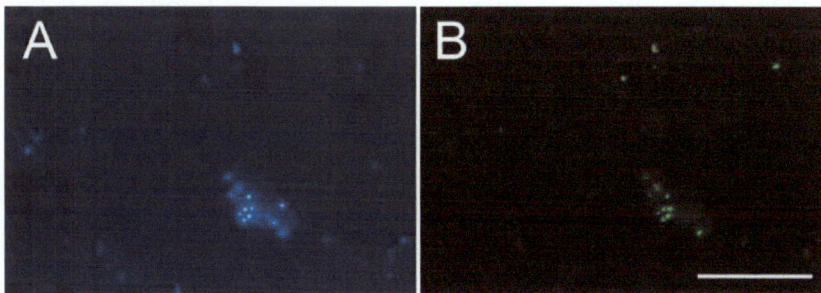

Fig. (4). Detection of bacteria in a marine sediment by quickHCR-FISH, which omits the permeabilization step.
(**A**) DAPI-stained cells. (**B**) Fluorescence signals of probes hybridized to target rRNA. Exposure time = 200 ms. Scale bar = 20 μm.

HIGHLY SENSITIVE FISH FOR TARGETING MRNA OR FUNCTIONAL GENES

The emergence of next-generation sequencing has become a driving force behind the rapid accumulation of genetic information from the natural world [47]. To fully grasp the dynamics of environmental microorganisms, information on mRNA and functional genes, in addition to rRNA, is indispensable [48, 49].

FISH, which visually detects microorganisms, has also expanded its applicability to the detection of low-abundance mRNA and functional genes with high sensitivity. These detection technologies can be roughly subdivided into two categories: one is composed of methods that increase fluorescence intensity by binding a large number of reporters to a small number of target molecules. Representative examples include recognition of individual genes (RING)-FISH, which involves polynucleotide probes [50], two-pass TSA-FISH, which employs the CARD system [51, 52], and GeneFISH, a combination of CARD and polynucleotide probes [53, 54]. The other category includes nucleic acid amplification (NAA), whereby the number of target molecules in a cell is increased *via* enzyme-catalyzed reactions. NAA includes *in situ* PCR [55], *in situ* loop-mediated isothermal amplification [56], *in situ* rolling circle amplification [57], cycling-primed *in situ* amplification -FISH [32], and others. Although these methods can achieve high sensitivity and visualize single-copy genes, they also come with their own challenges. It is necessary to ensure high cell penetrability to introduce substances with a high molecular weight for techniques that utilize enzymes or polynucleotide probes. Furthermore, enzyme-based methods have an increased risk of false positives in samples with endogenous activity as seen for CARD-FISH.

Until 2010, the application of these two detection technologies had been reported equally in the literature; however, the former category, which includes the CARD system, has been predominantly used in recent years for environmental microbiology cases. This is because the methods that rely on increasing fluorescence intensity are capable of detecting mRNA and functional genes with high sensitivity and stable detection rates. In the following section, we will describe two representative methods that are based on the CARD system: two-pass TSA-FISH and GeneFISH.

Two-pass TSA-FISH

This type of FISH is a method for achieving ultra-high sensitivity by taking the TSA reaction included in CARD-FISH and repeating it twice (Fig. **2B**). During the first TSA reaction, antigen-labeled tyramide is deposited in the cells, followed by incubation with an HRP-labeled antibody that binds the antigen. High fluorescence intensity is then achieved by deposition of fluorescently labeled tyramide for the second TSA reaction. Kubota *et al.* succeeded in detecting *mcr* mRNA of methanogenic archaea by developing this two-pass TSA-FISH employing oligonucleotide probes [27]. We speculated that with the high fluorescence intensity of two-pass TSA-FISH, it would be possible to detect single-copy genes and to apply the method to the gene detection technology. As a result, we reported a technique that detects genes by means of oligonucleotide probes [51] and polynucleotide probes [52]. The advantages of using oligonucleotide probes include the ability to design probes from existing PCR primers and a high phylogenetic classification ability; however, the detection rate is as low as 20%. Conversely, employing polynucleotide probes can provide high detection rates, up to 90%, yet it is more difficult to distinguish mismatches in comparison with oligonucleotide probes. Furthermore, due to the increased length of polynucleotide probes, they offer fewer options for designing specific probes that allow for discrimination of closely related microbial groups. Thus, selecting appropriate probe lengths depends on the purpose of the study.

GeneFISH

GeneFISH is a functional gene detection technology that combines CARD and polyribonucleotide probes [54]. GeneFISH achieves high detection rates and is based on a protocol that can be performed in combination with rRNA-targeting FISH. In recent years, GeneFISH has been employed alongside super-hig--resolution microscopy, where gene localization in the cell and spatial distribution of rRNA can be examined [53]. Thus, FISH analysis can now quantify gene copy numbers per cell with improved intracellular localization and the possibility of multiplexing.

APPLICATION OF FISH IN MARINE MICROBIOLOGY

The FISH technique can visually detect target microorganisms, clarify their abundance and niche, and confirm spatial exchange with other microorganisms. FISH has also contributed to the many findings that greatly advanced the field of marine microbiology and has helped determine the number of microorganisms in the ocean. With the introduction of high-sensitivity FISH methods, it became possible to detect oligotrophic bacteria, which have been previously overlooked [58, 59], leading to the correction of information regarding their abundance [60]. Even for the discovery of uncultured microorganisms, FISH is necessary for clarifying morphological features. For marine anammox bacteria, the morphology, abundance, niche, and other characteristics have been discovered by FISH, which greatly contributed to further ecological clarification [61]. For anaerobic methanotrophic archaea, FISH has revealed that anaerobic methanotrophic archaea are partnered with sulfate-reducing bacteria; this finding has led to estimation of the anaerobic methane oxidation reaction pathway [62]. Moreover, FISH has been used in combination with other technologies. One of the most popular combinations involves nanoscale secondary ion mass spectrometry (NanoSIMS). NanoSIMS investigates the atoms incorporated into the cell and can estimate the metabolism of target microorganisms. This method is increasingly being applied to marine bacteria and is used for elucidation of carbon and nitrogen cycles in oceans and specific niches [63]. Additionally, FISH and NanoSIMS are carried out to study the activity and biogeochemical effects of marine viruses [64]. Another example is bio-orthogonal noncanonical amino acid tagging (BONCAT)-FISH, which has been reported to positively label proteins produced by bacteria to evaluate their metabolic activity. BONCAT-FISH has also been applied to the characterization of extreme-environment microbiomes, such as methane-producing bacteria at deep-sea locations [65].

Microbial ecology is entering an era of utilization of a large amount of genetic information accumulated by metagenomic analysis and next-generation sequencing analysis. High-sensitivity FISH reveals which microorganisms are derived from these large amounts of genetic information by visual detection. In addition, FISH can be used in combination with a technology to estimate metabolism, and this strategy makes it possible to clarify the ecology of environmental microorganisms without cultivation. The FISH technology is under continuous development as mentioned in the chapter. In the future, the ecology of microorganisms at single-cell resolution will become clearer as sensitivity and resolution of the available methods increase.

CONSENT FOR PUBLICATION

Not applicable.

CONFLICT OF INTEREST

The author confirms that this chapter contents have no conflict of interest.

ACKNOWLEDGEMENTS

Declare none.

REFERENCES

[1]   Amann R, Fuchs BM. Single-cell identification in microbial communities by improved fluorescence *in situ* hybridization techniques. Nat Rev Microbiol 2008; 6(5): 339-48.
        [http://dx.doi.org/10.1038/nrmicro1888] [PMID: 18414500]

[2]   Amann RI, Ludwig W, Schleifer KH. Phylogenetic identification and *in situ* detection of individual microbial cells without cultivation. Microbiol Rev 1995; 59(1): 143-69.
        [PMID: 7535888]

[3]   Amann RI, Binder BJ, Olson RJ, Chisholm SW, Devereux R, Stahl DA. Combination of 16S rRNA-targeted oligonucleotide probes with flow cytometry for analyzing mixed microbial populations. Appl Environ Microbiol 1990; 56(6): 1919-25.
        [PMID: 2200342]

[4]   Manz W, Amann RI, Ludwig W, Wagner M, Schleifer KH. Phylogenetic oligodeoxynucleotide probes for the major subclasses of Proteobacteria: problems and solutions. Syst Appl Microbiol 1992; 15: 593-600.

[http://dx.doi.org/10.1016/S0723-2020(11)80121-9]

[5]   Manz W, Amann R, Ludwig W, Vancanneyt M, Schleifer KH. Application of a suite of 16S rRNA-specific oligonucleotide probes designed to investigate bacteria of the phylum cytophaga-flavobacte--bacteroides in the natural environment. Microbiology 1996; 142(Pt 5): 1097-106.
        [http://dx.doi.org/10.1099/13500872-142-5-1097] [PMID: 8704951]

[6]   Morris RM, Rappé MS, Connon SA, *et al.* SAR11 clade dominates ocean surface bacterioplankton communities. Nature 2002; 420(6917): 806-10.
        [http://dx.doi.org/10.1038/nature01240] [PMID: 12490947]

[7]   Neef A. Anwendung der *in situ* Einzelzell-Identifizierung von Bakterien zur populationsanalyse in komplexen mikrobiellen biozönosen, PhD thesis, Technische Universität München 1997.

[8]   Neef A, Amann R, Schlesner H, Schleifer KH. Monitoring a widespread bacterial group: *in situ* detection of planctomycetes with 16S rRNA-targeted probes. Microbiology 1998; 144(Pt 12): 3257-66.
        [http://dx.doi.org/10.1099/00221287-144-12-3257] [PMID: 9884217]

[9]   Manz W, Eisenbrecher M, Neu TR, Szewzyk U. Abundance and spatial organization of Gram-negative sulfate-reducing bacteria in activated sludge investigated by *in situ* probing with specific 16S rRNA targeted oligonucleotides. FEMS Microbiol Ecol 1998; 25: 43-61.
        [http://dx.doi.org/10.1111/j.1574-6941.1998.tb00459.x]

[10]   Stahl DA, Amann RI. Development and application of nucleic acid probes Nucleic Acid Techniques in Bacterial Systematics. Chichester, UK: JohnWiley & Sons Ltd 1991; pp. 205-42.

[11]   Wallner G, Amann R, Beisker W. Optimizing fluorescent *in situ* hybridization with rRNA-targeted oligonucleotide probes for flow cytometric identification of microorganisms. Cytometry 1993; 14(2): 136-43.
        [http://dx.doi.org/10.1002/cyto.990140205] [PMID: 7679962]

[12]   Loy A, Horn M, Wagner M. probeBase: an online resource for rRNA-targeted oligonucleotide probes. Nucleic Acids Res 2003; 31(1): 514-6.
        [http://dx.doi.org/10.1093/nar/gkg016] [PMID: 12520066]

[13]   Loy A, Maixner F, Wagner M, Horn M. probeBase--an online resource for rRNA-targeted oligonucleotide probes: new features 2007. Nucleic Acids Res 2007; 35(Database issue): D800-4.
        [http://dx.doi.org/10.1093/nar/gkl856] [PMID: 17099228]

[14]   Ludwig W, Strunk O, Westram R, *et al.* ARB: a software environment for sequence data. Nucleic Acids Res 2004; 32(4): 1363-71.
        [http://dx.doi.org/10.1093/nar/gkh293] [PMID: 14985472]

[15]   Thavarajah R, Mudimbaimannar VK, Elizabeth J, Rao UK, Ranganathan K. Chemical and physical basics of routine formaldehyde fixation. J Oral Maxillofac Pathol 2012; 16(3): 400-5.
        [http://dx.doi.org/10.4103/0973-029X.102496] [PMID: 23248474]

[16]   Frickmann H, Zautner AE, Moter A, *et al.* Fluorescence *in situ* hybridization (FISH) in the microbiological diagnostic routine laboratory: a review. Crit Rev Microbiol 2017; 43(3): 263-93.
        [http://dx.doi.org/10.3109/1040841X.2016.1169990] [PMID: 28129707]

[17]   Schramm A, Fuchs BM, Nielsen JL, Tonolla M, Stahl DA. Fluorescence *in situ* hybridization of 16S rRNA gene clones (Clone-FISH) for probe validation and screening of clone libraries. Environ Microbiol 2002; 4(11): 713-20.
        [http://dx.doi.org/10.1046/j.1462-2920.2002.00364.x] [PMID: 12460279]

[18]   Yilmaz LS, Parnerkar S, Noguera DR. mathFISH, a web tool that uses thermodynamics-based mathematical models for in silico evaluation of oligonucleotide probes for fluorescence *in situ* hybridization. Appl Environ Microbiol 2011; 77(3): 1118-22.
        [http://dx.doi.org/10.1128/AEM.01733-10] [PMID: 21148691]

[19]   Valm AM, Mark Welch JL, Rieken CW, *et al.* Systems-level analysis of microbial community

organization through combinatorial labeling and spectral imaging. Proc Natl Acad Sci USA 2011; 108(10): 4152-7.
[http://dx.doi.org/10.1073/pnas.1101134108] [PMID: 21325608]

[20]   Kvist T, Ahring BK, Lasken RS, Westermann P. Specific single-cell isolation and genomic amplification of uncultured microorganisms. Appl Microbiol Biotechnol 2007; 74(4): 926-35.
[http://dx.doi.org/10.1007/s00253-006-0725-7] [PMID: 17109170]

[21]   DeLong EF, Wickham GS, Pace NR. Phylogenetic stains: ribosomal RNA-based probes for the identification of single cells. Science 1989; 243(4896): 1360-3.
[http://dx.doi.org/10.1126/science.2466341] [PMID: 2466341]

[22]   Hoshino T, Yilmaz LS, Noguera DR, Daims H, Wagner M. Quantification of target molecules needed to detect microorganisms by fluorescence *in situ* hybridization (FISH) and catalyzed reporter deposition-FISH. Appl Environ Microbiol 2008; 74(16): 5068-77.
[http://dx.doi.org/10.1128/AEM.00208-08] [PMID: 18552182]

[23]   Mary I, Heywood JL, Fuchs BM, *et al.* SAR11 dominance among metabolically active low nucleic acid bacterioplankton in surface waters along an Atlantic meridional transect. Aquat Microb Ecol 2006; 45: 107-13.
[http://dx.doi.org/10.3354/ame045107]

[24]   Stoecker K, Dorninger C, Daims H, Wagner M. Double labeling of oligonucleotide probes for fluorescence *in situ* hybridization (DOPE-FISH) improves signal intensity and increases rRNA accessibility. Appl Environ Microbiol 2010; 76(3): 922-6.
[http://dx.doi.org/10.1128/AEM.02456-09] [PMID: 19966029]

[25]   Fuchs BM, Wallner G, Beisker W, Schwippl I, Ludwig W, Amann R. Flow cytometric analysis of the *in situ* accessibility of *Escherichia coli* 16S rRNA for fluorescently labeled oligonucleotide probes. Appl Environ Microbiol 1998; 64(12): 4973-82.
[PMID: 9835591]

[26]   Trebesius K, Amann R, Ludwig W, Mühlegger K, Schleifer KH. Identification of whole fixed bacterial cells with nonradioactive 23S rRNA-targeted polynucleotide probes. Appl Environ Microbiol 1994; 60(9): 3228-35.
[PMID: 16349377]

[27]   Kubota K. CARD-FISH for environmental microorganisms: technical advancement and future applications. Microbes Environ 2013; 28(1): 3-12.
[http://dx.doi.org/10.1264/jsme2.ME12107] [PMID: 23124765]

[28]   Yilmaz LS, Noguera DR. Mechanistic approach to the problem of hybridization efficiency in fluorescent *in situ* hybridization. Appl Environ Microbiol 2004; 70(12): 7126-39.
[http://dx.doi.org/10.1128/AEM.70.12.7126-7139.2004] [PMID: 15574909]

[29]   Loy A, Arnold R, Tischler P, Rattei T, Wagner M, Horn M. probeCheck--a central resource for evaluating oligonucleotide probe coverage and specificity. Environ Microbiol 2008; 10(10): 2894-8.
[http://dx.doi.org/10.1111/j.1462-2920.2008.01706.x] [PMID: 18647333]

[30]   Perry-O'Keefe H, Rigby S, Oliveira K, *et al.* Identification of indicator microorganisms using a standardized PNA FISH method. J Microbiol Methods 2001; 47(3): 281-92.
[http://dx.doi.org/10.1016/S0167-7012(01)00303-7] [PMID: 11714518]

[31]   Kubota K, Ohashi A, Imachi H, Harada H. Improved *in situ* hybridization efficiency with locked-nucleic-acid-incorporated DNA probes. Appl Environ Microbiol 2006; 72(8): 5311-7.
[http://dx.doi.org/10.1128/AEM.03039-05] [PMID: 16885281]

[32]   Kenzaka T, Tamaki S, Yamaguchi N, Tani K, Nasu M. Recognition of individual genes in diverse microorganisms by cycling primed *in situ* amplification. Appl Environ Microbiol 2005; 71(11): 7236-44.
[http://dx.doi.org/10.1128/AEM.71.11.7236-7244.2005] [PMID: 16269764]

[33] Yamaguchi T, Kawakami S, Hatamoto M, *et al. In situ* DNA-hybridization chain reaction (HCR): A facilitated *in situ* hybridization chain reaction system for the detection of environmental microorganisms. Environ Microbiol 2015; 17: 2532-41.
[http://dx.doi.org/10.1111/1462-2920.12745] [PMID: 25523128]

[34] Yamaguchi T, Fuchs BM, Amann R, *et al.* Rapid and sensitive identification of marine bacteria by an improved *in situ* DNA hybridization chain reaction (quickHCR-FISH). Syst Appl Microbiol 2015; 38(6): 400-5.
[http://dx.doi.org/10.1016/j.syapm.2015.06.007] [PMID: 26215142]

[35] Hao M, Tashiro T, Kato M, *et al.* Population dynamics of Crenarchaeota and Euryarchaeota in the mixing front of river and marine waters. Microbes Environ 2010; 25(2): 126-32.
[http://dx.doi.org/10.1264/jsme2.ME10106] [PMID: 21576863]

[36] Ishii K, Mussmann M, MacGregor BJ, Amann R. An improved fluorescence *in situ* hybridization protocol for the identification of bacteria and archaea in marine sediments. FEMS Microbiol Ecol 2004; 50(3): 203-13.
[http://dx.doi.org/10.1016/j.femsec.2004.06.015] [PMID: 19712361]

[37] Pernthaler A, Pernthaler J, Amann R. Fluorescence *in situ* hybridization and catalyzed reporter deposition for the identification of marine bacteria. Appl Environ Microbiol 2002; 68(6): 3094-101.
[http://dx.doi.org/10.1128/AEM.68.6.3094-3101.2002] [PMID: 12039771]

[38] Pavlekovic M, Schmid MC, Schmider-Poignee N, *et al.* Optimization of three FISH procedures for *in situ* detection of anaerobic ammonium oxidizing bacteria in biological wastewater treatment. J Microbiol Methods 2009; 78(2): 119-26.
[http://dx.doi.org/10.1016/j.mimet.2009.04.003] [PMID: 19389431]

[39] Barsotti O, Morrier JJ, Freney J, *et al.* Achromopeptidase for rapid lysis of oral anaerobic gram-positive rods. Oral Microbiol Immunol 1988; 3(2): 86-8.
[http://dx.doi.org/10.1111/j.1399-302X.1988.tb00088.x] [PMID: 3268757]

[40] Chassy BM, Giuffrida A. Method for the lysis of Gram-positive, asporogenous bacteria with lysozyme. Appl Environ Microbiol 1980; 39(1): 153-8.
[PMID: 6986847]

[41] Ezaki T, Suzuki S. Achromopeptidase for lysis of anaerobic gram-positive cocci. J Clin Microbiol 1982; 16(5): 844-6.
[PMID: 6759529]

[42] Woebken D, Fuchs BM, Kuypers MM, Amann R. Potential interactions of particle-associated anammox bacteria with bacterial and archaeal partners in the Namibian upwelling system. Appl Environ Microbiol 2007; 73(14): 4648-57.
[http://dx.doi.org/10.1128/AEM.02774-06] [PMID: 17526789]

[43] Pernthaler A, Pernthaler J, Schattenhofer M, Amann R. Identification of DNA-synthesizing bacterial cells in coastal North Sea plankton. Appl Environ Microbiol 2002; 68(11): 5728-36.
[http://dx.doi.org/10.1128/AEM.68.11.5728-5736.2002] [PMID: 12406771]

[44] Sekar R, Fuchs BM, Amann R, Pernthaler J. Flow sorting of marine bacterioplankton after fluorescence *in situ* hybridization. Appl Environ Microbiol 2004; 70(10): 6210-9.
[http://dx.doi.org/10.1128/AEM.70.10.6210-6219.2004] [PMID: 15466568]

[45] Dirks RM, Pierce NA. Triggered amplification by hybridization chain reaction. Proc Natl Acad Sci USA 2004; 101(43): 15275-8.
[http://dx.doi.org/10.1073/pnas.0407024101] [PMID: 15492210]

[46] Niu S, Jiang Y, Zhang S. Fluorescence detection for DNA using hybridization chain reaction with enzyme-amplification. Chem Commun (Camb) 2010; 46(18): 3089-91.
[http://dx.doi.org/10.1039/c000166j] [PMID: 20424746]

[47] Singer E, Bushnell B, Coleman-Derr D, *et al.* High-resolution phylogenetic microbial community

profiling. ISME J 2016; 10(8): 2020-32.
[http://dx.doi.org/10.1038/ismej.2015.249] [PMID: 26859772]

[48]   Danhorn T, Young CR, DeLong EF. Comparison of large-insert, small-insert and pyrosequencing libraries for metagenomic analysis. ISME J 2012; 6(11): 2056-66.
[http://dx.doi.org/10.1038/ismej.2012.35] [PMID: 22534608]

[49]   Lal A, Seshasayee ASN. The impact of next-generation sequencing technology on bacterial genomics A systems theoretic approach to systems and synthetic biology II: analysis and design of cellular systems. Dordrecht, Netherlands: Springer 2014; pp. 31-58.

[50]   Zwirglmaier K, Ludwig W, Schleifer KH. Recognition of individual genes in a single bacterial cell by fluorescence *in situ* hybridization--RING-FISH. Mol Microbiol 2004; 51(1): 89-96.
[http://dx.doi.org/10.1046/j.1365-2958.2003.03834.x] [PMID: 14651613]

[51]   Kawakami S, Kubota K, Imachi H, Yamaguchi T, Harada H, Ohashi A. Detection of single copy genes by two-pass tyramide signal amplification fluorescence *in situ* hybridization (Two-Pass TSA-FISH) with single oligonucleotide probes. Microbes Environ 2010; 25(1): 15-21.
[http://dx.doi.org/10.1264/jsme2.ME09180] [PMID: 21576847]

[52]   Kawakami S, Hasegawa T, Imachi H, *et al.* Detection of single-copy functional genes in prokaryotic cells by two-pass TSA-FISH with polynucleotide probes. J Microbiol Methods 2012; 88(2): 218-23.
[http://dx.doi.org/10.1016/j.mimet.2011.11.014] [PMID: 22172287]

[53]   Barrero-Canosa J, Moraru C, Zeugner L, Fuchs BM, Amann R. Direct-geneFISH: a simplified protocol for the simultaneous detection and quantification of genes and rRNA in microorganisms. Environ Microbiol 2017; 19(1): 70-82.
[http://dx.doi.org/10.1111/1462-2920.13432] [PMID: 27348074]

[54]   Moraru C, Lam P, Fuchs BM, Kuypers MM, Amann R. GeneFISH-an *in situ* technique for linking gene presence and cell identity in environmental microorganisms. Environ Microbiol 2010; 12(11): 3057-73.
[http://dx.doi.org/10.1111/j.1462-2920.2010.02281.x] [PMID: 20629705]

[55]   Hodson RE, Dustman WA, Garg RP, Moran MA. *in situ* PCR for visualization of microscale distribution of specific genes and gene products in prokaryotic communities. Appl Environ Microbiol 1995; 61(11): 4074-82.
[PMID: 8526521]

[56]   Maruyama F, Kenzaka T, Yamaguchi N, Tani K, Nasu M. Visualization and enumeration of bacteria carrying a specific gene sequence by *in situ* rolling circle amplification. Appl Environ Microbiol 2005; 71(12): 7933-40.
[http://dx.doi.org/10.1128/AEM.71.12.7933-7940.2005] [PMID: 16332770]

[57]   Hoshino T, Schramm A. Detection of denitrification genes by *in situ* rolling circle amplification-fluorescence *in situ* hybridization to link metabolic potential with identity inside bacterial cells. Environ Microbiol 2010; 12(9): 2508-17.
[http://dx.doi.org/10.1111/j.1462-2920.2010.02224.x] [PMID: 20406291]

[58]   Novas FE, Pol D. New evidence on deinonychosaurian dinosaurs from the Late Cretaceous of Patagonia. Nature 2005; 433(7028): 858-61.
[http://dx.doi.org/10.1038/nature03285] [PMID: 15729340]

[59]   Schippers A, Neretin LN, Kallmeyer J, *et al.* Prokaryotic cells of the deep sub-seafloor biosphere identified as living bacteria. Nature 2005; 433(7028): 861-4.
[http://dx.doi.org/10.1038/nature03302] [PMID: 15729341]

[60]   Schippers A, Kock D, Höft C, Köweker G, Siegert M. Quantification of microbial communities in subsurface marine sediments of the Black Sea and off Namibia. Front Microbiol 2012; 3: 16.
[http://dx.doi.org/10.3389/fmicb.2012.00016] [PMID: 22319518]

[61]   Strous M, Fuerst JA, Kramer EH, *et al.* Missing lithotroph identified as new planctomycete. Nature

1999; 400(6743): 446-9.
[http://dx.doi.org/10.1038/22749] [PMID: 10440372]

[62]   Orphan VJ, House CH, Hinrichs KU, McKeegan KD, DeLong EF. Methane-consuming archaea revealed by directly coupled isotopic and phylogenetic analysis. Science 2001; 293(5529): 484-7.
[http://dx.doi.org/10.1126/science.1061338] [PMID: 11463914]

[63]   Gao D, Huang X, Tao Y. A critical review of NanoSIMS in analysis of microbial metabolic activities at single-cell level. Crit Rev Biotechnol 2016; 36(5): 884-90.
[http://dx.doi.org/10.3109/07388551.2015.1057550] [PMID: 26177334]

[64]   Pasulka AL, Thamatrakoln K, Kopf SH, *et al.* Interrogating marine virus-host interactions and elemental transfer with BONCAT and nanoSIMS-based methods. Environ Microbiol 2018; 20(2): 671-92.
[http://dx.doi.org/10.1111/1462-2920.13996] [PMID: 29159966]

[65]   Hatzenpichler R, Scheller S, Tavormina PL, Babin BM, Tirrell DA, Orphan VJ. *in situ* visualization of newly synthesized proteins in environmental microbes using amino acid tagging and click chemistry. Environ Microbiol 2014; 16(8): 2568-90.
[http://dx.doi.org/10.1111/1462-2920.12436] [PMID: 24571640]

Part 3: Interaction Analysis between Artificial Materials and Micro/Macro-Organisms in Marine Environments

CHAPTER 16

Biofilm on Materials' Surfaces in Marine Environments

Hideyuki Kanematsu[1,*] and **Dana M. Barry**[2]

[1] *Department of Materials Science and Engineering, National Institute of Technology (KOSEN), Suzuka College, Japan*

[2] *Department of Electrical and Computer Engineering, Clarkson University, Potsdam, New York State, 13699, the USA/SUNY Canton, Canton, New York State, 13617, USA*

Abstract: Biofilm is inhomogeneous, thin film-like matter formed on materials' surfaces. It is actually a hydrogel and composed of water (about 80% depending on the formation stages), external polysaccharides (EPS) and bacteria themselves. When materials are immersed into marine environments, biofilms form by bacterial activity. However, the process contains many steps. The formation and growth capability of biofilms are determined by the combination of environmental factors (bacteria, biota, temperatures, *etc.*) and materials. In this chapter, the authors describe the general biofilm formation process and discuss how various materials affect biofilm formation and growth.

Keywords: Biofilms, EPS, Bacteria, Metals, Ceramics, Polymers, Hydrophilicity, Hydrophobicity, Water repellency, Van der Waals forces, Electric double layer, Zeta potential.

INTRODUCTION

What is Biofilm?

Biofilm is actually a hydrogel-like matter formed on materials' surfaces. It contains lots of water, polymeric substance and bacteria [1 - 6]. First, we would like to begin this chapter with definitions of important technical terms - what is biofouling? Biofouling is a process where organisms first attach to materials' surfaces and then cause deterioration of the materials in various ways. For example, ships (with this problem) in marine environments might sail slower and slower making their journeys become longer and longer. Other examples include

---

* **Corresponding author Hideyuki Kanematsu:** Department of Materials Science and Engineering, NIT, Suzuka College, Suzuka, Japan; Tel/FAX; +81-59-368-1848; E-mail: kanemats@mse.suzuka-ct.ac.jp

Toshiyuki Takahashi (Ed.)

the deterioration of sea vessels and ports by rust. These phenomena could be attributed to various degrees of biofouling.

Biofouling is classified into two main categories – microfouling and macrofouling [7 - 10]. Microfouling is composed of various phenomena including materials' degradation by the attachment of bacteria on them. The main product of microfouling is biofilms. Biofilms are composed of water and bacteria, along with their excreted exopolymeric substances (or extracellular polysaccharides that are usually called EPS). Water is the main component. Therefore, biofilms are a sort of water film. However, it is very sticky, since it contains organic compounds derived from bacteria. The schematic concept of biofilms is shown in Fig. (**1**).

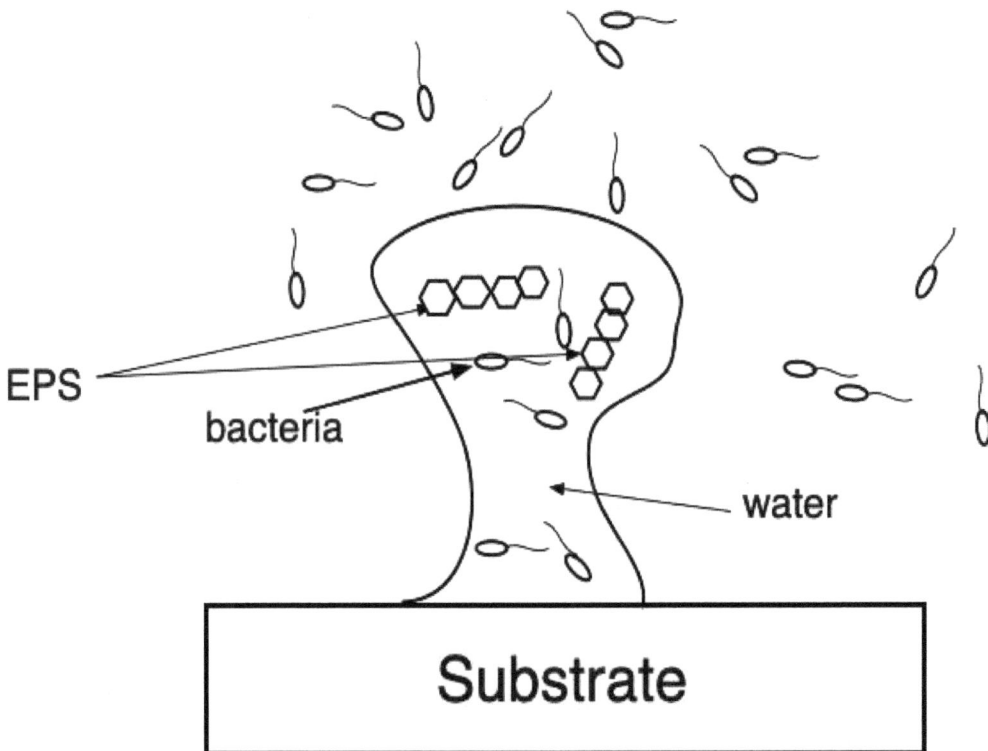

Fig. (1). Schematic illustration for biofilms.

Bacteria tend to attach to materials' surfaces, since carbon compounds (a form of their needed nutrition) generally exist on them. Various carbon compounds form very thin film-like matter on materials' surfaces. They are called "conditioning films". There are many kinds of carbon compounds constituting conditioning films. Particularly in marine environments, accumulated carbohydrates are often mentioned for the components. The thickness of conditioning film is very less, in the order of nanometers [11 - 15].

Generally, bacteria tend to attach to materials' surfaces because the conditioning film could be the nutrition for them. This is the first step for biofilm formation. In marine environments, there are many kinds of organic compounds. In particular, accumulated carbohydrates derived from organisms exist abundantly. Of course, other carbon compounds also exist. Therefore, bacteria try to attach to surfaces for their needed nutrition. This is the driving force for bacteria to attach to materials' surfaces (Fig. **2**).

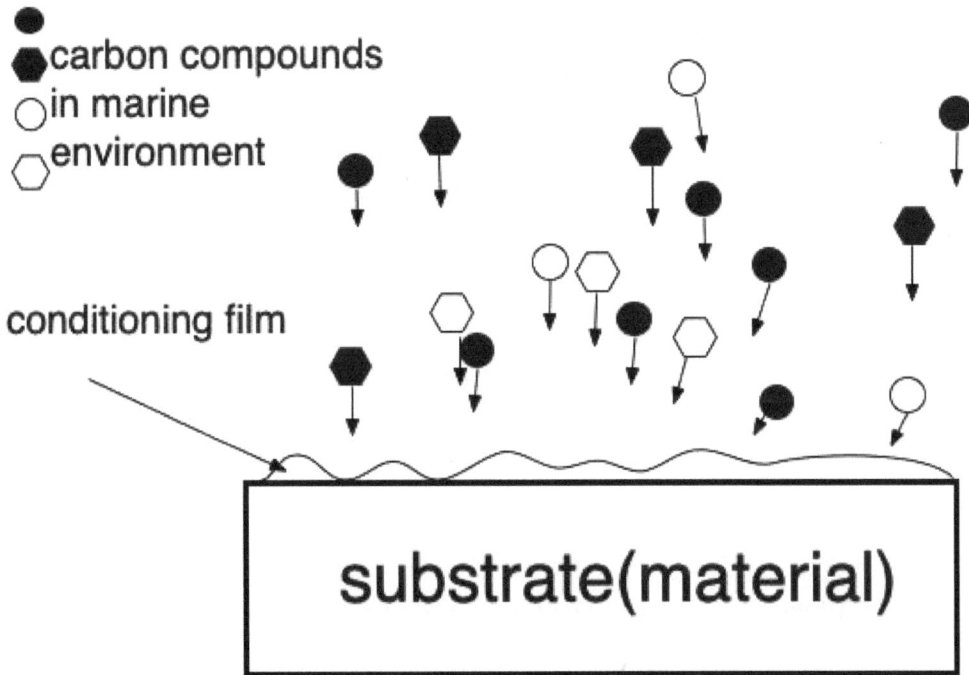

Fig. (2). Conditioning film on material in a marine environment.

The diagram above shows how bacteria tend to attach to materials' surfaces. However, materials' surfaces generally have potential barriers. Since a material's surface is an interface, both sides of it have potential differences [16, 17]. Therefore, bacteria must overcome the potential barrier in order to attach to a surface. Generally, an electric double layer forms on a material's surface. However, approaching bacteria also have their electric double layer. When the two double layers overlap, a repulsion force is produced (Fig. **3**). However, other various forces (depending on the type of material) would work and affect the balance in regards to biofilm formation on that particular material. From this viewpoint, it can be said that the type of material used is one of the important factors for biofilm formation and growth.

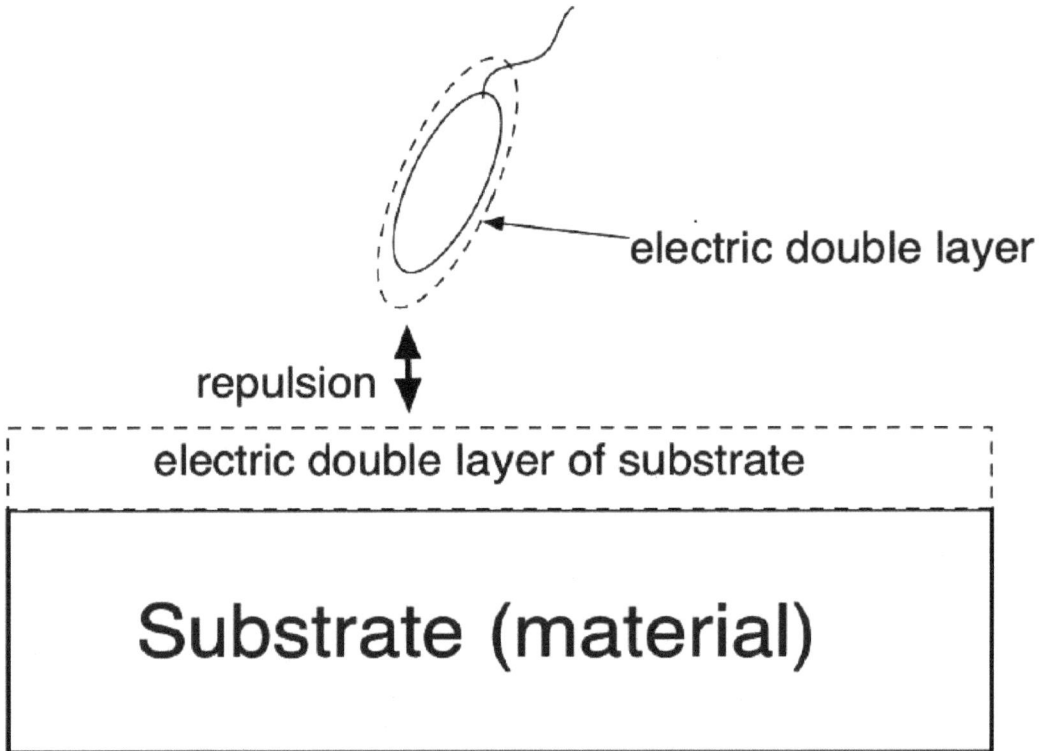

Fig. (3). The overlapping of electric double layers between the substrate and bacteria.

Other influential forces might be the electrostatic force between ionic matters, van der Waals forces, *etc*. The question is that which force will work and how will the balance be established? This will be explained below. However, bacteria eventually overcome the potential barrier and attach to materials' surfaces. The attached bacteria might detach from materials' surfaces. Such an attachment and detachment process might continue for a certain period. Then the attachment would dominate over the detachment process and the number of bacteria would increase.

When the number of bacteria attached to a material's surface exceeds a certain threshold value, bacteria on the material's surface would simultaneously begin to excrete polysaccharides. Then those bacteria would be surrounded by sticky EPS and biofilm would form at this point.

According to our experimental experiences, the situation where bacteria exist separately would not lead to biofilm formation. On the other hand, if there is a high concentration of bacteria in a local area then the excretion of polysaccharides would occur (Fig. **4**).

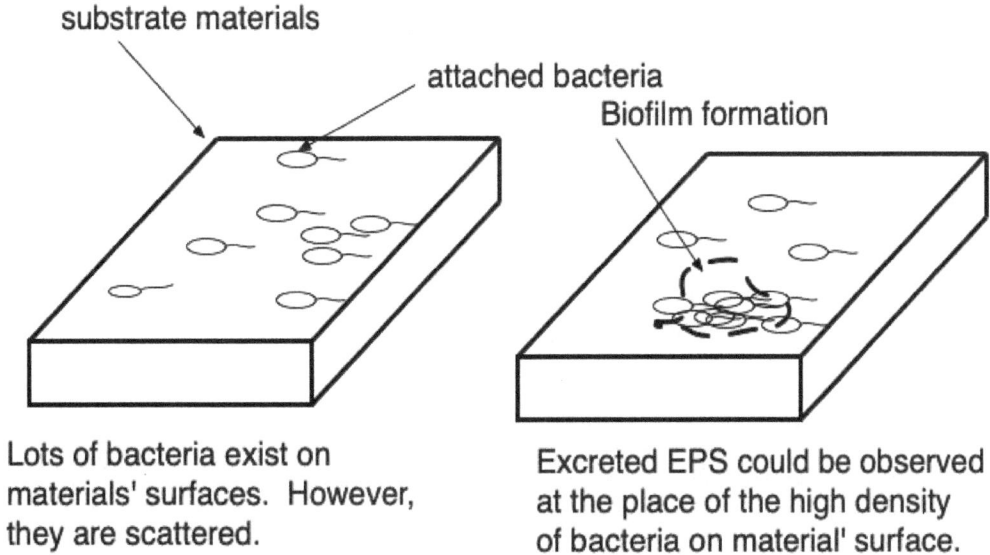

Fig. (4). Bacterial attachment and the excretion of EPS.

The excretion of EPS, mainly polysaccharides, would occur through a signal deduction process of proteins called quorum sensing [18 - 26]. The density of the signal proteins generally increases with the number of bacteria. (However, the local density seems to be important.) The high density of signal matters (called "auto inducers") would make it possible for them to enter bacterial cells through a transporter (the channel of proteins on bacterial cells). They would stimulate DNA through the cascade reactions among proteins. Then polysaccharides would be produced and excreted outside of the bacterial cells. Some characteristic signal matters are already known [26, 27]. For example, *E. coli*'s is AHL (N-acyl-L-homoserine lactone). It is also well known as an auto inducer for many other gram-negative bacteria. On the other hand, gram-positive bacteria such as *Staphylococcus epidermidis*' auto inducers are peptides. When auto inducers are affected in various ways, the biofilm formation and growth process have changed. This is already a point of countermeasure to control biofilm formation. Many countermeasures against biofilms have been proposed in a similar way. However, when we focus on substrate materials, there are not so many proposals about it. The reason could be attributed to the lack of knowledge and information about the correlation between bacteria and metals and between EPS and materials. A discussion is provided about the use of metals, ceramics, and polymers as substrates.

METALLIC MATERIALS

When metallic materials are immersed into marine environments, the electric double layer is immediately produced on materials' surfaces, as shown in Fig. (**3**). In addition, metal/metallic ions equilibrium would be established, as shown in Fig. (**5**).

The metallic ion could react with proteins around the bacterial cells, the bacterial cells' outer membranes, EPS of biofilms, *etc*. Therefore, the following effects would be possible.

#1: The contact of metallic ions with proteins scattered around bacterial cells.

Many kinds of proteins exist around bacterial cells including auto-inducers. Due to the electrostatic forces between metallic ions and them, they would react and bind with each other. Also, signal deduction processes might be changed to some extent.

#2: The contact of metallic ions with proteins on outer membranes of bacterial cells. There are many channels made of proteins on bacterial cells' outer membranes. They are generally called "transporters" and work as a channel for matter to enter or exit cells. When metallic ions react with these proteins and attach to them, the metabolism might be changed. In such a case, metallic ions would affect the bacterial growth.

#3: The reaction of metallic ions with EPS in biofilms. Metallic ions outside or inside biofilms might react with EPS. Sometimes metallic ions are the center of a bridge between huge polymer molecules (the center of cross-linkages). In such a case, metallic ions would affect the biofilm formation and growth processes.

CERAMIC MATERIALS

From the viewpoint of biofilms, ceramics can be classified into two types: bio-inert and bioactive [28]. As examples for the former, we could mention alumina, zirconia, silicon nitrides, and so on. The latter's concrete examples are carbon materials, glasses and calcium phosphates such as hydroxyapatite. Why is such a classification available and what does it mean for biofilms? It is mainly related to the interaction of energy between substrate material and bacteria/EPS.

Marine environment

$$M \leftrightarrow M^{n+} + ne^{-}$$

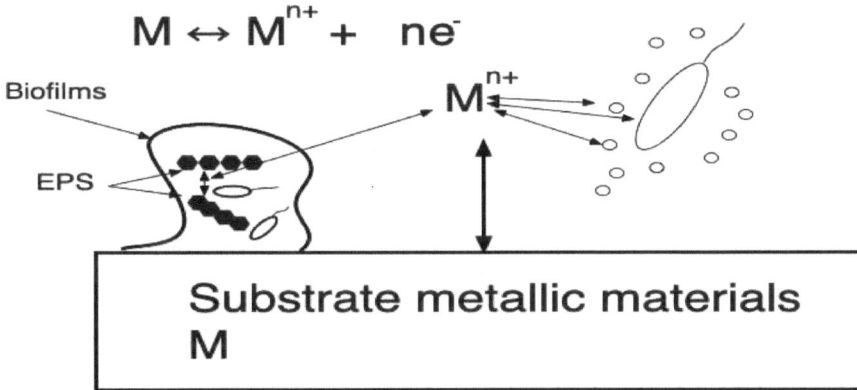

Fig. (5). Metallic substrate and biofilms in a marine environment.

Ceramics are usually constituted by ionic bonding and covalent ones. The difference of electronegativity between constituent atoms determines the type of bonding for the ceramic materials. From this viewpoint, bio-inert ceramics would have stronger covalent characteristics. On the other hand, bioactive ceramics have stronger ionic characteristics. In other words, ionic matter could interact with polymeric substances due to the crystal structures for bioactive ceramics.

Fig. (6) shows a schematic illustration for the interaction between ceramics and biofilms.

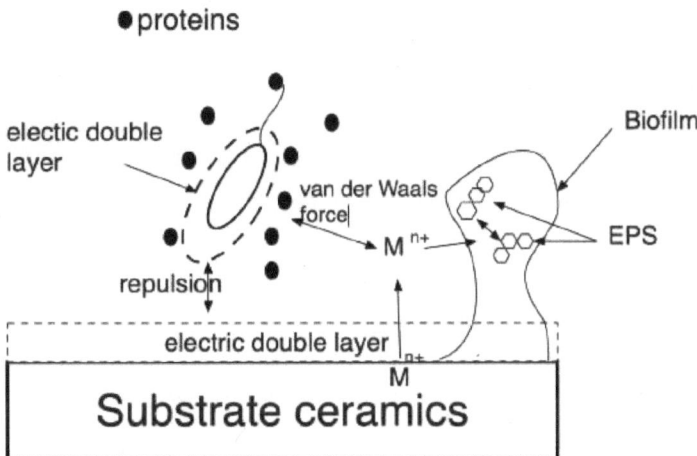

Fig. (6). The schematic relation between a ceramic substrate and biofilms.

When the specimen is bio-inert ceramics, then van der Waals forces (between polymeric substances from bacteria, on bacterial membranes, or in EPS and ionic components of ceramics) do not work very well. Therefore, strong bonding would not be produced. On the other hand, van der Waals forces would work for bioactive ceramics in various ways. Therefore, biofilms could form on bioactive ceramics.

POLYMERIC SUBSTANCES

Biofilm formation capability for polymeric substances is qualitatively evaluated by polarity. Concretely speaking, these substances are classified into hydrophilic and hydrophobic groups. Hydroxyl, amino, and carboxyl groups contribute to the hydrophilicity of polymeric substances. When the surface of polymeric materials is hydrophilic, water molecules adsorb to substrates due to their polar characteristics. When the tendency for hydrophilicity weakens, hydrophobicity increases. In this case, water molecules would be repelled by the substrate's surface and a sort of "water void" would form. Then bacteria could find a space to occupy at the vicinity of the surface (Fig. **7**). Therefore, it is said very often that hydrophobic surfaces of materials would be beneficial for bacteria to form biofilm. However, it is not always true and some contradictory results have been obtained so far [5]. There are several reasons for the confusion.

As shown in Fig. (**8**), the biofilm formation process is composed of multi steps, as already explained. In addition, at each step, different balances of forces among substrates, bacteria and EPS seem to be established. When we come to think about interactions between polymers, there are three important points.

One of them is the conditioning films. The absorbed compounds are carbon compounds such as hydrated carbon, proteins, *etc*. Therefore, van der Waals forces must work between those compounds and substrates. When the substrate is polarized locally, the van der Waals force must work. The van der Waals forces between carbon compounds (as conditioning film) and polarized polymer would balance the repulsion force due to the overlapping of the electric double layer (relating to zeta potential).

The second possibility is the attachment process of bacteria to polymer surfaces explained in Fig. (**7**). However, EPS would be produced after biofilm formation begins. Then we have to take the interaction between EPS and the polymer surfaces into consideration. Polytetrafluoroethylene (PTFE) is well known for its high water repellency. Actually, van der Waals forces on the material's surface are very small. However, it does not attract hydrophobic bacteria. It rather controls bacterial attachment and growth. The reason might be attributed to its contribution to other steps in the overall process.

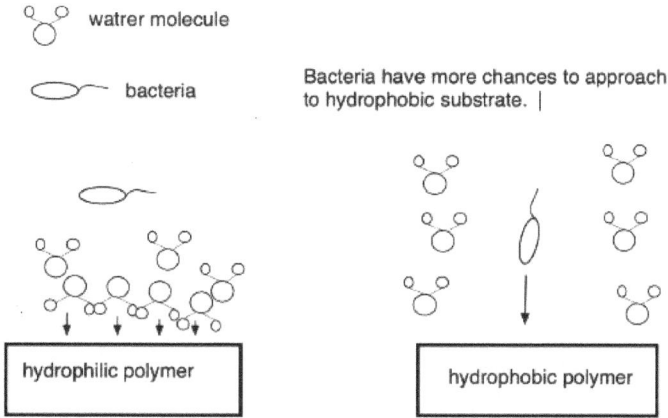

Fig. (7). Hydrophilicity/hydrophobicity of substrates and biofilms.

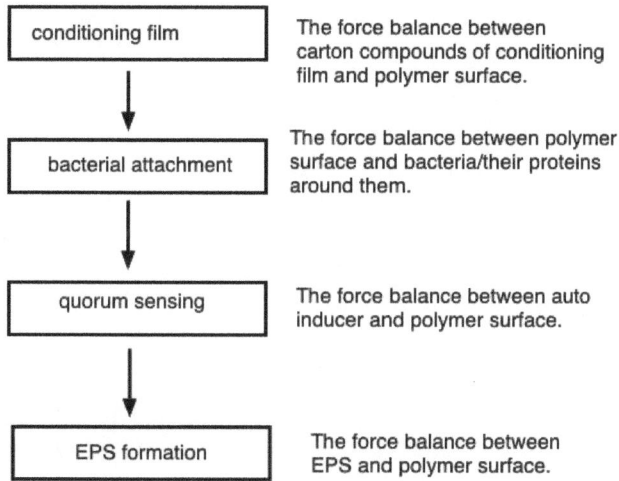

Fig. (8). The possibilities of van der Waals forces on polymer surfaces during the biofilm formation process.

CONSENT FOR PUBLICATION

Not applicable.

CONFLICT OF INTEREST

The authors confirm that this chapter contents have no conflict of interest.

ACKNOWLEDGEMENTS

Declare none.

REFFERENCES

[1]   Costerton JW, Cheng KJ, Geesey GG, *et al.* Bacterial biofilms in nature and disease. Annu Rev Microbiol 1987; 41: 435-64.
[http://dx.doi.org/10.1146/annurev.mi.41.100187.002251] [PMID: 3318676]

[2]   Flemming H-C, Geesey GG. Biofouling and biocorrosion in industrial water systems. Proceedings of the International Workshop on Industrial Biofouling and Biocorrosion, Stuttgart. 1-204.
[http://dx.doi.org/10.1007/978-3-642-76543-8]

[3]   Flemming H-C, Wingender J. The biofilm matrix. Nat Rev Microbiol 2010; 8(9): 623-33.
[http://dx.doi.org/10.1038/nrmicro2415] [PMID: 20676145]

[4]   Percival SL, Malic S, Cruz H, Williams DW. Introduction to biofilms. Biofilms and Veterinary Medicine. Berlin, Heidelberg: Springer-Verlag 2011; pp. 41-68.
[http://dx.doi.org/10.1007/978-3-642-21289-5_2]

[5]   Lewandowski Z, Beyenal H. Fundamentals of Biofilm Research, Second Edition (Ed, CRC Press), Boca Raton, London, New York. 2014; p. 642.

[6]   Kanematsu H, Barry DM. Biofilm and Materials Science. New York, the USA: Springer 2015; pp. 1-196.

[7]   Little BJ, Wagner PA. Succession in microfouling.Fouling Organisms of the Indian Ocean: Biology and Control Technology. New Delhi: Oxford and IBH 1997; pp. 105-34.

[8]   Dobretsov S, Railkin A. Correlative connections between marine maicro-and macrofouling. Biol Mora (Vladivost) 1994; 20: 115-9.

[9]   Muthukrishnan T, Abed RM, Dobretsov S, Kidd B, Finnie AA. Long-term microfouling on commercial biocidal fouling control coatings. Biofouling 2014; 30(10): 1155-64.
[http://dx.doi.org/10.1080/08927014.2014.972951] [PMID: 25390938]

[10] Dobretsov S, Dahms H-U, Tsoi MY, Qian P-Y. Chemical control of epibiosis by Hong Kong sponges: the effect of sponge extracts on micro-and macrofouling communities. Mar Ecol Prog Ser 2005; 297: 119-29.
[http://dx.doi.org/10.3354/meps297119]

[11] Lorite GS, Rodrigues CM, de Souza AA, Kranz C, Mizaikoff B, Cotta MA. The role of conditioning film formation and surface chemical changes on *Xylella fastidiosa* adhesion and biofilm evolution. J Colloid Interface Sci 2011; 359(1): 289-95.
[http://dx.doi.org/10.1016/j.jcis.2011.03.066] [PMID: 21486669]

[12] Kanematsu H, Barry DM. Conditioning Films. Biofilm and Materials Science. 1st ed. New York, USA: Springer 2015; pp. 9-15.

[13] Doiron K, Beaulieu L, St-Louis R, Lemarchand K. Reduction of bacterial biofilm formation using marine natural antimicrobial peptides. Coll Surf B Biointerf 2018; 167: 524-30.
[http://dx.doi.org/10.1016/j.colsurfb.2018.04.051] [PMID: 29729630]

[14] Hohmann S, Kögel S, Brunner Y, *et al.* Surface acoustic wave (saw) resonators for monitoring conditioning film formation. Sensors (Basel) 2015; 15(5): 11873-88.
[http://dx.doi.org/10.3390/s150511873] [PMID: 26007735]

[15] Francius G, El Zein R, Mathieu L, Gosselin F, Maul A, Block JC. Nano-exploration of organic conditioning film formed on polymeric surfaces exposed to drinking water. Water Res 2017; 109: 155-63.
[http://dx.doi.org/10.1016/ j.watres.2016.11.038] [PMID: 27883920]

[16]   Parsons R. The electrical double layer: recent experimental and theoretical developments. Chem Rev 1990; 90: 813-26.
[http://dx.doi.org/10.1021/cr00103a008]

[17]   Bockris JOM, Reddy AK. Modern electrochemistry 2B: electrodics in chemistry, engineering, biology and environmental science (Ed., Springer Science & Business Media) 2000; pp. 1539-2053.

[18]   Solano C, Echeverz M, Lasa I. Biofilm dispersion and quorum sensing. Curr Opin Microbiol 2014; 18: 96-104.
[http://dx.doi.org/10.1016/j.mib.2014.02.008] [PMID: 24657330]

[19]   Ikegai H. Genomics Approach. Biofilm and Material Science. New York, the USA: Springer 2015; pp. 53-60.

[20]   Whiteley M, Diggle SP, Greenberg EP. Progress in and promise of bacterial quorum sensing research. Nature 2017; 551(7680): 313-20.
[http://dx.doi.org/10.1038/nature24624] [PMID: 29144467]

[21]   Kim MK, Zhao A, Wang A, *et al.* Surface-attached molecules control *Staphylococcus aureus* quorum sensing and biofilm development. Nat Microbiol 2017; 2: 17080.
[http://dx.doi.org/10.1038/nmicrobiol.2017.80] [PMID: 28530651]

[22]   Jemielita M, Wingreen NS, Bassler BL. Quorum sensing controls *Vibrio cholerae* multicellular aggregate formation. eLife 2018; 7e42057
[http://dx.doi.org/10.7554/eLife.42057] [PMID: 30582742]

[23]   Morales-Soto N, Cao T, Baig NF, Kramer KM, Bohn PW, Shrout JD. Surface-growing communities of *Pseudomonas aeruginosa* exhibit distinct alkyl quinolone signatures. Microbiol Insights 2018; 11: 1178636118817738.
[http://dx.doi.org/10.1177/1178636118817738] [PMID: 30573968]

[24]   Singh N, Rajwade J, Paknikar KM. Transcriptome analysis of silver nanoparticles treated *Staphylococcus aureus* reveals potential targets for biofilm inhibition. Coll Surf B Biointerf 2019; 175: 487-97.
[http://dx.doi.org/10.1016/j.colsurfb.2018.12.032] [PMID: 30572157]

[25]   Heo YM, Kim K, Ryu SM, *et al.* Diversity and ecology of marine algicolous arthrinium species as a source of bioactive natural products. Mar Drugs 2018; 16(12): 508.
[http://dx.doi.org/10.3390/md16120508] [PMID: 30558255]

[26]   Sato M, Nakayama J. Quorum sensing of gram-positive bacteria and its inhibitors. Japan Society for Lactic Acid Bacteria 2010; 21: 95-106.

[27]   Dong W, Zhu J, Guo X, *et al.* Characterization of AiiK, an AHL lactonase, from Kurthia huakui LAM0618^T and its application in quorum quenching on *Pseudomonas aeruginosa* PAO1. Sci Rep 2018; 8(1): 6013.
[http://dx.doi.org/10.1038/s41598-018-24507-8] [PMID: 29662232]

[28]   Billotte WC. Ceramic Biomaterials. Biomaterials. Boca Raton, FL, the USA: CRC Press 2007; pp. 23-57.

CHAPTER 17

Effect of Metallic Nanoparticle-Dispersed Silane-Based Polymer Coatings on Anti-Biofilm Formation

Akiko Ogawa[1,*] and **Katsuhiko Sano**[2]

[1] *Department of Chemistry and Biochemistry, National Institute of Technology (KOSEN), Suzuka College, Suzuka, Mie, Japan*

[2] *D & D Corporation, Yokkaichi, Mie, Japan*

Abstract: Regulating biofouling is important for the maintenance of marine vessels, bridges, and other constructions. Marine organisms must also be protected. Here we propose a silane-based polymer coating that has features of both polysiloxanes and silicates: nontoxicity, weatherability, and a self-cleaning effect as an anti-biofouling method. We focus on the antibacterial activity of some metals and metallic nanoparticles applied to the silane-based coating that can remove biofouling at the late stage of biofilm formation and may be effective at the early stage. Silver, nickel, or copper nanoparticles dispersed to a silane-based polymer coating exhibit anti-biofouling behavior, whereas tungsten or molybdenum nanoparticles enhance biofouling.

Keywords: Biofilm, Coating, Metallic nanoparticles, Silane-based polymer.

INTRODUCTION

Marine biofouling usually presents economic challenges for ships, power plants, harbors, and bridges. In ships, it impedes movement, imposes a power penalty (>85%) and deteriorates coatings [1]. Marine biofouling also causes scaling and metallic corrosion in industrial equipment such as the pipes of heat exchangers and cooling systems in power plants [2], resulting in reduced power generating capability. Cleaning (removing) biofouling is essential for maintaining adequate propulsion in ships and efficient performance in power plants and prolonging the lifespans of marine constructions. However, cleaning biofouling is laborious and expensive. It has been estimated that shutting down a 235 MW power station due

---

* **Corresponding author Akiko Ogawa:** Department of Chemistry and Biochemistry, National Institute of Technology (KOSEN), Suzuka College, Suzuka, Mie, Japan; Tel: +81-59-368-1768; Fax: +81-59-368-1820; E-mail: ogawa@chem.suzuka-ct.ac.jp

Toshiyuki Takahashi (Ed.)

to biofouling costs US$100,000 per day [3] and the US Navy spends US$180–260 million per year to combat the effects of biofouling on its ships [4].

A historical antifouling method for ship hulls involves the application of heavy metals such as lead, arsenic, and mercury, and their organic derivatives. However, these substances present high environmental risks to marine life and have been banned [5]. Tributyltin (TBT) was used with self-polishing copolymer techniques, but it, too, has been banned worldwide since 2003 after bioaccumulation of TBT was linked to severe deformities in shellfish, duck, seals, and fish [5]. TBT-free alternatives that pose fewer environmental risks are therefore in demand. Recently, several eco-friendly antifouling coatings have been proposed, including combinations of alternative heavy metals such as copper and zinc with self-polishing copolymers that release biocides from the polymer backbone by hydrolysis and erode the surface resulting in polishing; booster biocides; foul release coatings; natural products deterring fouling derived from marine organisms; and electrochemical controls [6, 7].

Biofouling is a biological process in which microorganisms such as bacteria and microalgae adhere to a surface covered with conditioning films, then proliferate and secrete extracellular polymeric substrates (EPS), resulting in a biofilm to which macro-organisms such as young mussels, barnacles, and macroalgae attach [8] (Fig. **1**). Regulating biofilm formation is the most promising method of preventing biofouling.

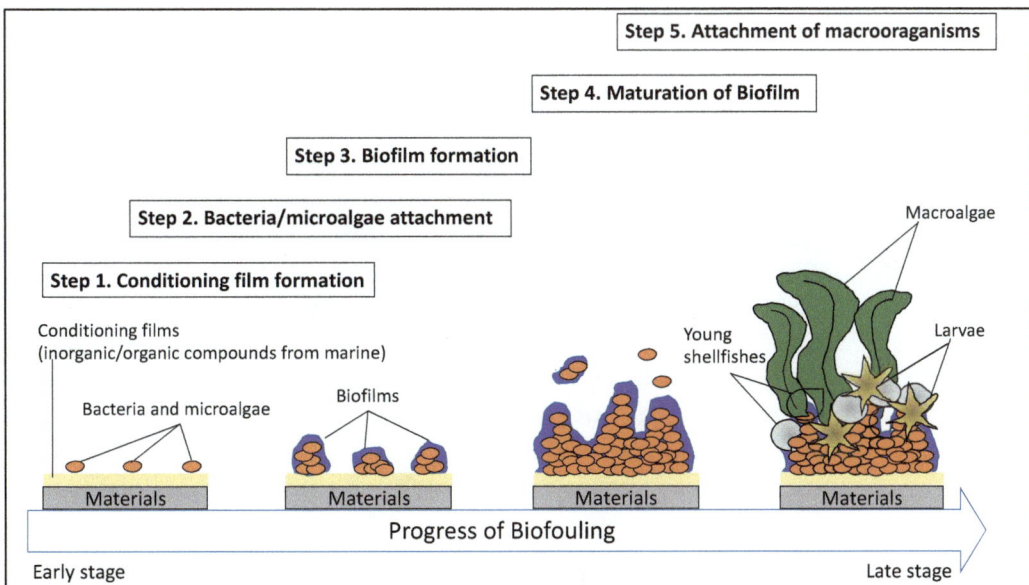

Fig. (1). Brief progression of biofouling.

In this section, we introduce an environmentally benign and effective anti-biofilm technology that is composed of metal nanoparticles and a silane-based polymer. Firstly, we introduce a biofilm and its formation process. Secondly, we explain the advantage of using a silane-based polymer as an anti-biofilm coating and the effect of several metals on anti-biofilm formation. Thirdly, we introduce successful applications of the metal nanoparticle (NP)–dispersed silane-based polymer coating to anti-biofilm formation in marine conditions.

Step 1: the materials are covered with conditioning films. Step 2: bacteria/microalgae are attached to the conditioning films. Step 3: the attached microorganisms gather one another, proliferate, and secrete EPS, resulting in forming biofilms. Step 4: The biofilms grow gradually in the mass and height, and when they are reached to threshold, the part of them are broken and some microorganisms are moved out. Step 5: macrooraganisms including larvae, young shellfishes and macroalgae, attach on the biofilm, then grow there.

Biofilms and Their Formation

Biofilms are composed of adhered bacteria and EPS produced by the bacteria as a survival mechanism. EPS are complexes of polysaccharides, proteins, glycolipids, glycoproteins, and extracellular DNA. Biofilm constituents form a highly hydrated matrix in which biofilm cells are gathered closely, embedded, and exchange information through the EPS components. The matrix not only supports the structure of the biofilm, but also provides biofilm cells with dissolved and particulate substances from the environment [9]. Conditioning films occur prior to biofilm formation. Conditioning films consist mainly of carbohydrates and other organic substrates such as proteins, polysaccharides, humic acids, lipids, and nucleic acids derived from metabolites of marine organisms [10, 11]. The development time of conditioning films is just 1 min after immersion, while biofilm formation generally takes 1 month [5]. Biofilm production involves three steps: (1) bacteria and microalgae reversibly attach to the surface of conditioning-film-covered materials (~ 24 h); (2) they irreversibly adhere there and proliferate to colonize (~1 week); and (3) they secrete EPS to interact with one another and promote survival (2 to 4 weeks). Once a biofilm is mature, biofilm cells are resistant to biocides, as much as three orders of magnitude less sensitive than planktonic cells. Biofilms also reduce susceptibility to metal toxicity, acid exposure, phagocytosis, and dehydration [12].

Silane-Based Polymers

Silane-based polymers are involved in polysiloxanes or silicones. Polysiloxanes consist of a siloxane backbone and alkyl groups (typically methyl groups) as side chains. Depending on the length of siloxane backbone, the kind of side chains, and crosslinking extent, polysiloxanes can exist in a range of states, from liquids to hard plastics [13]. Polysiloxanes are chemically and thermally stable, unreactive, hydrophobic, and provide electrical insulation [13, 14]. They are also resistance to ultraviolet (UV) light. UV light is separated into UV-A (315-400 nm), UV-B (280-315 nm), and UV-C (<280 nm). Most UV-B and all UV-C radiation are absorbed by the ozone layer, while UV-A reaches the surface of the earth. The Si-O bond energy is 444 kJ/mol [15], which is higher than UV-A (300-380 kJ/mol) or UV-B energy (380-430 kJ/mol). Therefore, a siloxane backbone is seldom degraded by sunlight. On the other hand, other organic polymers, such as epoxy resins and polyvinyl chloride resins, consist of a C-C or C-O backbone. With bond energies of 356 kJ/mol and 329 kJ/mol, respectively, C-C and C-O bonds are easily degraded by sunlight. This gives polysiloxanes an advantage as a coating in marine environments.

Polysiloxanes also exhibit relatively low surface tension as liquids and low free energy as solids (20 mJ/m^2) in comparison with other organic polymers such as polyurethane (approximately 50 mJ/m^2) [16] and acrylic resins ($\sim$ 40 mJ/m^2) [17]. If a biofilm forms on the surface of a polysiloxane, the matured (thick) biofilm will be easily removed because of the low surface energy.

However, the low surface energy also causes detachment of the coating from basal materials. To strengthen polysiloxane adhesion, inserting a silicate structure can be useful. In general, a silicate consists of an anionic SiO_2 or SiO_4 grouping, one or more cationic ions, and a silicate resin composed of one or more siloxane backbones and some alkoxy terminal groups (−OR). In the presence of water, the alkoxy groups can hydrolyze, polymerize, and dealcoholize each other, resulting in siloxane bonds. Ideally, all the alkoxy groups can be converted to siloxane bonds and a silica dioxide will be created. In practice, some alkoxy groups remain and an adamant silicate resin is more likely. Adamant silicate resin experiences high internal stress caused by a decrease in the volume of the resin during de-alcohol condensation. As a result, the resin tends to crack easily and detach from basal materials. However, a silicate resin exhibits higher adhesion than silicone resins. Therefore, hybridizing polysiloxanes and silicates present a promising solution to creating an antifouling coating that is highly adhesive, non-toxic, and mechanically strong.

We propose a hybridized silane-based polymer coating with desirable properties of both a polysiloxane and a silicate, *i.e.*, it could adhere to a basal material more strongly than a silicate while exhibiting the low surface energy of a polysiloxane. The polymer coating was synthesized from an alkylalkoxysilane oligomer (average molecular weight 360, D & D Co.) and N-2-aminomethyl-3-aminopropyl trimethoxysilane (KBM-603, molecular weight 222, Shin-Etsu Chemical). The alkylalkoxysilane oligomer is composed of siloxane main backbones and methoxy $(-CH_3O)$/methyl$(-CH_3)$/phenyl$(-C_6H_5)$ groups as randomly connected side chains [18]. The ratio of these side chain groups was balanced with hardening, solidity, and flexibility to exert the best performance as a coating. A methyl group and a phenyl group are related to solidity and flexibility, respectively. A methoxy group is involved in dealcoholized polymerization by hydrolysis. KBM-603 is an aminoalcoxysilane composed of a siloxane core, three methoxy groups, and a primary amino group. In general, alkylalkoxysilane needs a catalyst such as dibutyltin acetate and titanium n-butoxide for curing. Without it, amino compounds can catalyze dealcoholization condensation of the alkylalkoxysilane oligomer [19]. Therefore, the alkylalkoxysilane oligomer can progress polymerization by dealcoholization following hydrolysis in the presence of KBM-603 and moisture in the air. The methoxy groups of KBM-603 can also be copolymerized with those of alkylalkoxysilane. Simultaneously, KBM-603 can polymerize without a catalyst using the moisture in the air, presenting a lower environmental burden than a silane-based polymer. Due to a comparatively flexible backbone and the unhomogeneous side groups of alkylalkoxysilane, the internal silane-based polymer has some free volumes that are flexible and available to accommodate several compounds, such as nanosized particles and colloids [20, 21]. The free volume can change shape and volume marginally, allowing water and some ions to transfer from one free volume to another. Wel and Adan describe a theory of transportation and equilibrium sorption of moisture in polymer films and organic coating [22]. Palza proposes that polymer/silver or copper-nanoparticle composites possess antimicrobial properties that allow packed metallic nanoparticles to dissolve and move through the free volume to the surface, where they are released [23].

Antimicrobial Metals and Their Compounds

Several metals are known antimicrobial agents. Excessive levels of essential metals, such as chromium (Cr), vanadium (V), manganese (Mn), iron (Fe), cobalt (Co), nickel (Ni), copper (Cu) and zinc (Zn), can be lethal to cells. Certain non-essential metals, such as silver (Ag), mercury (Hg) and tellurium (Te), have a strong microbicidal effect on most bacteria at extremely low concentrations. Since ancient times, Egyptians, Phoenicians, Greeks, and Romans have used silver and copper to preserve food and protect water from infection. Ag and its compounds have long been applied in fields such as medicine and food products as antimicrobial agents, and Cu and its compounds have also been known as antimicrobial agents [24]. Ag and Cu can inhibit biofilm formation [25, 26]. Today, antimicrobial metal compounds are applied widely in industry, agriculture, and healthcare. Metals can selectively inhibit metabolic pathways of bacteria and interact with bactericidal activity through different mechanisms in mammalian cells. Metal compounds can therefore be effective tools to protect materials from undesirable bacterial actions and biofilm formation [24, 27].

Three feasible determinants of antibacterial toxicity are available: donor atom selectivity, reduction potential, and speciation. Antibacterial mechanisms of metal toxicity depend on metal species. Cu(II), Ag(I), Hg(II), Cd(II), Zn(II), Te(IV), Pb(II), and Ni(II) can lead to protein dysfunction. Ag(I), Cu(II), and Cd(II) can also lead to impaired membrane function. Fe(II) or Fe(III) produce reactive oxygen species (ROS) that are damaging to cellular functions and disrupt DNA strands. Fe(III), Cu(II), Cr(VI), As(III), and Te(IV) can produce ROS via intermediate sulfur radical chemistry and disrupt antioxidants. Ag(II), Cd(II), Co(II), Zn(II), Cr(VI), As(III), and Te(IV) can disrupt cellular thiol-disulfide exchange enzymes, which results in oxidative stress and creation of ROS. Several metals exert an influence on cellular functions in ionized state.

Nano-scale metallic particles are relatively easier to produce and they have been widely applied by various industries. Nanoparticles (NPs) range in diameter from 1 to 100 nm. Generally, NPs increase the surface area more than that of the corresponding bulk volume, which facilitates chemical reactions and makes it possible to dramatically reduce the usage and the cost of the metals. Both Ag-NPs and Cu-NPs have been used widely in industrial and medical fields [24]. However, NPs possess high cellular toxicities for living organisms, including animals. They easily penetrate epithelial and endothelial barriers into the lymph and blood streams, which can carry the nanoparticles to other organs, such as the brain, heart, and bone marrow, where NPs accumulate and cause several disorders. From there they can be transferred into the cells, causing DNA lesions, mutations, and cell death [28]. Consequently, controlling the release of NPs or ionizing them is a critical issue to ensure the safety of animals and other marine organisms.

APPLICATION OF METALLIC NANOPARTICLES DISPERSED SILANE-BASED POLYMER COATINGS TO MARINE ENVIRONMENT

We proposed that antifouling coatings based on metallic nanoparticle–dispersed silane-based polymers would exhibit synergistic effects due to combining metallic NPs with silane-based polymers (Fig. **2**). With the help of several metals with antimicrobial properties, a silane-based polymer enables late-phase (matured) biofilms to detach from a surface spontaneously due to low surface energy. These polymers were prepared as follows. Two precursors (alkylalkoxysilane oligomer and KBM-603) were mixed in a 250-mL polypropylene bottle containing zirconia beads (1 mm diameter) for 30 min by a paint shaker (Toyoseiki Co., Tokyo, Japan) in a nitrogen gas atmosphere. Simultaneously, metal nanoparticles were added to the polypropylene bottle. After shaking, the mixture was filtered through a nylon mesh #110 (NBC Meshtech Inc., Hino, Japan) to remove the zirconia beads and any residues. The filtrate was loaded into an air-spray apparatus then sprayed on the surface of basal materials. Here, the weight ratio of the alkylalkoxysilane oligomer to KBM-603 was set at 80:20 to harden the polymer for approximately 2 h at ambient temperatures. We prepared 5 varieties of coatings: Ag-NPs, Cu-NPs, Ni-NPs, tungsten (W)-NPs, and molybdenum (Mo)-NPs were dispersed in silane-based polymer coatings. These NPs were purchased from Sigma-Aldrich (St. Louis, MO, USA) and the size of all NPs as 100 nm. A soda-lime glass was used as the basal material because of its chemical stability, which permitted evaluation of the effect of the hybrid coatings on anti-biofouling. We tested the ability of the biofilm formations because biofouling begins during biofilm formation, and we evaluated the biofilms quantitatively and qualitatively by comparing the degree of surface configuration (vertical interval and sea-island structure) and by Raman spectroscopic analysis, respectively. A biofilm grows higher in a vertical direction according to maturation, which can be detected in the difference of the vertical intervals. Formed biofilms usually assume a sea-island configuration because they are scattered across the surface of materials at a relatively early phase in the biofilm-forming condition [29]. Raman spectroscopy can detect polysaccharides, proteins, lipids, and nucleotides, which are the main components of EPS derived from biofilms. Some researchers have applied Raman spectroscopy to biofilm identification [30 - 36].

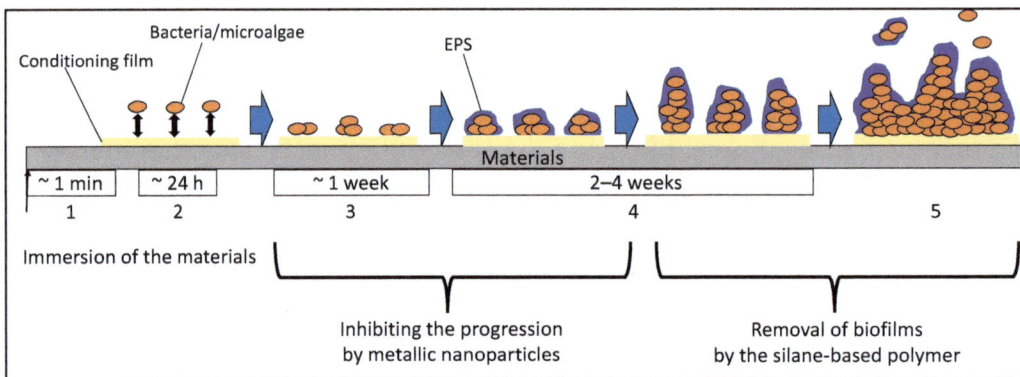

Fig. (2). Time course of biofilm formation and the expected effects of metallic nanoparticles and the silane-based polymer on regulating biofilm formation.

1–5 indicate the process of biofilm formation as follows. 1: conditioning film formation. 2: reversible attachment of bacteria/microalgae. 3: irreversible adhesion of bacteria/microalgae and colonization. 4: EPS secretion and growing the biofilms. 5: biofilm maturation.

First, we investigated the effects of Ag-NPs dispersed in a silane-based polymer-coated coupon and Cu-NPs dispersed in a silane-based polymer-coated coupon (these basal plates were 10 mm square and 1 mm thick) using a closed laboratory biofilm reactor (LBR) containing natural seawater [37]. Ag-NPs and Cu-NPs were dispersed at 0.1 mol% or 5 mol%. All coupons were anchored in the LBR and soaked in circulating seawater for 3 days. The seawater, which was obtained from one part of Ise Bay (34°30.4533′N, 136°48.8626′E), was analyzed for bacterial flora by next-generation sequencing. The analysis showed the seawater exhibited the features of surface seawater and possessed the ability of biofilm formation because *Rhodobacteraceae* (key members of the microbial community of the initial biofilm formed in eastern Mediterranean coastal waters) [38] and a relatively dominant bacterial family found in marine biofilms [39], were abundant (16%).

After a 3-day incubation period, a non-dispersed silane-based polymer-coated coupon (silane-based polymer) was almost covered with biofilm (Fig. **3**). We considered that 3 days was a suitable period to evaluate the progress of biofilm formation because silane-based polymers appeared to be covered with mature biofilms. In comparison with a silane-based polymer, at a low NP concentration (0.1 mol%), a Cu-NP–dispersed silane-based polymer-coated coupon (Cu-polymer) moderately inhibited biofilm formation, and an Ag-NP–dispersed silane-based polymer-coated coupon (Ag-polymer) strongly inhibited biofilm formation. At a high NP concentration (5 mol%), the Cu-polymer more effectively inhibited biofilm formation than did a low concentration. On the other hand, the Ag-polymer did not inhibit biofilm formation less than the low concentration. Both Ag- and Cu-polymers were effective at inhibiting or delaying biofilm formation in marine conditions; the effect of Cu-NPs on anti-biofouling increased in a dose-dependent manner. However, Ag-NPs affect anti-biofouling only at low concentrations. Other researchers have reported that Ag-NPs affect anti-biofilm activity in a dose-dependent manner [40 - 42] but the concentrations are all lower than 0.1 mol%. When we observed the surface of the Ag-polymer at 5 mol%, the color was light gray, which may indicate silver aggregation. This suggests that Ag-NPs gathered in a mass-like state and that they lost some advantages of NPs (high chemical activity and easily ionization). Improvement of the dispersing silver and polymer-coating process at high concentrations is therefore required. On the other hand, the 0.1 mol% Ag-polymer possessed more effective anti-biofilm activity than the 0.1 mol% Cu-polymer, and the Cu-polymer will be able to inhibit biofilm formation at a wide range of NP concentrations. Considering marine environmental anti-biofouling applications, Ag-polymer and Cu-polymer can not only provide a low environmental impact but also serve as an economically useful method to protect the materials, because:

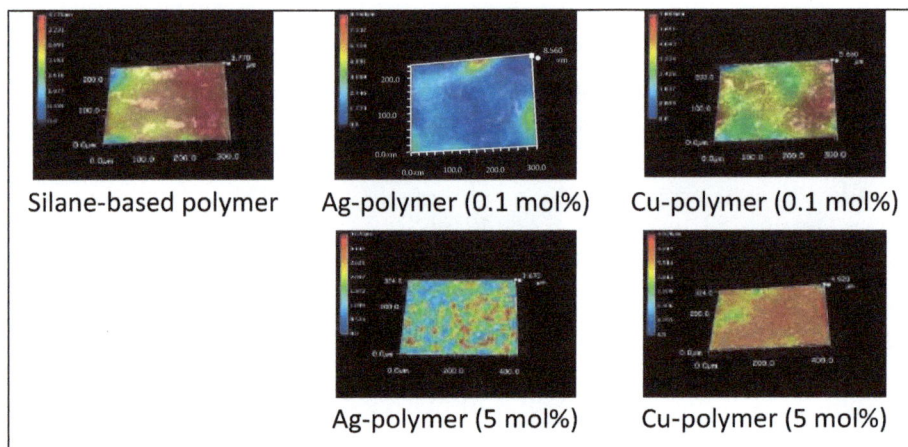

Fig. (3). Microscopic images of the surface of coupons after immersion test.

1. NPs are gradually changed to metallic ions and released from the polymer for a long period in comparison with non-stabilized NPs, and
2. both low Ag-NP concentrations and using Cu-NP can lead to cost savings.

Next, we investigated the effects of other metal-NP–dispersed silane-based polymer-coated coupons on anti-biofilm formation [43]. We selected Ni, W, and Mo because these metals had been rarely reported with respect to anti-biofouling or the effect on biofilm formation. Ni and Mo are known as common components of stainless steels applied in marine construction materials and ships [44, 45], while W is a heavy and hard metal that is heat-resistant and serves as an alternative to lead [46, 47]. Here, the concentration of NPs in the coating was 0.1 mol%, the Ag-polymer was used as an example of an anti-biofilm coating, and the biofilm formation test was performed in an open LBR-circulated tap water containing resident laboratory microbiota. To quantify the amount of biofilm formed on the surface of each coupon, the biofilms were stained with crystal violet dye, which bound to negatively charged compounds such as polysaccharides and nucleic acids that were components of EPS [48, 49]. Intensity and color were measured by a colorimeter. After a 5-day incubation test, a W-NP–dispersed silane-based polymer-coated coupon (W-polymer) was stained the densest, and a Mo-NP–dispersed silane-based polymer-coated coupon (Mo-polymer) was stained denser than the silane-based polymer (Fig. **4**). On the other hand, a Ni-NP–dispersed silane-based polymer-coated coupon (Ni-polymer) and Ag-polymer were stained lighter than the silane-based polymer. All coupons were identified by the specific Raman peaks derived from EPS components. These results show that the Ni-polymer effectively inhibited biofilm formation the same as the Ag-polymer but the W-polymer and Mo-polymer accelerated it.

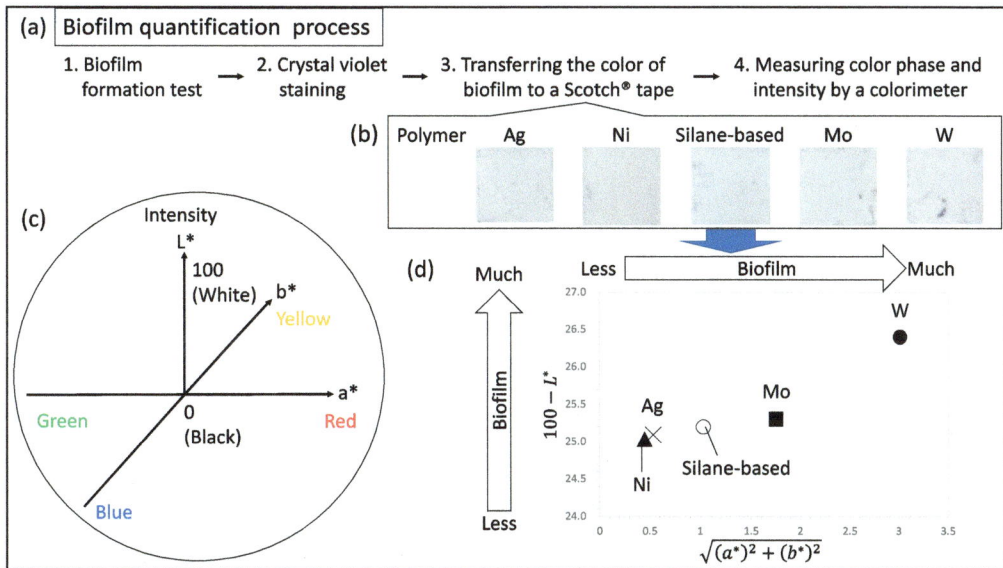

Fig. (4). The result of biofilm quantification by crystal violet staining and a colorimeter.

(**a**): Brief biofilm quantification process. (**b**): Digital photo images of crystal violet-stained biofilms transferred to Scoth® tape. (**c**): Color phase and intensity. Crystal violet dye has a blue-violet color that indicates a red-blue phase. (d): The color phase value and intensity value of formed biofilm on each coupon. The bigger the values are, the better the amount of biofilm is.

CONCLUDING REMARKS

In this chapter, we proposed a silane-based polymer coating for anti-biofouling in a marine environment. This coating offers the advantages of nontoxicity, weather resistance (especially UV resistance), low surface energy, and simple construction. We applied metallic NPs (0.1 mol%) to the silane-based polymer in expectation of an additional anti-biofouling effect. The Ag-, Cu-, and Ni-polymer can inhibit or delay the biofilm formation. Conversely, W- and Mo-polymer accelerates it. A combination of metallic NPs and the silane-based polymer coating can enhance the effect of anti-biofouling under marine environment, when the concentration and the kind of metal NPs are appropriate. Metallic NPs inhibit bacterial growth, metabolism, and proliferation at the early stages of biofilm formation. Even if the biofilm continues growing and matures, the surface of silane-based polymer coatings can easily remove the biofilm in the late stage of biofilm formation. Metallic NP–dispersed silane-based coatings are therefore an effective and marine-friendly anti-biofouling method.

CONSENT FOR PUBLICATION

Not applicable.

CONFLICT OF INTEREST

The authors confirm that this chapter contents have no conflict of interest.

ACKNOWLEDEMENTS

The authors would like to thank Enago (www.enago.jp) for English language review.

REFFERENCES

[1]   Bressy C, Lejars MN. Marine fouling: an overview. J Ocean Technol 2014; 9: 19-28.

[2]   Bixler GD, Bhushan B. Biofouling: lessons from nature. Philos Trans- Royal Soc, Math Phys Eng Sci 2012; 370(1967): 2381-417.
        [http://dx.doi.org/10.1098/rsta.2011.0502] [PMID: 22509063]

[3]   Venkatesan R, Sriyutha Murthy P. Macrofouling Control in Power Plants. Marine and industrial biofouling. Springer-Verlag Berlin Heidelberg 2009; pp. 265-92.
        [http://dx.doi.org/10.1007/978-3-540-69796-1_14]

[4]   Holm ER. Barnacles and biofouling. Integr Comp Biol 2012; 52(3): 348-55.
        [http://dx.doi.org/10.1093/icb/ics042] [PMID: 22508866]

[5]   Chambers LD, Stokes KR, Walsh FC, Wood RJK. Modern approaches to marine antifouling coatings. Surf Coat Tech 2006; 201: 3642-62.
        [http://dx.doi.org/10.1016/j.surfcoat.2006.08.129]

[6]   Chambers LD, Wharton JA, Wood RJK, Walsh FC, Stokes KR. Techniques for the measurement of natural product incorporation into an antifouling coating. Prog Org Coat 2014; 77: 473-84.
        [http://dx.doi.org/10.1016/j.porgcoat.2013.11.013]

[7]   Nurioglu AG, Esteves ACC, de With G. Non-toxic, non-biocide-release antifouling coatings based on molecular structure design for marine applications. J Mater Chem B Mater Biol Med 2015; 3: 6547-70.
        [http://dx.doi.org/10.1039/C5TB00232J]

[8]   Cao B, Shi L, Brown RN, *et al.* Extracellular polymeric substances from Shewanella sp. HRCR-1 biofilms: characterization by infrared spectroscopy and proteomics. Environ Microbiol 2011; 13(4): 1018-31.
        [http://dx.doi.org/10.1111/j.1462-2920.2010.02407.x] [PMID: 21251176]

[9]   Flemming HC, Neu TR, Wozniak DJ. The EPS matrix: the "house of biofilm cells". J Bacteriol 2007; 189(22): 7945-7.
        [http://dx.doi.org/10.1128/JB.00858-07] [PMID: 17675377]

[10]   Siboni N, Lidor M, Kramarsky-Winter E, Kushmaro A. Conditioning film and initial biofilm formation on ceramics tiles in the marine environment. FEMS Microbiol Lett 2007; 274(1): 24-9.
        [http://dx.doi.org/10.1111/j.1574-6968.2007.00809.x] [PMID: 17578524]

[11]   Kanematsu H, Barry DM. Conditioning films. Biofilm and Materials Science. Springer International Publishing: Switzerland 2015; pp. 9-15.

[12]   Otter JA, Vickery K, Walker JT, *et al.* Surface-attached cells, biofilms and biocide susceptibility: implications for hospital cleaning and disinfection. J Hosp Infect 2015; 89(1): 16-27.

[http://dx.doi.org/10.1016/j.jhin.2014.09.008] [PMID: 25447198]

[13]   Hacker MC, Mikos AG. Synthetic polymers.Principles of Regenerative Medicine. 2nd ed. San Diego: Academic Press 2011; pp. 587-622.
[http://dx.doi.org/10.1016/B978-0-12-381422-7.10033-1]

[14]   Hill RG. 10 - Biomedical polymers.Biomaterials, Artificial Organs and Tissue Engineering. Woodhead Publishing 2005; pp. 97-106.
[http://dx.doi.org/10.1533/9781845690861.2.97]

[15]   Ahmad S, Gupta AP, Sharmin E, Alan M, Pandey SK. Synthesis, characterization and development of high performance siloxane-modified epoxy paints. Prog Org Coat 2005; 54: 248-55.
[http://dx.doi.org/10.1016/j.porgcoat.2005.06.013]

[16]   Barquins M, Cognard J. Adhesion characteristics of gold surfaces. Cold Bulletin 1986; 19: 82-6.

[17]   Serrano-Granger C, Cerero-Lapiedra R, Campo-Trapero J, Del Río-Highsmith J. *In vitro* study of the adherence of *Candida albicans* to acrylic resins: relationship to surface energy. Int J Prosthodont 2005; 18(5): 392-8.
[PMID: 16220804]

[18]   Sugiyama Y, Komori T, Yanagihara T, *et al.* Sealer. Japan Patent 3816354 2006 June; 16

[19]   Mark JE, Schaefer DW, Lin G. The polysiloxanes. New York, U.S.: Oxford University Press 2015.

[20]   Kondo T, Yoshii K, Horie K, Itoh M. Photoprobe study of siloxane polymers. 3. Local free volume of polymethylsilsesquioxane probed by photoisomerization of azobenzene. Macromolecules 2000; 33: 3650-8.
[http://dx.doi.org/10.1021/ma991765y]

[21]   White RP, Lipson JE. Polymer free volume and its connection to the glass transition. Macromolecules 2016; 49: 3987-4007.
[http://dx.doi.org/10.1021/acs.macromol.6b00215]

[22]   ] Wel van der G K, Adan O C G. Moisture transport and equilibrium in organic coatings. Heron 2000; 45: 125-52.

[23]   Palza H. Antimicrobial polymers with metal nanoparticles. Int J Mol Sci 2015; 16(1): 2099-116.
[http://dx.doi.org/10.3390/ijms16012099] [PMID: 25607734]

[24]   Lemire JA, Harrison JJ, Turner RJ. Antimicrobial activity of metals: mechanisms, molecular targets and applications. Nat Rev Microbiol 2013; 11(6): 371-84.
[http://dx.doi.org/10.1038/nrmicro3028] [PMID: 23669886]

[25]   Shih HY, Lin YE. Efficacy of copper-silver ionization in controlling biofilm- and plankton-associated waterborne pathogens. Appl Environ Microbiol 2010; 76(6): 2032-5.
[http://dx.doi.org/10.1128/AEM.02174-09] [PMID: 20080997]

[26]   Kawakami H, Kittaka K, Sato Y, Kikuchi Y. Bacterial attachment and initiation of biofilms on the surface of copper containing stainless steel. ISIJ Int 2010; 90: 133-8.
[http://dx.doi.org/10.2355/isijinternational.50.133]

[27]   Gold K, Slay B, Knackstedt M, Gaharwar AK. Antimicrobial activity of metal and metal-oxide based nanoparticles. Adv Ther 2018; 11700033
[http://dx.doi.org/10.1002/adtp.201700033]

[28]   Sukhanova A, Bozrova S, Sokolov P, Berestovoy M, Karaulov A, Nabiev I. Dependence of nanoparticle toxicity on their physical and chemical properties. Nanoscale Res Lett 2018; 13(1): 44-64.
[http://dx.doi.org/10.1186/s11671-018-2457-x] [PMID: 29417375]

[29]   Barry DM, Kanematsu H. Cooling water. Biofilm and Materials Science. Cham, Switzerland: Springer 2015; pp. 79-83.

[30]   Sandt C, Smith-Palmer T, Pink J, Brennan L, Pink D. Confocal Raman microspectroscopy as a tool for studying the chemical heterogeneities of biofilms *in situ.* J Appl Microbiol 2007; 103(5): 1808-20.
[http://dx.doi.org/10.1111/j.1365-2672.2007.03413.x] [PMID: 17953591]

[31]   Ivleva NP, Wagner M, Horn H, Niessner R, Haisch C. Towards a nondestructive chemical characterization of biofilm matrix by Raman microscopy. Anal Bioanal Chem 2009; 393(1): 197-206.
[http://dx.doi.org/10.1007/s00216-008-2470-5] [PMID: 18979092]

[32]   Samek O, Al-Marashi JFM, Telle HH. The potential of Raman spectroscopy for the identification of biofilm formation by *Staphylococcus epidermidis.* Laser Phys Lett 2010; 7: 378-83.
[http://dx.doi.org/10.1002/lapl.200910154]

[33]   Chao Y, Zhang T. Surface-enhanced Raman scattering (SERS) revealing chemical variation during biofilm formation: from initial attachment to mature biofilm. Anal Bioanal Chem 2012; 404(5): 1465-75.
[http://dx.doi.org/10.1007/s00216-012-6225-y] [PMID: 22820905]

[34]   Sharma G, Prakash A. Combined use of Fourier transform infrared and Raman spectroscopy to study planktonic and biofilm cells of *Cronobacter sakazakii.* J Microbiol Biotechnol Food Sci 2014; 3: 310-4.

[35]   Sano K, Kanematsu H, Kogo T, Hirai N, Tanaka T. Corrosion and biofilm for a composite coated iron observed by FTIR-ATR and Raman spectroscopy. Trans IMF. 139-45.
[http://dx.doi.org/10.1080/00202967.2016.1167315]

[36]   Kanematsu H, Sato M, Shindo K, *et al.* Biofilm formation behaviors on graphene by *E. coli* and *S. epidermidis.* ECS Trans 2017; 80: 1167-75.
[http://dx.doi.org/10.1149/08010.1167ecst]

[37]   Ogawa A, Kanematsu H, Sano K, *et al.* Effect of silver or copper nanoparticles-dispersed silane coatings on biofilm formation in cooling water systems. Materials (Basel) 2016; 9(8): 632-50.
[http://dx.doi.org/10.3390/ma9080632] [PMID: 28773758]

[38]   Elifantz H, Horn G, Ayon M, Cohen Y, Minz D. Rhodobacteraceae are the key members of the microbial community of the initial biofilm formed in Eastern Mediterranean coastal seawater. FEMS Microbiol Ecol 2013; 85(2): 348-57.
[http://dx.doi.org/10.1111/1574-6941.12122] [PMID: 23551015]

[39]   Kviatkovski I, Minz D. A member of the Rhodobacteraceae promotes initial biofilm formation via the secretion of extracellular factor(s). Aquat Microb Ecol 2015; 75: 155-67.
[http://dx.doi.org/10.3354/ame01754]

[40]   Kalishwaralal K, BarathManiKanth S, Pandian SR, Deepak V, Gurunathan S. Silver nanoparticles impede the biofilm formation by Pseudomonas aeruginosa and Staphylococcus epidermidis. Coll Surf B Biointerfa 2010; 79(2): 340-4.
[http://dx.doi.org/10.1016/j.colsurfb.2010.04.014] [PMID: 20493674]

[41]   Gurunathan S, Han JW, Kwon D-N, Kim J-H. Enhanced antibacterial and anti-biofilm activities of silver nanoparticles against Gram-negative and Gram-positive bacteria. Nanoscale Res Lett 2014; 9(1): 373-89.
[http://dx.doi.org/10.1186/1556-276X-9-373] [PMID: 25136281]

[42]   Hashimoto M, Yanagiuchi H, Kitagawa H, Yamaguchi S, Honda Y, Imazato S. Effect of metal nanoparticles on biofilm formation of *Streptococcus mutans.* Nanomedicine (Lond) 2017; 9: 61-8.

[43]   Ogawa A, Kiyohara T, Kobayashi Y, Sano K, Kanematsu H. Nickel, molybdenum, and tungsten nanoparticle-dispersed alkylalkoxysilane polymer for biomaterial coating: evaluation of effects on bacterial biofilm formation and biosafety. Biomedical Research and Clinical Practice 2017; 2: 1-7.
[http://dx.doi.org/10.15761/BRCP.1000138]

[44]   www.ssina.com/overview/alloyelements_intro.html

[45]   www.jssa.gr.jp/english/

[46]   www.nittan.co.jp/en/tech/course/index.html

[47]   www.itia.info/a-substitute-for-lead.html

[48]   Peeters E, Nelis HJ, Coenye T. Comparison of multiple methods for quantification of microbial biofilms grown in microtiter plates. J Microbiol Methods 2008; 72(2): 157-65.
[http://dx.doi.org/10.1016/j.mimet.2007.11.010] [PMID: 18155789]

[49]   Pantanella F, Valenti P, Natalizi T, Passeri D, Berlutti F. Analytical techniques to study microbial biofilm on abiotic surfaces: pros and cons of the main techniques currently in use. Ann Ig 2013; 25(1): 31-42.
[PMID: 23435778]

CHAPTER 18

Chemical Characteristics of Steelmaking Slag in Aqueous Environments

Seiji Yokoyama[*]

Department of Mechanical Engineering, Toyohashi University of Technology, Toyohashi, Japan

Abstract: The oxidizing slag discarded from the oxidation process in an electric arc furnace was analyzed using the tank leaching test based on the Japanese industrial standards. The dissolved concentrations of environmentally regulated substances from the oxidizing slag were lower than those stipulated in the environmental quality standards. Therefore, this slag could be used as soil. Ca, K, P, Mg, S, Fe, Mn, Zn, Cu, B and Mo, which are essential elements for plants, were dissolved in the slag. Si was also dissolved in the slag. The slag-containing aqueous solution exhibited buffering action, and its pH was approximately 8.5. This value was smaller than those of mortar and limestone solutions. The slag-free eluate also exhibited buffering action. The Fe(II) ion was dissolved in the slag, and therefore, it could reduce the Cr(VI) ion.

Keywords: Buffering Action, Cr(VI), Essential Nutrient, Environmental Quality Standard, Environmentally Regulated Substance, Leaching Test, pH.

INTRODUCTION

Approximately 36.6 million ton of slag has been produced in 2017 in Japan as ironmaking and steelmaking by-product [1]. The amounts of slags produced in blast furnaces, converters (basic oxygen furnaces) and electric arc furnaces (EAFs) were approximately 22.8, 10.9, and 2.7 million ton, respectively [1]. Slags have been roughly classified into blast furnace (BF) and steelmaking slags. Blast furnace slag is produced under reducing atmosphere and comes into contact with carbon-saturated molten iron. Therefore, the metal oxides in the slag that are reduced by the carbon in molten pig iron do not remain in the slag, and dissolve into the molten iron. In addition, most environmentally regulated substances are also transferred from the BF slag into the molten iron. Blast furnace slag has been reused as cement, road bed, and concrete aggregates, and its recycling ratio is approximately 100%.

---

[*] **Corresponding author Seiji Yokoyama**: Department of Mechanical Engineering, Toyohashi University of Technology, Toyohashi, Japan; Tel: +81-532-44-6696; Fax: +81-532-44-6690; E-mail: yokoyama@tf.me.tut.ac.jp

Toshiyuki Takahashi (Ed.)

Most converter and EAF slags, which are classified as steelmaking slag, have been reused as road beds or civil engineering materials. However, approximately 0.21 million ton of steelmaking slag is discarded into final disposal sites, and its recycling ratio is approximately 98.5%. In Japan, the decrease in public works decreased the demand for road bed materials, and therefore waste disposal became expensive. In addition, the final disposal space itself decreased. These problems have been accelerating the research into developing new slag applications [2]. In contrast with BF slag, steelmaking slag is produced in oxidizing atmosphere. Therefore, most elements dissolved in molten steel are oxidized and are transferred from the molten steel into the steelmaking slag. Accordingly, most steelmaking slag includes iron oxide as well as various other oxides. Consequently, if molten steel includes environmentally regulated substances, steelmaking slag could also include environmentally regulated substances as oxides, which is one of the reasons steelmaking slag is not reused.

The EAF steelmaking process usually consists of oxidizing and reducing steps. The EAF reducing slag, which is discarded from the reduction process, is similar to the BF slag, while the EAF oxidizing slag, which is discarded from the oxidation process, is similar to the steelmaking slag discarded from converters. Generally, various iron scraps are used as raw materials for EAF steelmaking. In addition, various steels are produced in an EAF, and therefore, the chemical composition of EAF slags varies. The chemical composition of an EAF oxidizing slag discarded from a typical steelmaking process is listed in Table **1** [3]. Fig. (**1**) illustrates the X-ray diffraction (XRD) pattern of an EAF oxidizing slag. The peaks of $2CaO \cdot SiO_2$, $FeO \cdot Al_2O_3$, FeO, and $MgFeAlO_4$ have been observed in the XRD spectrum of this slag along with many unidentified peaks.

Table 1. Chemical composition of electric arc furnace oxidizing slag (wt%).

FeO	SiO_2	CaO	Al_2O_3	MgO	MnO	CuO	ZnO	Cr_2O_3	NiO	CuO
35.1	19.2	20.8	15.2	4.1	5.1	1.8	0.071	0.43	0.028	0.025

LEACHING TEST AND DISSOLUTION OF ENVIRONMENTALLY REGULATED SUBSTANCES

When using or discarding wastes and slag in the soil or seawater, leaching tests must be performed. If the wastes do not pass the leaching test, they must be discarded at final disposal sites. Leaching tests exists for each country [4], and are not intended to measure the chemical content of a material, but to determine the amount of dissolved material under certain conditions.

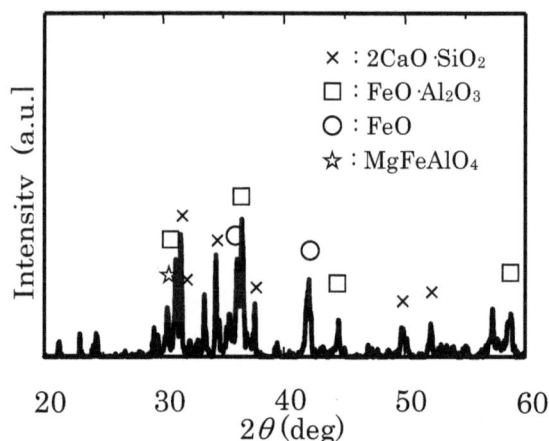

Fig. (1). X-ray diffraction pattern of electric arc furnace oxidizing slag.

Two leaching test methods are used in Japan: the shaking [5] and the tank leaching method [6]. The former is more commonly used, and it typically applies to samples featuring particles smaller than 2 mm in size. The latter is for samples which we will actually use in soil, as a rule. For soil disposal, the initial pH of the aqueous solution should be 5.8–6.3 and the ratio of the sample weight (g) to the aqueous solution volume (mL) should be 1/10. For seawater disposal, the pH should be 7.8–8.3, and the ratio 3/100. The details of the leaching test should be referred to. Both leaching test are performed for 6 h.

DISSOLVED CHEMICAL COMPONENTS

Table **2** summarizes the chemical components of solutions formed when 100 g of EAF oxidizing slag was dissolved in 1 L of water with different initial pH values using the tank leaching test [3]. The initial pH was adjusted by adding hydrochloric acid or sodium hydroxide to the solution. In addition, Table **2** includes the Japanese environmental quality standards (EQSs) for soil, which are the same as those for groundwater, seawater, and water pollution. The concentrations of environmentally regulated substances in the aqueous solution with the initial pH of 6 should be compared with the EQSs for soil. As the dissolved concentrations were lower than the EQSs for soil, this slag could be used on and under the ground. Here, Cr(VI), was not analyzed, however, because the total Cr concentration was lower than the corresponding EQSs, the concentration of Cr(VI) was also smaller than the EQSs of Cr(VI).

Table 2. Dissolved environmentally regulated substances and their corresponding environmental quality standards (EQSs) (mg/L).

		Initial pH				EQS		
		4	6	8	10	Soil[*1]	Marine[*2]	Water[*3]
Environmentally Regulated Substances	Total As				ND[*4]	0.01	0.1	0.01
	Total B	0.77	0.28	0.07	0.08	1		1[*8]
	Total Be	ND	ND	ND	ND		2.5	
	Total Cd	ND	ND	ND	ND	0.01	0.1	0.01
	Total Cr	0.024	ND	ND	ND		2	
	Cr(VI)	-	ND	-	-	0.05	0.5	0.05
	Total Cu	0.003	ND	0.004	0.006		3	
	Total Pb	ND	ND	ND	ND	0.01	0.1	0.01
	Hg	ND	ND	ND	ND	0.0005	0.005	0.0005
	Total Ni	0.001	ND	ND	ND		1.2	
	Total Se	ND	0.003	ND	ND	0.01	0.1	0.01
	Total V	0.004	0.01	0.003	0.004		1.5	
	Total Zn	0.006	0.014	0.012	ND		2	0.03[*5], 0.02[*5], 0.01[*7]
	F-	0.2	0.5	0.2	0.2	0.8	15	0.8[*8]
	Total CN	ND	ND	ND	ND	not detectable	1	not detectable

[*1]: EQSs for soil pollution. [*2]: Law on prevention of marine pollution and maritime disasters. [*3]: human health EQSs for water pollutants. [*4]: ND - not detected. [*5]: Habitable River or lake for (EQSs for conservation of living environment). [*7]: Habitable coastal water (EQSs for conservation of living environment). [*8]: Habitable coastal water that requires conservation in particular for nidus and nursery ground (EQSs for conservation of living environment).

Habitable coastal water that requires conservation in particular for nidus and nursery ground (EQSs for conservation of living environment).

The concentrations of environmentally regulated substances in the solution with the initial pH of 8 were lower than the EQSs for marine ecosystems. However, the slag weight to solution volume ratio for this leaching test was different from the regulated one. The annual average EQS values from the water pollution law are also listed in Table **2**. Because the water pollution law provides no leaching test data, these EQSs are only listed as reference. The concentration of dissolved Zn was the only one that was larger than the regulated value. Since the steelmaking slag has been produced at 1900 K and above and was cooled in air, it was considered that the organic materials regulated by the EQSs were decomposed and oxidized at high temperature. The safety of this slag was confirmed using serial batch dissolution testing [7].

The concentrations of the slag solution components, except for the environmentally regulated ones, are also listed in Table 2. The concentrations of Al, Ca, Fe, Mg, and Si will be described later. In addition to the substances listed in Tables 2 and 3, Sb, Bi, Co, Li W, Co, Ag, Tl, Sn, and Ti were also analyzed. However, the concentrations of these components were lower than the detection limits of the inductively coupled plasma (ICP) spectrometer used for analyzing. Generally, the concentrations listed in Table 3 decreased as the initial pH of the solution increased.

Table 3. Concentration of slag components except for environmentally regulated substances (mg/L).

Substance	Initial pH				Substance	Initial pH			
	4	**6**	**8**	**10**		**4**	**6**	**8**	**10**
Total Ba	0.13	0.094	0.075	0.047	Total Sr	0.11	0.074	0.07	0.056
Total Mn	0.004	ND	0.003	0.002	Total S	0.7	1.4	0.5	ND
Total Mo	0.002	0.007	0.002	0.002	Total W	0.027	0.056	0.03	0.03
Total K	3.8	4.8	3.4	0.5	Cl⁻	6	6	4	ND
Total Na	1.5	3.6	1.5	2.5					

ND: Not detected.

MAJOR DISSOLVED COMPONENTS

Fig. (2) illustrates the changes in the concentrations of Al, Ca, Fe, Mg, and Si with time [3]. The tank leaching test, which was described above, was performed for 48 h. The concentrations of Ca, Mg, and Si increased with time, and the concentrations of Al, Ca, Fe, Mg, and Si increased as the initial solution pH decreased. By comparison, the concentrations of Al and Fe were the highest 6 h after the slag was added to water, and decreased thereafter. As described in one of the following sections, the pH changed through the leaching test. Al and Fe, which dissolved during the early stage of leaching, precipitated in the solution as hydroxides ($Al(OH)_3$ and $Fe(OH)_2$) and iron oxide (FeO) as the pH decreased.

Several elements are essential for plant growth, and they can be classified as macronutrients (N, K, Ca, Mg, P, and S) and micronutrients (Cl, Fe, B, Mn, Zn, Cu, Mo, and Ni). Of these, Ca, K, P, Mg, S, Fe, Mn, Zn, Cu, B, and Mo dissolve in water, as listed in Table 3) and illustrated in Fig. (2). While Si is not included on the list of essential elements, it is essential for plants in the Gramineae family and the phytoplankton of the Bacillariophyceae family. However, Al, which is toxic for plants, also dissolves in water. While both useful and harmful elements dissolve in water, their dissolved concentrations determine whether they are useful or harmful.

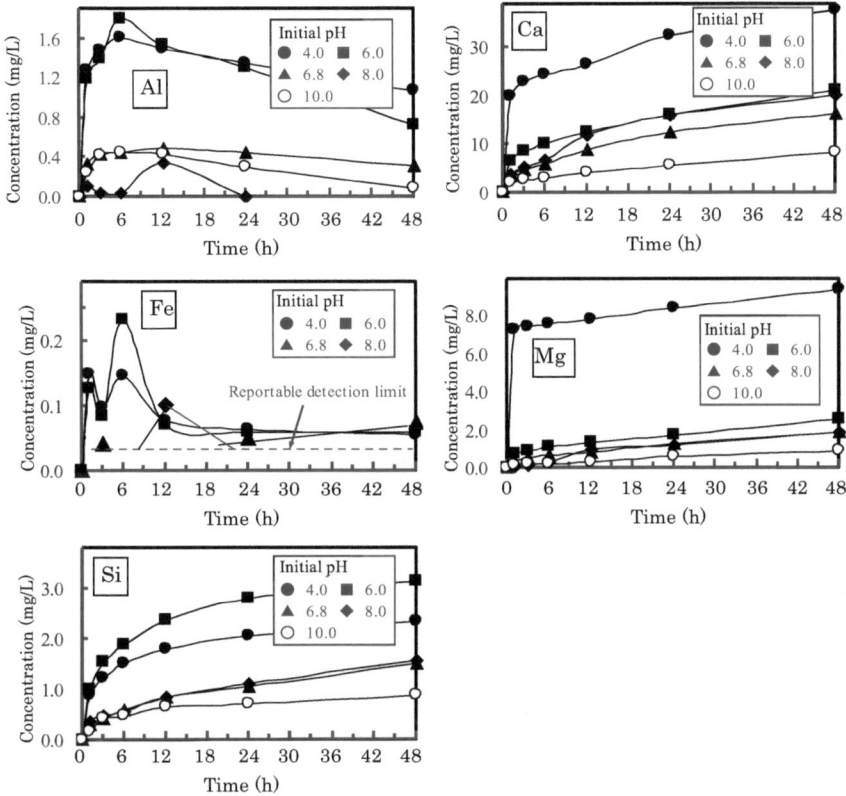

Fig. (2). Changes in concentrations of Al, Ca, Fe, Mg, and Si with time [3].
EAF oxidizing slag (100 g) was added to water (1 L), and the initial pH was adjusted by adding hydrochloric acid or sodium hydroxide. Slag particle size was 1—2 mm.

PH BEHAVIOR

Fig. (3) illustrates the changes in pH with time during the tank leaching test. When the pH of the initial solution was below 7, the pH steeply increased immediately after the slag was added to the solution, and then it gradually decreased with time. When the pH of the initial solution was above 7, the pH monotonously decreased with time after the slag was added to it. Generally, these pH values appeared to converge to approximately 8.5.

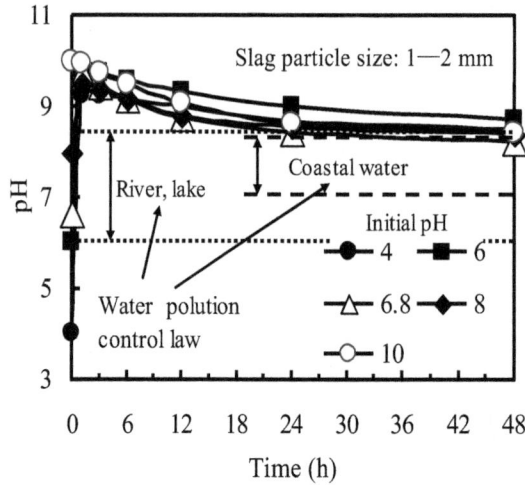

Fig. (3). Changes in pH with time [3].
EAF oxidizing slag (100 g) was added to water (1 L), and the initial pH was adjusted by adding hydrochloric acid or sodium hydroxide.

Fig. (4) depicts the changes in pH with time for the EAF oxidizing slag discarded from the oxidizing process of stainless steelmaking [8]. The chemical composition of this slag is listed in Table 4. By comparison with the slag in Table 1, this slag contained lower FeO and higher SiO_2 amounts. For these experiments, the slag was added to 1 L pure water. The pH values of the solutions obtained by adding 0.1, 1, and 10 g slag to water steeply increased immediately after the slag was added to water and then gradually increased with time. The pH of the solution obtained by adding 100 g of slag to 1 L of water steeply increased immediately after the slag was added to water and then decreased with time. The pH changes observed for these slag solutions were almost the same as those of the solutions of EAF oxidizing slag discarded from normal steelmaking processes presented in Fig. (3).

Table 4. Chemical composition of electric arc furnace oxidizing slag from stainless steelmaking (wt%).

FeO	CaO	SiO_2	Al_2O_3	MgO	MnO	Cr_2O_3	CaS	ZnO	CuO	NiO
0.75	34.2	45.9	5.6	7.98	4.24	0.68	0.34	0.01	0.24	0.06

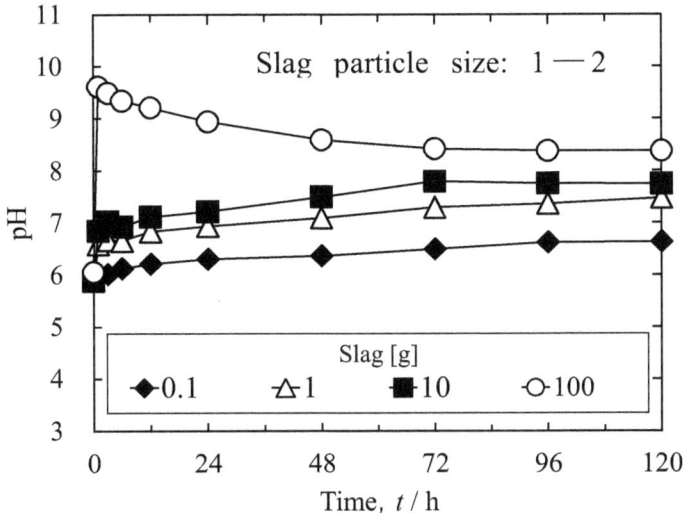

Fig. (4). Changes in pH with time [8].
Different amounts of EAF oxidizing slag discharged from oxidizing process of stainless steelmaking were added to 1 L pure water.

Fig. (**5**) illustrates the changes in pH with time for EAF oxidizing slag, mortar, and limestone [9]. The initial aqueous solution was prepared by adding artificial acid rain (pH of 3.5) to water until the pH of the solution was 4.5. The artificial acid rain was prepared using pure water, sulfuric and nitric acid solutions, sodium hydroxide, and so on based on the Japanese Industrial Standard (JIS) H 8502 [10]. The slag was EAF oxidizing slag, and its composition is summarized in Table **1**. Mortar was prepared using Portland cement, standard sand, and water, based on JIS A 6206 [11]. Limestone is a natural rock. The samples were ground to obtain particles 1–2 mm in size. The chemical compositions of the slag, mortar, and limestone are listed in Tables **1** and **5**. Immediately after the mortar and limestone were added to the initial solutions, the pH steeply increased, and then gradually decreased with time, as illustrated in Figs. (**3 and 4**).

Table 5. Chemical compositions of mortar and limestone (wt%).

Mortar	CaO	SiO_2	Al_2O_3	Fe_2O_3	Na_2O	K_2O
	32.2	60.3	2.8	1.75	0.31	0.11
Limestone	$CaCO_3$	SiO_2	Al_2O_3	Fe_2O_3	MgO	
	99.5	0.07	0.01	0.02	0.39	

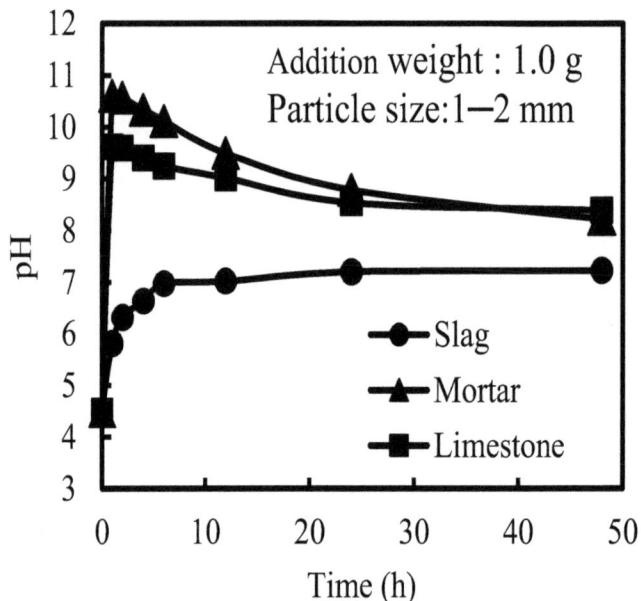

Fig. (5). Changes in pH with time [9].
Aqueous solution with initial pH of 4.5 was prepared by adding artificial acid rain (pH of 3.5) to water. Subsequently, 1.0 g slag, mortar, and limestone were separately added to 500 mL solution.

Limestone is used to improve the acidity of lakes and ponds [12]. In this study it was revealed that mortar and limestone could significantly improve water acidity. However, it may be said that they are too strong. Compared with mortar and limestone, the ability of the EAF slag to improve the acidity of lakes and ponds is weak. Nevertheless, adding any of these substances to water must be carefully monitored. Depending on the added slag mass and solution volume, the slag can neutralize the pH of water pH. While it appears that the formation of surface complexes on the slag could influence the pH behavior, only eluates without such complexes present buffering action [13].

REACTION BETWEEN EAF OXIDIZING SLAG AND Cr(VI) IN AQUEOUS SOLUTION

Cr(VI) compounds are genotoxic carcinogens. Therefore, the concentration of Cr(VI) in aqueous solutions is environmentally regulated. Because EAF slag contains iron(II) oxide (FeO), Fe^{2+} ions that are dissolved from EAF slag can reduce Cr(VI) as follows:

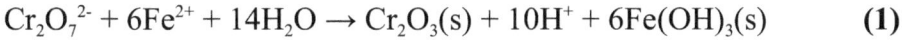

$$Cr_2O_7^{2-} + 6Fe^{2+} + 14H_2O \rightarrow Cr_2O_3(s) + 10H^+ + 6Fe(OH)_3(s) \qquad (1)$$

and

$$Cr_2O_7^{2-} + 6FeO(s) + 2H^+ + 8H_2O \rightarrow Cr_2O_3(s) + 6Fe(OH)_3(s), \qquad (2)$$

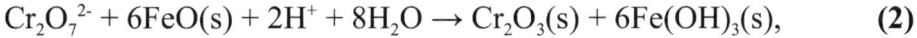

where (s) indicates solid phase.

Fig. (**6**) illustrates the changes in the concentration of Cr(VI) and total concentration of Cr(III) and Cr(VI), Cr_t, with time, during the reduction of Cr(VI).

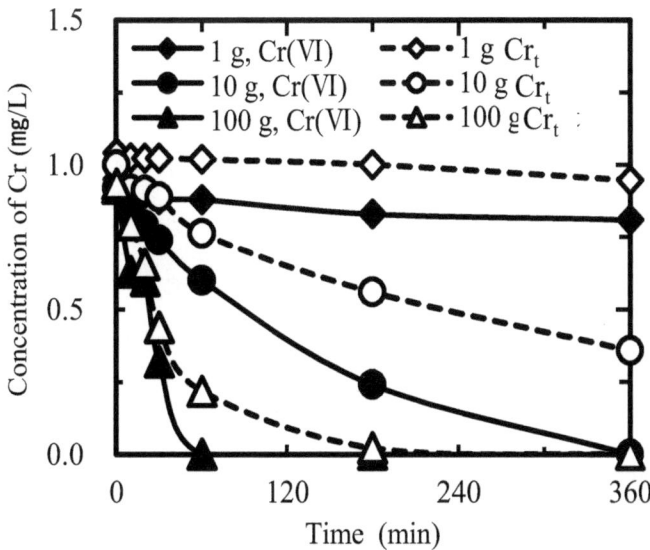

Fig. (6). Changes in concentration of Cr with time [14].
The masses represent the masses of added slag, and Cr_t is the total concentration of Cr(VI) and Cr(III) ions. Slag particle size is 1–2 mm. Carbon dioxide gas is introduced into the solution during the experiment.

In this experiment, EAF slag was added to aqueous solutions where the Cr(VI) concentration was adjusted by adding potassium dichromate ($K_2Cr_2O_7$) [14]. The total Cr and Cr(VI) concentrations were analyzed using an ICP spectrometer and the 1,5-diphenylcarbazide spectrometric method, respectively. In addition, carbon dioxide gas was introduced into the solution to enhance the dissolution of the Fe ions from the slag, because carbonic acid reduced the pH of the solution. Generally, the concentration of both Cr(VI) and total Cr decreased with time after the addition of EAF oxidizing slag. In addition, the reduction rates of Cr(VI) and total Cr increased as the mass of added slag increased.

Fig. (**7**) depicts the relationship between the dissolved Fe concentration and time. Here, the concentration of Fe refers to the Fe(II) ion, because the Fe(III) ion could not dissolve in this pH range. Generally, the concentration of Fe was almost constant in the early stage of dissolution, and then it increased with time. This tendency became remarkable as the mass of added slag increased. After the concentration of Cr(VI) decreased sufficiently, the concentration of Fe increased. The concentration of Mn increased with time from the beginning of the dissolution process, and this was similar to the dissolution behavior of Ca, as illustrated in Fig. (**2**). Therefore, the Mn ions did not contribute to the reduction of Cr(VI).

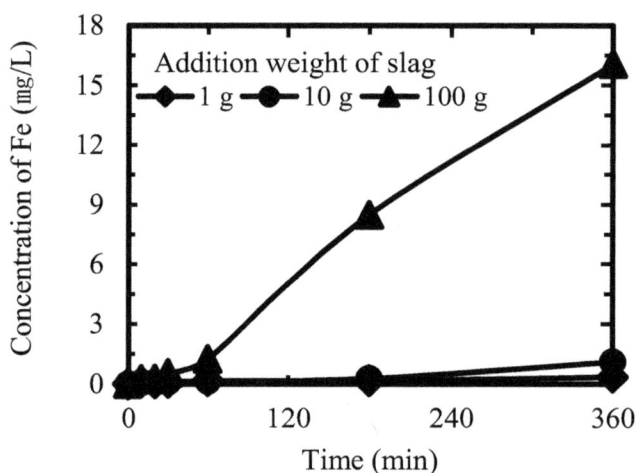

Fig. (7). Changes in concentration of dissolved Fe with time. Slag particle size is 1–2 mm.

If carbon dioxide gas was not introduced into the solution, the concentrations of total Cr and Cr(VI) decreased imperceptibly. This indicated that the reduction of Cr(VI) ions according to Eq. (2) hardly occurred. The reaction between Cr(VI) and slag has been reported in the literature [15, 16]. Presently, rain is mostly acid, and acid rain dissolves the Fe(II) ions from slag, and reduces Cr(VI) ions.

CONSENT FOR PUBLICATION

Not applicable.

CONFLICT OF INTEREST

The author confirms that this chapter contents have no conflict of interest.

ACKNOWLEDGEMENTS

Declare none.

REFERENCES

[1]   Nippon Slag Association. Tekko Suragu Toukeinennpyou 2017.

[2]   Ito K. Steelmaking Slag for Fertilizer Usage. Nippon Steel and Sumitomo Metal Tech Rep 2014; 399: 132-8.

[3]   Yokoyama S, Suzuki A, Izaki M, Umemoto M. Elution behavior of electric arc furnace oxidizing slag into fresh water. Tetsu To Hagane 2009; 95: 434-43.
[http://dx.doi.org/10.2355/tetsutohagane.95.434]

[4]   Yokoyama S. Dissolution assay, Corrosion Control and Surface Finishing, H Kanamatsu, Dana. Springer Japan 2016; pp. 153-67.
[http://dx.doi.org/10.1007/978-4-431-55957-3_15]

[5]   Ministry of Environment of Japan, Notification No.19.

[6]   JIS K 0058-1:2005. Test methods for chemicals in slags −Part 1 : Leaching test method

[7]   Yokoyama S, Suzuki A, Nor NHBM, *et al.* Serial batch elution of electric arc furnace oxidizing slag discharged from normal steelmaking process into fresh water. ISIJ Int 2010; 50: 636-8.
[http://dx.doi.org/10.2355/isijinternational.50.630]

[8]   Yokoyama S, Shimomura T, Hisyamudin MNN, Takahashi T, Izaki M. Influence of amount of oxidizing slag discharged from stainless steelmaking process of electric arc furnace on elution behavior into fresh water. J Phys Conf Ser 2012; 352012051
[http://dx.doi.org/10.1088/1742-6596/352/1/012051]

[9]   Latif Muhammad Syahiran Naim bin A, Yokoyama S. Application of EAF oxidizing slag for acid rain, CAMP-ISIJ 31, 2108, PS-59

[10]   JIS H. 8502: 1999, Methods of corrosion resistance test for metallic coatings.

[11]   JIS A. 6206: 2013, Ground granulated blast-furnace slag for concrete.

[12]   Lewis C J, Boynton R S. Acid Neutralization with lime for environmental control and manufacturing process, National Lime Association, 216.

[13]   Takahashi T, Yokoyama S. Bioassay of components eluted from electric arc furnace steel slag using microalgae chlorella. ISIJ Int 2016; 56: 1495-503.
[http://dx.doi.org/10.2355/isijinternational.ISIJINT-2015-539]

[14]   Yokoyama S, Okazaki K, Sasano J, Izaki M. Removal of hexavalent chromium in carbonic acid solution by oxidizing slag discharged from steelmaking process in electric arc furnace. AIP Conf Proc 2014; 1585: 181-9.
[http://dx.doi.org/10.1063/1.4866639]

[15]   Erdem M, Altundoğan HS, Turan MD, Tümen F. Hexavalent chromium removal by ferrochromium slag. J Hazard Mater 2005; 126(1-3): 176-82.
[http://dx.doi.org/10.1016/j.jhazmat.2005.06.017] [PMID: 16098660]

[16]   Han C, Jiao Y, Wu Q, Yang W, Yang H, Xue X. Kinetics and mechanism of hexavalent chromium removal by basic oxygen furnace slag. J Environ Sci (China) 2016; 46: 63-71.
[http://dx.doi.org/10.1016/j.jes.2015.09.024] [PMID: 27521937]

CHAPTER 19

Utilization of Steelmaking Slag in Sea Areas

Ryo Inoue[*]

Graduate School of International Resource Sciences, Akita University, Akita, Japan

Abstract: In recent years, many studies have been conducted on the use of steelmaking slag to improve marine environments. By burying steelmaking slag alone or in a mixture of steelmaking slag and soil (or sand) containing organic materials in marine environments, the growth of marine plants can be promoted and rocky-shore denudation can be repaired. Furthermore, it is known that the occurrence of blue tide can be prevented. These effects are understood to be the result of a steady supply of iron, which is a major component of steelmaking slag. This chapter outlines the generation and conventional applications of ironmaking and steelmaking slags, which are by-products of the steel industry. Author then highlight previous studies on the environmental protection of sea areas using steelmaking slag and discuss novel applications of steelmaking slag.

Keywords: Blue tide, Environmental protection, Iron, Rocky-shore denudation, Steelmaking slag.

INTRODUCTION

Types and Amounts of Ironmaking and Steelmaking Slags

Based on economic growth in China, India, and South America, steel production in these countries continues to increase. The annual worldwide crude steel production increased from 850 million tons in 2000 to 1.79 billion ton in 2018 [1]. China's crude steel production has increased dramatically since 1998 and China accounted for 52% of the world's annual crude steel production in 2018. In contrast, Japan's annual crude steel production volume has been relatively stable at approximately 100 million tons since 1972. Japan has been the world's second-largest crude steel producer since 2000. However, Japanese crude steel production volume during the fiscal year of 2018 was 104 million tons (5.8% of the world's annual crude steel production), which was overtaken by India, leaving Japan ranked third in the world [1].

---

[*] **Corresponding author Ryo Inoue:** Graduate School of International Resource Sciences, Akita University, Akita, Japan; Tel/Fax: +81-18-889-2948; E-mail: ryo@gipc.akita-u.ac.jp

Toshiyuki Takahashi (Ed.)

Steel production processes can be roughly divided into the blast furnace converter process and electric arc furnace process. The former uses a blast furnace to create pig iron (highly carbon-rich iron melt) using coke (from which water, coal tar, pitch, *etc.* have been previously removed by roasting) [2]. In the converter, the carbon in the pig iron is removed by blowing oxygen through the mixture to create molten steel (iron melt with low carbon concentration) [3]. The latter produces molten steel in an electric arc furnace using iron scrap as the main raw material [3]. This process accounts for 25% of the crude steel production in Japan [1].

Because impurities, such as phosphorus and sulfur, in steel products negatively affect the mechanical properties of steel materials, refining processes for removing impurities are indispensable in both the blast furnace-converter and electric arc furnace processes [3]. In these refining processes, lime (CaO) is typically added as a flux and oxide slag containing impurities is discharged as a by-product. A generation chart for iron and steel slags is presented in Fig. (**1**) [4].

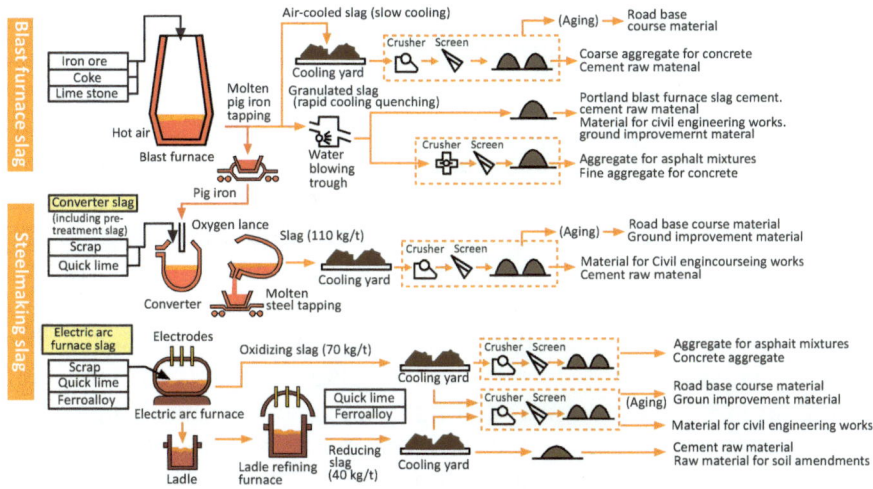

Fig. (1). Iron and steel slag production in different production processes [4].

Iron and steel slags are classified based on their production processes, as shown in Fig. (**2**).

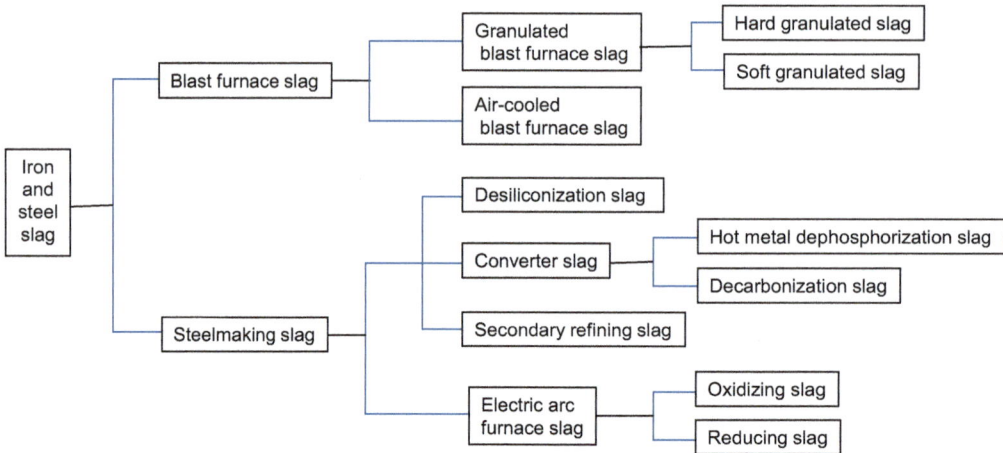

Fig. (2). Types of iron and steel slags.

The slag discharged from a blast furnace is called "blast furnace slag". Its non-iron components, which are contained in the iron ore, mainly include SiO_2 and Al_2O_3. The ash in the limestone and coke are melted together and discharged as they separate from the hot metal (molten pig iron). Representative chemical compositions of slag [4] are provided in Table **1**. Blast furnace slag is divided into slowly cooled slag and quenched slag according to the cooling rate from the molten state. The latter is obtained by injecting pressurized water into the molten slag flow or by injecting the molten slag into water to quench and granulate the slag. This is called "granulated slag" [4]. By manipulating the production conditions of granulated blast furnace slag, it is possible to create fine particles with a small amount of pores (hard granulated slag) or light particles with many pores (soft granulated slag) [4]. Granulated slag is vitreous and contains crystallization energy based on rapid cooling. It is known to have high chemical reactivity and a latent hydraulic property (the property of forming a hydrate and hardening in the presence of an alkaline aqueous solution [5]). The amount of blast furnace slag discharged is approximately 300 kg per ton of pig iron in Japan. The total amount of slag produced in 2018 was 22.7 million tons, with 83% being granulated slag, while the amount of pig iron produced in 2018 was 75.9 million tons [6].

The slag discharged from a converter is called "converter slag" and it is formed through the dissolution of lime, magnesium oxide, and iron oxide, which are added as auxiliary materials, and P_2O_5, which is formed from the phosphorus contained in hot metal during oxidation refining [3]. Because the dephosphorization of hot metal tends to progress thermodynamically as temperature decreases [3], it is common in Japan for dephosphorization to be performed in a hot metal pre-treatment furnace with a low operating temperature prior to converter operation. Therefore, decarburization is the main job of the converter. Hot metal dephosphorization slag and converter decarburization slag are generically referred to as converter slag. Again, representative chemical compositions are provided in Table **1** [4, 7]. The amount of converter slag produced in Japan is approximately 120 to 130 kg per ton of crude steel. The total amount produced in 2018 was 11.0 million tons, while the amount of crude steel produced by blast furnace-converter process was 76.9 million tons in 2018 [6].

Table 1. Examples of iron and steel slag compositions [4].

| | Blast Furnace Slag | Converter Slag | | Electric Arc Furnace Slag | | Ordinary Portland Cement |
		Dephosphorization Slag [4, 7]	Decarburization Slag	Oxidizing Slag	Reducing Slag	
CaO	41.7	37.3-52.2	45.8	22.8	55.1	64.2
SiO_2	33.8	14.1-18.3	11.0	12.1	18.8	22.0
Total Fe	0.4	12.2-26.1	17.4	29.5	0.3	3.0
MgO	7.4	2.6-2.8	6.5	4.8	7.3	1.5
Al_2O_3	13.4	2.6-3.4	1.9	6.8	16.5	5.5
S	0.8	0.03-0.05	0.06	0.2	0.4	2.0
P_2O_5	<0.1	2.9-3.8	1.7	0.3	0.1	-
MnO	0.3	4.3-5.6	5.3	7.9	1.0	-

The slag discharged from an electric arc furnace includes oxidizing slag formed during the oxidation smelting stage (decarburization, dephosphorization, and dehydrogenation) and reducing slag generated during the reduction smelting stage (deoxidation and desulfurization) [4]. Representative chemical compositions of these slags are provided in Table **1** [4]. Japan produces approximately 70 kg of oxidizing slag and 40 kg of reducing slag per ton of crude steel [4]. The total amount produced in 2018 was 2.79 million tons, while the amount of crude steel produced by electric arc furnace process was 26.0 million tons in 2018 [6].

Applications of Ironmaking and Steelmaking Slags

Hard granulated slag from the blast furnace process is typically used as a fine aggregate for concrete. Soft granulated slag is used as a raw material for cement, civil engineering, and fertilizers. Fig. (**3**) presents a pie chart representing blast furnace slag utilization in 2018 [6]. One can see that 83% of blast furnace slag is granulated slag and that 89% of this slag is used as a raw material for cement.

Blast furnace cement, which is a mixture of a fine powder of blast furnace granulated slag and Portland cement, is classified into three types of A, B, and C according to the blending amount of granulated blast furnace slag based on the Japanese Industrial Standard JIS-R-5211. Blast furnace cement B, whose blending amount of granulated blast furnace slag powder is between 30% and 60%, is the most popular type in Japan [8]. Blast furnace cement has a high energy-saving efficiency of 40% or more in terms of cement production energy during the firing process and provides a 41% reduction in CO_2 emissions compared to pure Portland cement [9]. In Fig. (**3**), shows that 71% of slowly air-cooled slag is used as a road construction material. In recent years, it has been reported that the chemical stability of concrete aggregates can be improved by reducing the porosity of slowly cooled blast furnace slag, which reduces its water absorption [10, 11].

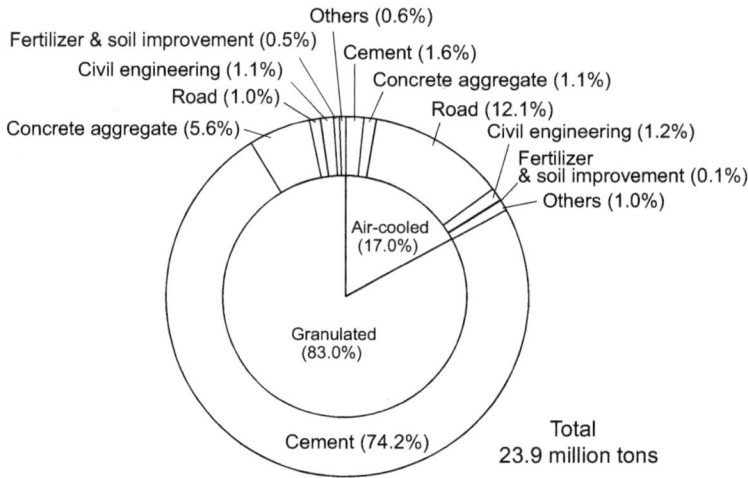

Fig. (3). Utilization of blast furnace slag in 2018 [6].

The amount of converter slag utilized in 2018 [6] is presented in Fig. (**4**). One can see that 22% of the slag was reused as refining flux, 33% as road construction material, and 30% as civil engineering material. Because converter slag has poor chemical reactivity, unlike the potential hydraulicity of blast furnace granulated slag, utilization with high added value is not performed. It is well-known that converter slag is rich in various minerals, including Ca, Mg, Mn, Fe, and P. Therefore, its use as a fertilizer/soil improvement material is a promising application for taking advantage of these minerals [12]. However, current use for this application is less than 0.9%.

When the environmental standards for water pollution [13] were revised in 1999 and those for soil contamination [14] were revised in 2001, fluorine and boron, which had previously required monitoring, were updated to standard items. However, ironmaking and steelmaking slags were excluded from the regulations defined in the report of Central Environment Council of Ministry of the Environment (December 26, 2000) [14, 15]. To improve the refining ability of steelmaking slag, slag basicity (CaO/SiO_2 mass concentration ratio) must be enhanced during operations [3]. However, when increasing the CaO concentration, steelmaking slag enters a solid-liquid coexistence state and the reactivity and fluidity of the slag deteriorate. In the past, because fluospar (CaF_2) has traditionally been used as a melting point depressant for converter slag, Inoue and Suito examined the fluorine content in the mineral phases of steelmaking slag [16] and improved the elution test procedure [17]. Their efforts aimed to establish technology for controlling fluorine dissolution from furnace slag [17 - 19], secondary refining slag [19, 20], and electric arc furnace slag [21]. Out of consideration for environmental protection, the steel industry has begun to use fluorine-free and boron-free fluxes for most refining processes when using steelmaking slag as a road construction material and civil engineering material [15, 22]. For the use of blast furnace slag and converter slag on land, safety metrics (Table **2**) [23] have been evaluated based on the dissolution tests defined in the soil environmental standards. Such slags have been shown to fall within the soil environmental standard values [4]. Environmental safety has also been evaluated based on bottom sediment standards under the Marine Pollution Control Law when using slag in sea areas [4].

Table 2. Environmental standards for inorganic substances [23].

Element	Environmental limit	
	In Eluate	**In Agricultural Soil**
Soil environmental standard (inorganic elements)		
Cadmium	0.01 mg/L	
Total Cyan	Not detected in solution	
Lead	0.01 mg/L	
Hexavalent chromium	0.05 mg/L	
Arsenic	0.01 mg/L	15 mg/kg soil
Total mercury	0.0005 mg/L	
Alkyl mercury	Not detected in solution	
Selenium	0.01 mg/L	
Fluorine	0.8 mg/L	
Boron	1 mg/L	

(Table 2) cont.....

Copper		125 mg/kg soil
Required monitoring items of public water area		
Nickel	—	
Molybdenum	0.07 mg/L	
Antimony	0.02 mg/L	
Total manganese	0.2 mg/L	
Uranium	0.002 mg/L	

Fig. **(5)** presents the amount of electric arc furnace slag used in 2018 in itemized form [6]. Similar to converter slag, the main applications are road construction material (48%) and civil engineering material (32%). Additionally, oxidizing slag also contains various minerals, including Ca, Mg, Mn, Fe, and P, but it may also contain Cr and Ni. Although meaningful research has been conducted on the suppression of Cr dissolution from slag [21] and suppression of oxidation to hexavalent chromium [24, 25], greater attention has been focused on the use of electric arc furnace slag as a road construction material.

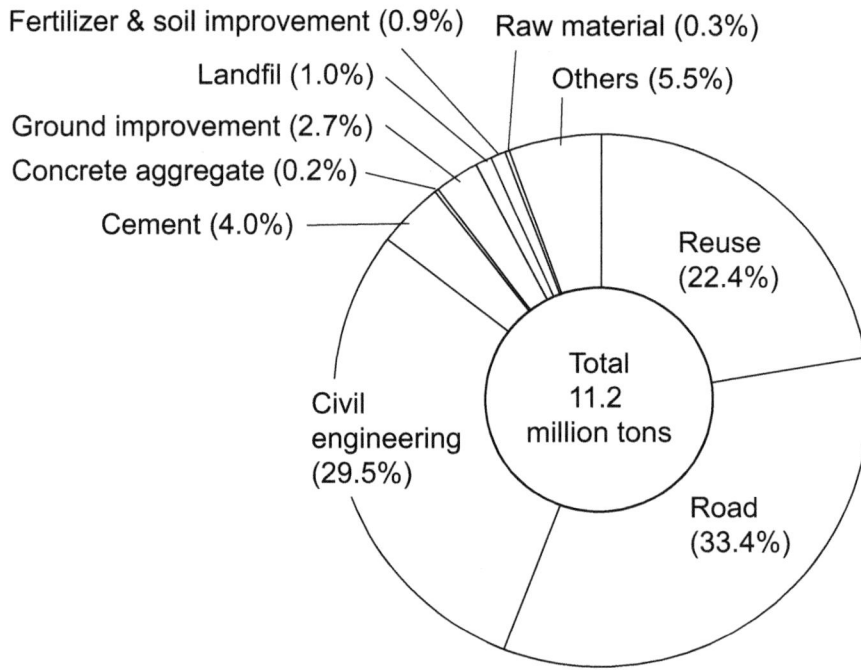

Fig. (4). Utilization of converter slag in 2018 [6].

NEW DEVELOPMENT OF APPLICATION OF STEELMAKING SLAG

Recently, based on a decrease in the number of public works, the use of steelmaking slag as a civil engineering and road material has started to reduce. Therefore, the use of slag in marine environments has emerged as a hot topic in the field of steelmaking slag. To use steelmaking slag as a seashore restorative material, the following three points have received significant attention:

1. the development of artificial stone (hydrated solid) using a mixture of steelmaking slag and granulated blast furnace slag powder;
2. development of solidified carbonate produced by blowing CO_2 into a mixture of steelmaking slag and granulated blast furnace slag powder; and
3. development of landfill materials mixed with clay and steelmaking slag (calcia-modified soil).

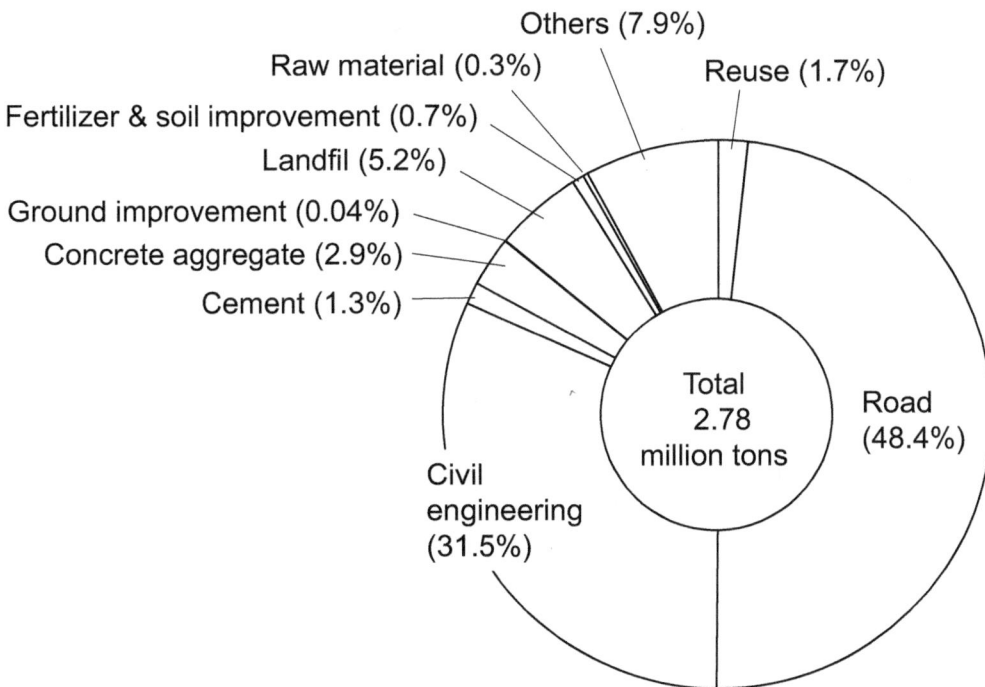

Fig. (5). Utilization of electric arc furnace slag in 2018 [6].

The solidification of steelmaking slag using the treatments described in points (1) and (2) [26, 27] makes it possible to increase the strength of slag blocks and suppress increases in the pH values of the water eluted from steelmaking slag.

While points (1) and (2) use steelmaking slag as a substitute for natural sand and natural stone as a marine environment remediation material, the goal of point (3) is to prevent rocky-shore denudation (the covering of rocky areas with calcareous algae causes a reduction in large seaweed [28 - 31]) and improve bottom sediments in closed sea areas [32, 33] using the minerals eluted from steelmaking slag, especially iron. Both of these goals have been considered in demonstration projects in the field of water environment improvement technology in closed sea areas implemented by the Ministry of the Environment [34, 35]. The effectiveness and safety of these methods have been confirmed.

ENVIRONMENTAL PROBLEMS IN SEA AREAS

Seaweed beds, where seaweed gathers to form communities in coastal areas, have the following functions [36]: (1) maintenance of biodiversity (habitats of fish and selfish); (2) carbon dioxide absorption through the photosynthesis of marine algae, which stores carbon in the ground and supplies oxygen; (3) the purification of seawater; (4) coastal protection through erosion control; and (5) creation of habitats with hydrophilicity that facilitate environmental learning based on diving, fishing, and observation.

Land forests absorb carbon dioxide and store carbon in wood and roots. This type of carbon storage is referred to as "green carbon." Coastal ecosystems (mangrove forests, seaweed beds, and salt marshes) absorb carbon dioxide and store it in the ground (seafloor sediment). This type of carbon storage is referred to as "blue carbon" [37]. According to a report compiled in October of 2009 by the United Nations Environment Program [38], the global amount of blue carbon is 1.5 times of the amount of green carbon and the amount of absorbed CO_2 in coastal areas (representing only 0.2% of the total sea surface) has reached 1.65 billion tons, which is comparable to annual emissions in Japan. However, this ecosystem declines by 2% to 7% every year. This rate of decline is considered to be greater than that of tropical rain forests. It has been hypothesized that reversing the decline of coastal ecosystems caused by land reclamation and development may be a powerful countermeasure against global warming. It has been simulated that CO_2 absorption by seaweed beds will increase by 33% to 48% with an area increase in seaweed bed area of 17% to 28% by 2030 [37].

Rocky-Shore Denudation

Rocky-shore denudation is a phenomenon in which rocks and bedrock on the bottom of the sea are covered by a calcareous algae called "abscissor coral algae." This coverage causes seaweed communities, such as kelp and sargassum, to decline and disappear [39]. This occurrence is thought to be related to an increase in sea water temperature and seaweed intake by herbivorous organisms, but one major factor is a lack of iron supply based on artificial environmental changes on land [28].

Many studies on the relationship between iron and algae growth have been conducted. Three major aspects of this relationship have been emphasized. The first is the effect of iron on the growth of phytoplankton. This aspect is widely known as "Martin's iron hypothesis" [40, 41] that a lack of iron is a controlling factor for algae growth. This is an assumption that the growth of phytoplankton is suppressed by a lack of iron, even when other nutrients, such as nitrogen and phosphorus, are present in sufficient quantities. To verify this iron deficiency theory, several international collaborative research projects dispersed ferrous sulfate in the North Pacific from June to August of 2001. In these projects, when the iron concentration increased to 3 to 4 nM/L (0.1 ppb), a large outbreak of phytoplankton was observed. This experimental finding revealed that a lack of iron ions impedes the production of phytoplankton [42 - 45].

The mechanism by which a lack of iron suppresses the growth of phytoplankton is considered to be similar to the growth inhibition mechanism of seaweed [46]. As shown in Fig. (**6**) [31], iron functions as an essential element in the production of enzymes and chromoprotein, which are indispensable to photosynthesis and nitrogen assimilation/metabolism, respectively, in plants and seaweed [28]. When an enzyme in seaweed operates during the absorption of nutrients, such as nitrogen and phosphorus, from seawater, iron is essential for smooth operation [47]. If iron has not been charged in seaweed previously, it is thought that seaweed has difficulty in absorbing nutrients. In the case of floating diatoms, which are a type of phytoplankton, the iron concentration range necessary for their growth has been reported as 6.6 to 66 nM/L (0.2 to 2 ppb) [48].

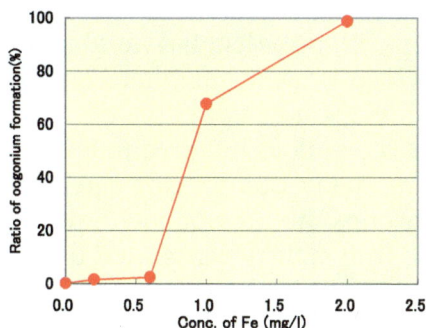

Fig. (6). Relationship between the concentration of ethylenediaminetetraacetic acid chelated iron and the oogonium formation of Laminaria [31].

The second aspect is that iron affects the growth of large algae. After Martin *et al.* [40, 41] presented studies on the effects of iron deficiency on the reduction of phytoplankton formation in the 1980s, Suzuki *et al.* [49] studied the influence of iron ions on the growth of kelp in Japan. Iron was considered to be an essential element not only for leaf growth, but also for the reproductive processes of gametophyte maturation and sporulation. Tsutsumi *et al.* [50] reported the effects of iron on kelp growth. In their experiments, it was shown that seawater with an added mixture of steelmaking slag and artificial humus soil significantly enhanced the growth of kelp compared to regular seawater. A demonstration experiment in Mashike-cho, Hokkaido confirmed that an iron ion concentration of at least 5 ppb is necessary for the fertilization and growth of kelp [50].

According to Takeda [46], the iron concentration in the coastal area of the Japan Seto Inland Sea is approximately 5 to 110 nM/L (0.3 to 6.2 ppb). Based on the required iron concentrations for large algae, there are some coastal zones that are clearly lacking in iron. It is believed that iron deficiency causes rocky-shore denudation in coastal zones where the dissolved iron concentration in seawater is less than 1 to 2 ppb. Therefore, supplying iron in such coastal areas is considered to be a promising method for eliminating rocky-shore denudation.

Regarding large algae other than kelp, Zhang *et al.* [51, 52] revealed that iron deficiency affects the coloration of nori. Ueki *et al.* [53] reported that a lack of iron ions turns the color of nori to a reddish brown.

The third aspect is the study of the iron supply routes to coastal areas. Kuma *et al.* [54] and Suzuki *et al.* [55] revealed that iron is supplied from land areas to coastal waters in the form of humic iron. Various theories have been proposed regarding the mechanism of iron supply from land areas to sea areas [47, 54, 56, 57]. When bacteria in the soil decompose fallen leaves from forests, humic substances, such as alkali-soluble humic acid and acid-soluble fulvic acid, are produced. The former is an acidic amorphous high-molecular-weight organic substance that forms a reddish brown to blackish brown precipitate under acidic conditions. It dissolves iron oxide particles in the soil to form divalent iron ions. The latter is an amorphous polymer organic acid that does not precipitate under acidic or alkaline conditions. Because it contains many complexing agents that have the strong chelating effects of carboxyl groups, it combines with divalent iron ions to form fulvic acid iron. The fulvic acid iron is then transported from the forest to rivers.

In rivers, divalent iron ions that are unreacted with fulvic acid will be exposed to air and oxidized into trivalent ions. The solubility of divalent and trivalent iron ions was reported by Pourbaix *et al.* in the form of an iron solubility-pH diagram [58]. Recently, Zhang *et al.* [59] derived solubility values using the thermodynamic calculation software FACT SAGE. Their results are presented in Fig. (7). One can see that although the solubility of divalent iron ions in the neutral pH range is high, when a divalent iron ion is oxidized into a trivalent ion, the trivalent iron ion transforms into $Fe(OH)_3$ and precipitates based on its low solubility of 1 to 10 nM/L (0.056 to 0.56 ppb). In contrast, iron ions bound by fulvic acid are extremely stable and reach the sea in an aqueous soluble form.

Because the iron concentration in the Seto Inland Sea is 0.3 to 6.2 ppb, which is greater than the potential solubility of ferric ions [46], it is understood that iron is chelated and transported to the sea [46, 60]. However, it has been pointed out that there is a possibility that the growth of seaweed is affected by water flow stagnation caused by dam construction and revetment work in recent years, which can prevent the iron supply from land areas from reaching sea areas [28, 57]. Suzuki *et al.* [61] studied the use of humic acid iron accumulated in dam bottoms for supplying iron. The embedding of steelmaking slag containing a large amounts of iron oxides directly into sea areas as an iron supplier has also been investigated [28 - 31, 50].

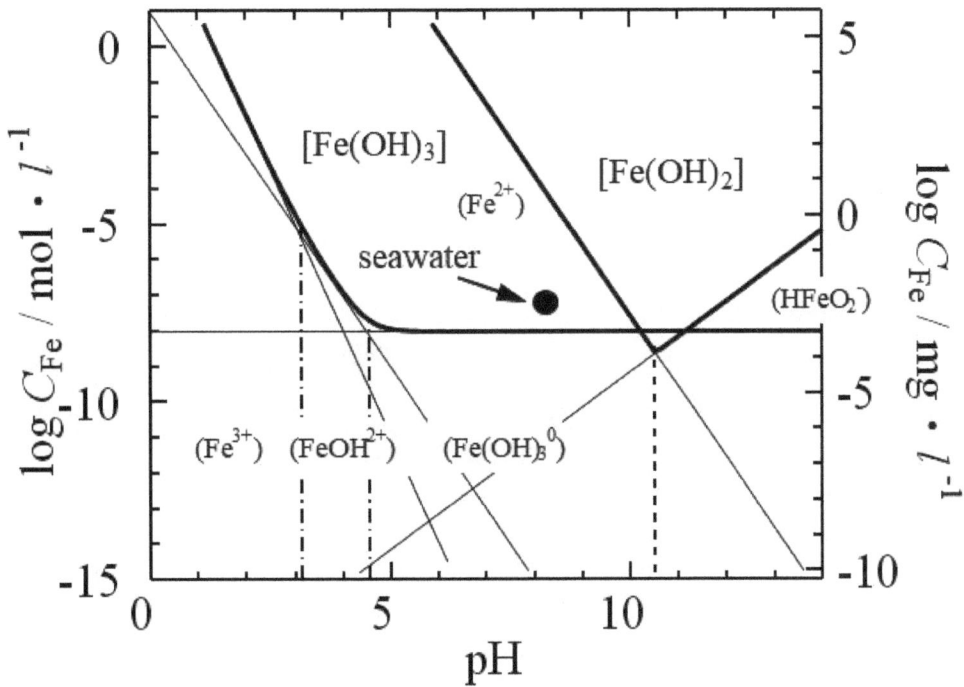

Fig. (7). Stability diagram for iron in seawater at 298 K [59].

Blue Tide

The occurrence of blue tide is a problem where sediments are deposited in the depressions in closed sea areas. The generation mechanism for blue tide can be explained as follows [62, 63]. First, the eutrophication of seawater leads to the growth of too much plankton. Dead plankton then fall to the seabed and are decomposed by bacteria. During this process, a large amount of oxygen in the sea areas where water stagnates is consumed. As a result, an oxygen-depleted water zone with extremely low amounts of dissolved oxygen is formed. Because anaerobic bacteria grow in anoxic water zones, sulfate-reducing bacteria, which are a type of anaerobic bacteria, generate large amounts of hydrosulfide. When water containing a large amount of hydrosulfide rises to the upper layer of seawater, the hydrosulfide is oxidized by oxygen in the vicinity of the surface layer of seawater to form fine particles of sulfur or sulfur oxide. These fine particles float in the seawater as colloids, reflecting sunlight and discoloring the seawater to milky blue or milky white. The occurrence of this blue tide causes various problems, including the deterioration of living environments for organisms and the generation of a foul odor [63].

During a period of economic growth in Japan, large-scale landfills were created in the coastal areas of the Three Great Bays (Tokyo Bay, Osaka Bay, and Ise Bay) and Seto Inland Sea. A large amount of soil and sand from the seabed was mined as a cheap landfill material. As a result, many submarine sediment mining sites were left in these areas [64 - 66]. These ruined sites (called "deep pits") have poor seawater circulation, which led to the deterioration of water quality in terms of poor oxygen levels and hydrosulfide generation, resulting in blue tide. In a study [67], it was reported that 90% of the total sulfide in Tokyo Bay existed in depressions generated by sediment mining. Blue tide contains a large amount of hydrosulfide, which is extremely toxic to most organisms, causing serious damage to ecosystems in tidal flats and shallow land areas [68]. The need for backfilling these depressions to restore coastal environments has been explored by the Central Environment Council and Transport Policy Council [69]. It has been reported that the sulfide concentrations in the sediments of the coastal areas of Japan are high in areas where a large amount of organic matter is accumulated in deep pits and under-farmed areas, leading to sulfide ion concentrations in excess of 20 ppm [70].

As improvement methods for seafloor environments in the deep pits generating blue tide, backfilling, sandblasting, oxygen supply *via* air bubbling, *etc.* have been carried out. However, a large amount of soil and sand is required to backfill deep pits and improve sediment quality. From the viewpoint of preserving the natural environment, the utilization of industrial by-products as a substitute for natural sand and crushed stone is a promising solution. The following two points were presented in a report by the Ministry of Land, Infrastructure, Transport, and Tourism [69]:

1. While analyzing the environmental influence of expanding the supply of sandblasting materials and promoting high utilization of coral sand and recycled materials, the maintenance/regeneration/creation of tidal flats, beaches, and seaweed beds must be promoted.
2. In the Three Great Bays, Seto Inland Sea, *etc.*, large-scale depression sites have been created by previous sediment collection projects for landfill and concrete aggregates, leading to low-oxygen sea areas and blue tide. To backfill these depressions, it is necessary to actively promote the expansion of procurement sources for backfill materials and the utilization of recycled materials, as well as soil and sand obtained from port construction.

As recycled materials for backfilling and sandblasting, clay [71], fired shell materials [72], granular coal ash and cement mixtures [73, 74], blast furnace granulated particles [75, 76], and massive steelmaking slag [77] have been examined. Considering supply, stability, quality, manufacturing and processing costs, and transportation costs, it may be appropriate to use massive steelmaking slag as a recycled material.

UTILIZATION OF STEELMAKING SLAG IN SEA AREAS

Because the repair and improvement of sea area environments are required to combat rocky-shore denudation and blue tide, many studies on the utilization of steelmaking slag have been conducted in recent years.

Counteracting Rocky-Shore Denudation

Since it was determined that iron has a significant effect on the growth of phytoplankton, several studies have been conducted focusing on steelmaking slag as an iron source. From 1999 to 2001, the CO_2 Fixation Research Group of Japan Iron and the Steel Institute of Japan was activated to enhance the proliferation of marine phytoplankton using steelmaking slag as a nutrient source. This was accomplished by combating CO_2 fixation by algae by using active ingredients, such as phosphorus, silicon, and iron, contained in steelmaking slag as nutrient sources for marine plankton growth. In this group, researchers studied engineering facilities and marine fisheries, and reported the elution behavior of steelmaking slag into seawater [78], effectiveness of steelmaking slag as an essential nutrient source for phytoplankton [48], characteristics of slag hydrated solids and bioadhesiveness in marine environments [79], and dissolution of phosphorus and silicon from steelmaking slag and their effects on the growth of natural phytoplankton communities [80].

Yamamoto *et al.* [28 - 30] reported a method for supplying iron to coastal sea areas by chelating iron ions eluted from slag using the humic acid in artificial humus soil. In this case, in the area of Mashike on the east coast of Hokkaido, where rocky-shore denudation was significant, a fertilization unit consisting of a steelmaking slag containing approximately 20 mass% of iron and an artificial humus soil with fermented waste wood chips was installed on the seashore Fig. (**8**) [29]. Iron components eluted from the steelmaking slag were complexed by the humic acid in the artificial humus soil and supplied to the seawater. Fig. (**9**) presents the iron concentration in the coastal area after installing the fertilization unit [30]. One can see that the iron concentration near the fertilization unit is greater than that far from the fertilization unit. Additionally, as shown in Fig. (**10**), the growth of kelp near the fertilization unit was verified [29]. Consequently, it was confirmed that supplying iron in this manner is an effective method for the generation of seaweed beds [29 - 31]. When steelmaking slag was mixed with artificial humic substances, it was shown to be superior in terms of iron dissolution rate and lifespan compared to steelmaking slag or artificial humic substances alone [28, 81].

Fig. (8). Photograph of fertilizer unit installation [29].

It can be concluded that these studies are very significant from the perspective of improving sea environments by providing a stable supply of chelated iron and effectively utilizing steelmaking slag. However, the production of artificial humus soil and steelmaking slag combined with artificial humus soil, as well as the installation of mixed materials in sea areas are costly; therefore, cost reduction is considered to be an important future issue. Matsunaga [82] pointed out that humic acid has the effect of suppressing the growth of calcareous algae. However, in the future, it should be determined if iron or humic acid is more effective or if both are necessary for the elimination of rocky-shore denudation.

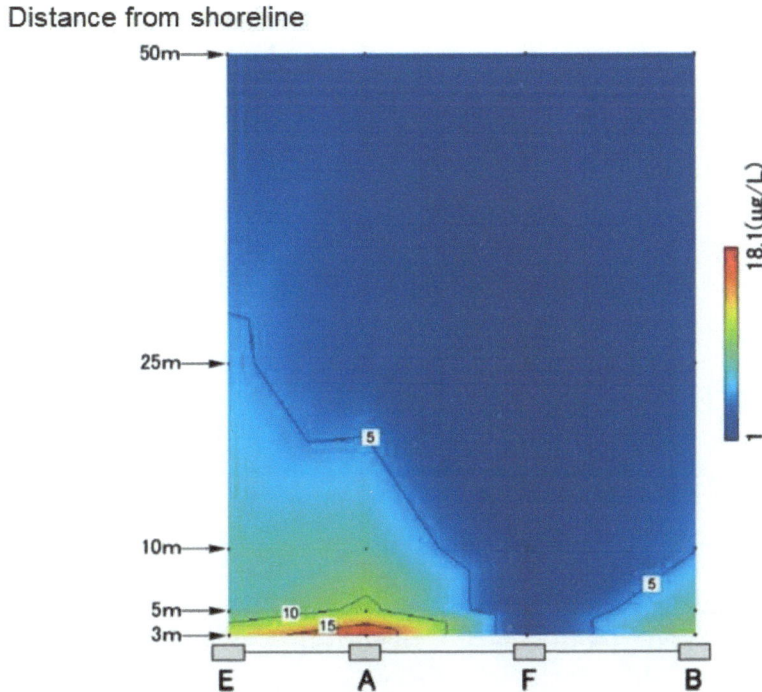

Fig. (9). Iron concentration in a coastal sea area after fertilizer unit installation [41].

Inoue *et al.* [7] determined that the amount of iron eluted increases significantly when reacting steelmaking slag with humic acid by developing a flow injection analysis method for iron determination up to ppb levels in eluate. They experimentally clarified that the formation of an iron complex with humic acid contains more trivalent iron ions than divalent iron ions. Zhang *et al.* [83, 84] focused on gluconic acid, which is contained in honey, wine, vinegar, plants, and fruits, and reported that the elution of elements in steelmaking slag is promoted by gluconic acid and that trivalent iron ions form more easily than divalent ions in a complex with gluconic acid.

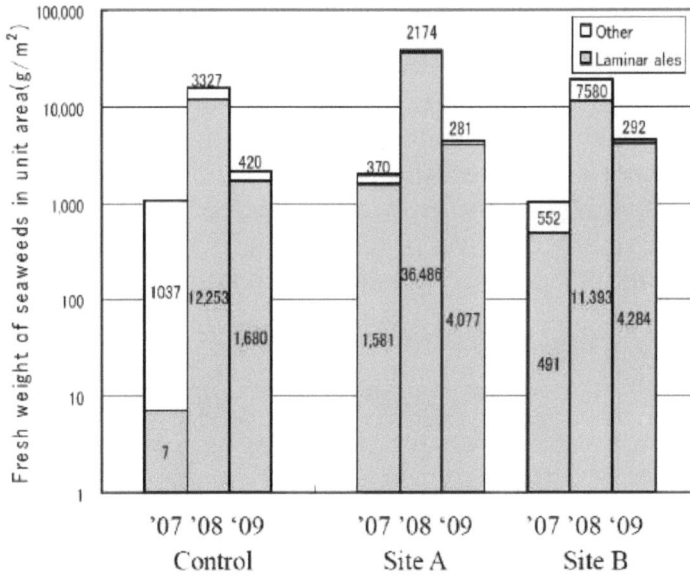

Fig. (10). Changes in the amount of seaweed after fertilizer unit installation [41].

In the sea area around Kawasaki City, Japan, the effects of a mixture of steelmaking slag and harbor clay on the growth of seaweed were demonstrated and evaluated [85]. When bottom soil and clay include organic acids, such as fulvic acid and humic acid [86], these organic acids complex the iron in steelmaking slag and dissolve it in seawater to encourage the growth of seaweed. The difference between this technology and the aforementioned method proposed by Yamamoto *et al.* [29 - 31] (use of artificial humus as a complex former) is in the use of clay. Clay collected from the sea bottom has little influence on the ecosystem. A mixed material of clay and steelmaking slag (Calcia-modified soil) was developed as a material for creating shallow land areas and backfilling after deep mining [87], but another possible use as an iron supplier is also conceivable. Zhang *et al.* [88] experimentally determined that element elution from steelmaking slag is promoted by clay. The utilization of both unused clay soil and steelmaking slag, which is inevitably generated during the steelmaking process, is considered to be socially significant from the perspective of effective utilization of resources. Although it has been reported that humic acid iron is abundant in dam bottom sediment [28, 89, 90], cost issues related to the transportation of sediment from dam bottoms to coastal areas are difficult to overcome

In addition to seaweed, Tanimoto *et al.* [91] also studied the effects of blast furnace slag and steelmaking slag on the growth of eelgrass. From 2010 to 2013, the following points were systematically investigated by a research group from the Iron and Steel Institute of Japan Innovative Program for Advanced Technology in a study on the utilization of steelmaking slag in sea areas: a) The effects of composition, size, and thermal history of steelmaking slag on the dissolution behavior of iron and other elements from steelmaking slag alone and steelmaking slag-clay mixtures [7, 92]; b) Mechanism of the growth promotion effect of marine plants caused by eluted elements [93]; and c) Quantitative evaluation of the CO_2 absorption effect of ocean plants [94].

Counteracting Blue Tide

Backfilling and Sand-Coating

To improve sediment environments in closed water areas, various studies have been conducted on the backfilling of deep pits and suppression of hydrogen sulfide generation *via* sand-coating using clay sand [71], fired shells [72], granulated materials of coal ash and cement [73, 74], blast furnace granulated sand [75, 76], massive steelmaking slag [77], *etc*. As a result of demonstration experiments using these materials, it has become clear that any of these materials can be useful under appropriate construction conditions. However, a large amount of landfill material is required to repair sea area depressions and a period of several years or more will be required for landfilling.

Among the various studies using steelmaking slag for the backfilling of deep pits and shallow land construction for counteracting blue tide, the Japan Iron and Steel Federation carried out a verification project on the evaluation of safety and environmental improvement effects based on the use of converter slag in sea areas [95] with assistance from the Ministry of Economy, Trade, and Industry from 2004 to 2007. In this project, the following advantages were identified when mixing steelmaking slag and soft clay: 1) improvement of soil strength, 2) suppression of high-pH water elution from slag, and 3) adsorption of sulfide on steelmaking slag. Miyata *et al.* [32] confirmed that sulfide generation from the bottoms of closed sea areas and oyster farms can be suppressed by using steelmaking slag.

Activity Control of Sulfate-Reducing Bacteria

It is known that sulfate-reducing bacteria attenuate their activity under high-pH conditions. Takahashi *et al.* [75] made pore water more alkaline using blast furnace sand and suppressed the generation of hydrogen sulfide from sediments. However, the possibility that the sulfur contained in blast furnace slag may be a source of hydrogen sulfide should be taken into consideration. Miki *et al.* [96] reported that the elution of alkaline components from steelmaking slag can suppress the activity of sulfate-reducing bacteria. For this method, it is necessary to monitor the environmental impact of alkalizing seafloor sediment.

Control of Hydrogen Sulfide Generation

It has been reported that steelmaking slag suppresses the generation of hydrogen sulfide in addition to being useful as a marine civil engineering material. Ito *et al.* [77] studied pH dependency on sulfide removal to clarify the mechanism of sulfide generation suppression from seafloor mud using steelmaking slag. As shown in Fig. (11), sulfide removal at a pH of 8.2 is 31.7 mgS·g^{-1}, which decreases with an increase in pH, approaching zero at a pH of 9.9. Ito *et al.* also reported that the removal of sulfide occurred in an area rich in calcium silicate. A research group from the Ministry of Economy, Trade, and Industry working on a joint project with the Iron and Steel Federation demonstrated that steelmaking slag suppresses the generation of hydrogen sulfide [95]. Miyata *et al.* [97] installed vessels containing massive steelmaking slag, granite blocks, and concrete blocks on the seafloor. By comparing sulfide concentrations in pore water two months after installation, it was determined that the steelmaking slag had the effect of suppressing hydrogen sulfide generation and that the redox potential for pore water of steelmaking slag is greater than that of granite. Taniguchi *et al.* [98] conducted experiments to reduce the hydrogen sulfide generated from seafloor sediment using steelmaking slag and natural stone. They reported that steelmaking slag has a greater effect in terms of suppressing hydrogen sulfide generation compared to natural stone.

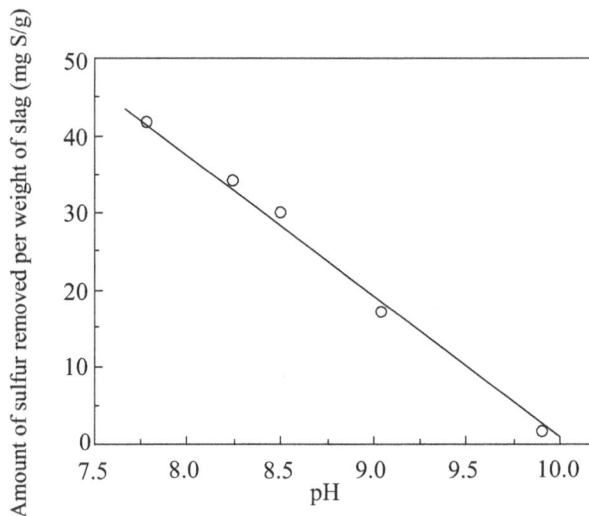

Fig. (11). Dependence of sulfide removal from seawater on pH level [77].

Effects of Iron on Suppression of Hydrogen Sulfide Generation

In a study on hydrogen sulfide generation suppression mechanisms using coal ash granules [99], it was reported that the iron in coal ash granules reacts with sulfide ions to suppress sulfide generation. When the oxides of Fe and Mn in coal ash granules reacted with hydrogen sulfide to form sulfides of Fe and Mn, it was determined that the oxidation of sulfide ions into S^0, $S_2O_3^{2-}$, and SO_4^{2-} is the main mechanism of hydrogen sulfide inhibition when shifting to an oxidized state [100]. However, regarding the involvement of metal oxides, it has also been suggested that manganese oxide contributes to the oxidation of hydrogen sulfide [101]. Schoonen *et al.* [102] reported that Fe ions and sulfide ions react to form an amorphous black precipitate (FeS), which rapidly transforms into S.

Numerous studies have been conducted on the reaction of hydrogen sulfide and iron ions in solution in the world. They can be summarized as follows:

1. In the reaction of Fe and H_2S, amorphous FeS is formed as a precursor, which reacts with S^0 to form FeS_2. Although FeS_2 is thermodynamically stable, FeS_2 is not generated by a direct reaction between Fe and S [98].
2. Sulfide is not accumulated in sediment until iron oxide or iron hydroxide is transformed into iron sulfide [103].
3. Iron oxide and iron hydroxide that react with hydrogen sulfide have significantly different reaction rates based on differences in their mineral phases [104]. Specifically, the reaction rates differ significantly depending on the concentrations of other metal ions and pH [105].
4. Hydrogen sulfide transforms into S^0, $S_2O_3^{2-}$, and SO_4^{2-} *via* inorganic polysulfide under oxidation conditions [100].
5. In tidal flats, iron sulfide concentrations increase in summer based on low oxygen levels. Iron sulfide decreases and iron oxide increases from late fall to spring as the oxidizing atmosphere recovers [106].
6. The accumulation of hydrogen sulfide in sediments depends on the balance between the generation and extinction of hydrogen sulfide through oxidation [107].
7. It is clear that almost no generation of hydrogen sulfide can be observed in natural tidal flats with high iron content in bottom mud. In contrast, tidal flats with low iron content exhibit the generation and accumulation of hydrogen sulfide [106 - 112].

Based on these studies, it can be concluded that hydrogen sulfide generated in a reducing atmosphere during summer will react with iron and become iron sulfide when sufficient iron levels exist in tidal flats and sediments. When the amount of oxygen dissolved in seawater increases to create an oxidizing atmosphere in winter, hydrogen sulfide is considered to transform into polysulfide, S^0, $S_2O_3^{2-}$, and SO_4^{2-}, and iron sulfide is converted to iron oxide. As a result, when there is sufficient iron, the generation of hydrogen sulfide will be suppressed throughout the year [106].

Hayashi *et al.* [85, 113, 114] and Miyata *et al.* [115, 116] conducted a demonstration test using steelmaking slag instead of granulated coal ash as an iron supplier. They successfully demonstrated the detoxification of hydrogen sulfide by the iron oxide in the slag.

SUMMARY

Summaries of previous studies and clarifications of remaining issues concerning (a) the effects of steelmaking slag on seaweed society formation for the prevention of rocky-shore denudation, (b) improvement of deep pit environments by suppressing hydrogen sulfide generation using steelmaking slag, and (c) elucidation of the hydrogen generation suppression mechanism are provided below.

Examination of Seaweed Bed Creation using Steelmaking Slag

1. Iron is essential for the growth of seaweed. In recent years, iron-deficient coastal areas have been observed, which is considered to be one of the main causes of rocky-shore denudation.
2. Iron is dissolved in the form of iron complexes in coastal waters.
3. When a mixture of steelmaking slag and artificial humus soil is added to iron-deficient coastal areas, iron is supplied in the form of humic acid iron and the rocky-shore denudation is remedied. However, cost issues related to the manufacturing, transportation, and construction of such a mixture remain.
4. Mixtures of steelmaking slag and clay are also being researched as a back-up material for shallow-field construction and deep excavation. When clays contain humic acid, the humic acid in the clay and iron in steelmaking slag supply humic acid-iron chelate.
5. Based on these previous researches, it should be possible to elute humic acid-iron chelate from a mixture of steelmaking slag and clay, which are inexpensive and readily available.

Elucidation of Hydrogen Sulfide Generation Control Mechanisms Based on Steelmaking Slag

1. Hydrogen sulfide generation suppression by coal ash granules is a result of iron absorbing and oxidizing hydrogen sulfide.
2. Iron in sediment forms iron sulfide in the summer based on low oxygen levels

and iron oxide in the winter based on high oxygen levels. In nature, iron cyclically contributes to hydrogen sulfide fixation. Therefore, by introducing steelmaking slag into seafloor sediment, there is a possibility that the generation of hydrogen sulfide can be suppressed over a long period of time.

CONSENT FOR PUBLICATION

Not applicable.

CONFLICT OF INTEREST

The author confirms that this chapter contents have no conflict of interest.

ACKNOWLEDGEMENTS

Declare none.

REFERENCES

[1]   Steel Statistics Handbook, Japan Iron and Steel Federation. Tokyo 2019.

[2]   Sakamoto N. University teaching text for steel engineering process, Ironmaking, JFE 21st Century Foundation. Tokyo 2007.

[3]   Nakato A. University teaching text for steel engineering process, Steelmaking, JFE 21st Century Foundation. Tokyo 2007.

[4]   Environmental materials—Iron and steel slag, 4th edition, Japan Iron and Steel Slag Association. Tokyo 2011.

[5]   Kasai Y, Kobatashi M. Admixtures for cement and concrete, Institute of Technology. Tokyo 1993.

[6]   Iron and steel slag statics annual report, Nippon Slag Association. December 2019: FS-169. http://www.Slg.Jp/statistics/report.html

[7]   Inoue R, Ueda S, Ariyama T, Miki T, Hayashi A. Symposium on "New function search for steelmaking slag and application technology development in the marine environment", Osaka, Japan Iron and Steel Institute, 1-8 September 22, 2011.

[8]   Utilization of iron and steel slag for blast furnace cement, 2018 edition) Nippon Slag Association. 2018: FS-162. 2018. http://www.slg.jp/pdf/fs-162.pdf

[9]   Fujii M. Effective Utilization of Iron and Steel Slags. Bull Iron Steel Inst Jpn (Ferrum) 2008; 13: 450-4.

[10]   Ta Y, Tobo H, Watanabe K. Development of manufacturing process of blast furnace slag coarse aggregate with low absorption. JFE Technical Report 2017; 40: 57-61.

[11]   Nakanishi K, Murata Y, Ta Y. Evaluation of concrete applicability for coarse aggregate of dence blast furnace slag with low water absorption ratio. JFE Technical Report 2017; 40: 62-8.

[12]   Goto I. An effective application of converter slag on the reconstruction of farmland struck by Tsunami. Bull Iron Steel Inst Jpn (Ferrum) 2012; 17: 554-9.

[13]   Shinya M. Change in measurement methods of environmental quality standards for water in Japan. Seikatsu Eisei 2011; 55: 46-58.

[14] Partial revision about environmental standard to affect pollution of soil (Aqueduct No. 44), Ministry of the Environment Government of Japan, 28 March. https://www.env.go.jp/water/dojo/law/h130328_44.pdf

[15] Iron and steel slag, ministry of land, infrastructure, transport and tourism. http://www.mlit.go.jp/kowan/recycle/2/07.pdf

[16] Inoue R, Suito H. Fluorine-containing mineral phases in ironmaking and steelmaking slags and their solubilities in aqueous solution. ISIJ Int 2002; 42: 785-93.
[http://dx.doi.org/10.2355/isijinternational.42.785]

[17] Suito H, Inoue R. Behavior of fluorine dissolution from hot metal pretreatment slag. Tetsu To Hagane 2002; 88: 340-6.
[http://dx.doi.org/10.2355/tetsutohagane1955.88.6_340]

[18] Inoue R, Suito H. Immobilization of fluorine dissolved from hot metal pretreatment slag. Tetsu To Hagane 2002; 88: 347-54.
[http://dx.doi.org/10.2355/tetsutohagane1955.88.6_347]

[19] Inoue R, Suito H. Influence of gypsum addition and hydrothermal treatment on dissolution behavior of fluorine in hot metal pretreatment slags. ISIJ Int 2002; 42: 930-7.
[http://dx.doi.org/10.2355/isijinternational.42.930]

[20] Suito H, Inoue R. Dissolution behavior and stabilization of fluorine in secondary refining slags. ISIJ Int 2002; 42: 921-9.
[http://dx.doi.org/10.2355/isijinternational.42.921]

[21] Suito H, Inoue R. Immobilization of fluorine and chromium in steelmaking slag and sludge. Sanyo Techn Rep 2003; 10: 9-18.

[22] Nakamura K, Yoneda M. Estimating grain size and ph effects on steel slag dissolution mechanisms. J Jpn Soc Mat Cycl Waste Manag 2014; 25: 25-35.
[http://dx.doi.org/10.3985/jjsmcwm.25.25]

[23] Soil Environmental Standards (Additional table), Ministry of the Environment Government of Japan. http://www.env.go.jp/kijun/dt1.html

[24] Sato Y, Takasaki Y, Shibayama A, Inoue R. Elution Behavior of Iron from Steelmaking Slag in Fluvial and Marine Environments. The 6th International Congress on the Science and Technology of Steelmaking (ICS2015). The Chinese Society for Metals, 918-920

[25] Inoue R, Sato Y, Takasaki Y, Shibayama A. Immobilization of hexavalent chromium in stainless steelmaking slag. The Tenth International Conference on Molten Slags, Fluxes, and Salts (Molten16). Seattle, USA 2016; 865-72
[http://dx.doi.org/10.1002/9781119333197.ch92]

[26] Inoue Y, Matsunaga H, Watanabe K. Properties of iron and steel slag hydrated matrix exposed in sea area for long period. JFE Tech Rep 2017; 40: 19-24.

[27] Yabuta K, Sugimoto K, Hayashi E, Takahama S, Genba K. Effects of seaweed epiphytic nad fish-gathering due to the construction by steel-making slag. JFE Tech Rep 2017; 40: 13-8.

[28] Yamamoto M, Hamasuna N, Fukushima M, *et al.* Recovery from Barren Ground by Supplying Slug and Humic Substances. J Jpn Inst Energy 2006; 85: 971-8.
[http://dx.doi.org/10.3775/jie.85.971]

[29] Kiso H, Tsutsumi N, Shibuya M, Nakagawa M. Real sea area experiment on growth of kelp *etc.* at the time of marine fertilization–development of seaweed bed construction technology using converter steelmaking slag (1)–. 20th Ocean Engineering Symposium. Tokyo. The Japan: Society of Naval Architects and Ocean Engineers 2008.

[30] Kato T, Aimoto M, Miki O, Nakagawa M. Study on distribution of fe concentration in the sea area in ocean fertilization test using steelmaking slag–development of seaweed bed construction technology

using converter steelmaking slag (2)–. 20th Ocean Engineering Symposium. Tokyo. The Japan: Society of Naval Architects and Ocean Engineers 2008.

[31] Tsutsumi N, Kato T, Honmura T, Nakagawa M. Effects of fertilization raw material components on kelp growth in water tank experiment–development of seaweed bed construction technology using converter steelmaking slag (3)–. 20th Ocean Engineering Symposium. Tokyo. The Japan: Society of Naval Architects and Ocean Engineers 2008.

[32] Miyata Y, Sato Y, Shimizu S, Oyamada K. Environmental Improvement in the Sea Bottom by Steelmaking Slag. JFE Tech Rep 2008; 19: 1-5.

[33] Miyata Y, Tanishiki K, Watanabe K, Yamamoto T, Urabe N. Improvement of marine environment using "marine stone™" made of steelmaking slag. JFE Tech Rep 2017; 40: 7-12.

[34] Demonstration Test Report "Environmental Improvement Technology for Coastal Area by Converter Steelmaking Slag Products", Fiscal 2009 Environmental Technology Demonstration Project "Technology for Improving Water Environment in Closed Sea Area", New Nippon Steel / JFE Steel, March 2010, Ministry of the Environment Government of Japan https://www.env.go.jp/policy/etv/pdf/list/h21/02_h_4%5B1%5D.pdf

[35] Demonstration Test Report " Seaweed bed construction and water quality improvement technology using steelmaking slag", Fiscal 2009 Environmental Technology Demonstration Project "Technology for Improving Water Environment in Closed Sea Area", JFE Steel, March 2010, Ministry of the Environment Government of Japan. https://www.env.go.jp/policy/etv/pdf/list/h21/02_h_3%5B1 %5D.pdf

[36] Tidal flats and seaweed beds, Ministry of the Environment Government of Japan. . http://www.env.go.jp/council/09water/y0917-06/mat03.pdf

[37] About blue carbon, Waterrfront Vitalization and Environment Research Foundation. http://www.wave.or.jp/bluecarbon/bluecarbon_about.pdf

[38] Nellemann C, Corcoran E, Duarte CM, *et al*. Blue Carbon—The role of healthy oceans in binding carbon, A rapid response assessment. United Nations Environment Programme 2009.

[39] The Current State of Algology at the Beginning of the 21st Century The 50th Anniversary Publication of the Japan Society of Algal Science, (Eds., Fujita, D., Hori, T., Ohno, M., Horiguchi, T.) . 2002; p. 102-105.

[40] Gordon RM, Martin JH, Knauer GA. Iron in north-east Pacific waters. Nature 1982; 299: 611-2. [http://dx.doi.org/10.1038/299611a0]

[41] Martin JH, Fitzwater SE. Iron-deficiency limits phytoplankton growth in the Northeast Pacific Subarctic. Nature 1988; 331: 341-3. [http://dx.doi.org/10.1038/331341a0]

[42] Obata H. Analytical chemistry of marine trace metals, emphasizing iron, in seawater. J Oceanogr 2003; 12: 449-60. [http://dx.doi.org/10.5928/kaiyou.12.449]

[43] Tsuda A, Takeda S. Mesoscale iron-enrichment experiments in the North Pacific, an introduction and summary. Jpn J Ecol 2005; 55: 514-9.

[44] Nishioka J. Speciation of iron and the role of iron for phytoplankton growth in the Subarctic North Pacific. J Oceanogr 2006; 15: 19-36.

[45] Boyd PW, Jickells T, Law CS, *et al*. Mesoscale iron enrichment experiments 1993-2005: synthesis and future directions. Science 2007; 315(5812): 612-7. [http://dx.doi.org/10.1126/science.1131669] [PMID: 17272712]

[46] Takeda S. Kaiyo monthly. 1996; 10: p. 66-77.

[47] Komai T. Basis of Humic Substance and Example of Application. Symposium on New function search of steelmaking slag and utilization technology development in marine environment. 49-78.

[48] Arita K, Umiguchi Y, Taniguchi A. Availability of steelmaking slag as a source of essential elements for phytoplankton. Tetsu To Hagane 2003; 89: 415-21.
[http://dx.doi.org/10.2355/tetsutohagane1955.89.4_415]

[49] Suzuki Y, Kuma K, Matsunaga K. Effect of iron on oogonium formation, growth rate and pigment synthesis of *Laminaria japonica* (Phaeophyta). Fish Sci 1994; 60: 373-8.
[http://dx.doi.org/10.2331/fishsci.60.373]

[50] Tsutsumi N, Kato T, Kitano Y. Availability of iron and steel slag for earth-friendly material. Bull Iron Steel Inst Jpn (Ferrum) 2012; 17: 539-49.

[51] Zhang J, Nagahama T, Ohwaki H, Ishibashi Y, Fujita Y, Yamazaki S. Analytical approach to the discoloration of edible laver "*nori*" in the Ariake Sea. Anal Sci 2004; 20(1): 37-43.
[http://dx.doi.org/10.2116/analsci.20.37] [PMID: 14753255]

[52] Zhang J, Sato T, Maruyama R, *et al.* Trace element deficiency, especially iron deficiency, causes the discoloration of sea laver, nori, in the Ariake sea. Bull Soc Sea Water Sci Jpn 2009; 63: 158-66.

[53] Ueki C, Murakami A, Kato T, Saga N, Motomura T. Effects of nutrient deprivation on photosynthetic pigments and ultrastructure of chloroplasts in *Porphyra yezoensis*. Nippon Suisan Gakkaishi 2010; 76: 375-82.
[http://dx.doi.org/10.2331/suisan.76.375]

[54] Kuma K, Nishioka J, Matsunaga K. Controls on iron(III) hydroxide solubility in seawater: The influence of pH and natural organic chelators. Limnol Oceanogr 1996; 41: 396-407.
[http://dx.doi.org/10.4319/lo.1996.41.3.0396]

[55] Suzuki A, Hamazaki A, Aratake H, Saito T. Effect of iron contained in dam bottom mud on growth and proliferation of seaweed *Yatsumatamoku*. J Water and Waste 2012; 54: 136-44.

[56] Nagao S. Material transport from land to ocean through rivers–characteristics of humic substances and complexation ability http://www.chikyu.ac.jp/AMORE/2003.4FS/nagaoAmur.pdf

[57] Hatayama S. Iron prevents global warming. Tokyo: Bungei Shunju 2008.

[58] Pourbaix M, de Zoubov N. Atlas D'Equilibres Electrochimiques, Gauthier-Villars & Cie,. Paris: Editeur-Imprimeur-Libraire 1963; pp. 307-21.

[59] Zhang X, Matsuura H, Tsukihashi F. Dissolution mechanism of various elements into seawater for recycling of steelmaking slag. ISIJ Int 2012; 52: 928-33.
[http://dx.doi.org/10.2355/isijinternational.52.928]

[60] Rose AL, Waite TD. Kinetics of iron complexation by dissolved natural organic matter in coastal waters. Mar Chem 2003; 84: 85-103.
[http://dx.doi.org/10.1016/S0304-4203(03)00113-0]

[61] Suzuki Y, Kuma K, Kudo I, Matsunaga K. Iron requirement of the brown macroalgae Laminaria japonica, Undaria pinnatifida (Phaeophyta) and the crustose coralline alga Lithophyllum yessoense (Rhodophyta), and their competition in the northern Japan Sea. Phycologia 1995; 34: 201-5.
[http://dx.doi.org/10.2216/i0031-8884-34-3-201.1]

[62] Marumo K, Yokota M. Review on aoshio and biological effects of hydrogen sulfide. Rep Mar Ecol Res Inst 2012; 15: 23-40.

[63] Blue tide, Fisheries Agency / Japan Fisheries Resource Conservation Association. http://www.fish-jfrca.jp/02/pdf/pamphlet/013.pdf

[64] Biology of Tokyo Bay. Numata M, Furoda T, Eds. Tsukiji Shokan, Tokyo 1997.

[65] Remaining quantity of depression after dredging in Tokyo Bay, first document of 2003 Tokyo Bay back sea area environment creation project, Chiba construction office of Kanto Regional Development Bureau of Ministry of Land, Infrastructure, Transport and Tourism. 2003.

[66] Naito R, Nakamura Y, Imamura H, Sato M. Under construction effects of geomorphological

restoration of subaqueous borrow pits and extraction of related research & development tasks. Proceedings of civil engineering in the ocean, 2016; 22: pp. 649-54.

[67]   Sasaki J, Isobe M, Watanabe A, Gomyo M. Study on the occurrence scale of blue tide in Tokyo Bay. Coast Eng J 1996; 43: 1111-5.

[68]   Iimura A, Kobayashi H, Ogura H. Occurrence of blue tide in Tokyo Bay, Annual report of Chiba Prefectural Environmental Research Center 2010.

[69]   Fundamental direction of the port environment policy in the future (report), Transport Policy Council, Ministry of Land, Infrastructure, Transport and Tourism 2005.

[70]   Asaoka S, Yamamoto T, Takahashi Y, Yamamoto H, Kim KH, Orimoto K. Development of an On-site Simplified Determination Method for Hydrogen Sulfide in Marine Sediment Pore Water Using a Shipboard Ion Electrode with Consideration of Hydrogen Sulfide Oxidation Rate. Interdisciplinary Studies on Environmental Chemistry-Environmental Pollution and Ecotoxicology 2012; 6: 345-52.

[71]   Discussion Material of the 2nd Ise Bay Renewal Meeting, Mikawa Section, Chubu Regional Development Bureau, October . 2011.

[72]   Terai A, Yamamoto F, Oohashi T, Toyohara H. Development of efficient precipitator composed of baked shell. Nippon Suisan Gakkaishi 2011; 77: 871-5.
       [http://dx.doi.org/10.2331/suisan.77.871]

[73]   Nakahara M, Hiraoka K, Yamamoto T, Ueshima H. Improvement of coastal marine sediments by capping with granulated coal ash. Mizu Kankyo Gakkaishi 2012; 35: 159-66.
       [http://dx.doi.org/10.2965/jswe.35.159]

[74]   Tamai K, Nishino H, Izuro Y, Hibino T, Suto A, Nishidoi M. Empirical study of sustainability for sediment environmental improvement through covering on seabed with fly ash beans. J Jpn Soc Civil Engs, Ser B3 (Ocean Engineering), 2012; 68: I_1145-50.

[75]   Takahashi T, Yabuta K. New Application of Iron and Steelmaking Slag. NKK Technical Report 2002; 178: 43-8.

[76]   Asaoka S, Yamamoto T. Blast furnace slag can effectively remediate coastal marine sediments affected by organic enrichment. Mar Pollut Bull 2010; 60(4): 573-8.
       [http://dx.doi.org/10.1016/j.marpolbul.2009.11.007] [PMID: 20003992]

[77]   Ito K, Nishijima W, Shoto E, Okada M. Control of sulfide and ammonium release from coastal bottom sediments by converter slag. Mizu Kankyo Gakkaishi 1997; 20: 670-3.
       [http://dx.doi.org/10.2965/jswe.20.670]

[78]   Miki T, Shitogiden K, Samada Y, Nagasaka T, Hino M. Consideration of dissolution behavior of elements in steelmaking slag based on their stability diagram in seawater. Tetsu To Hagane 2003; 89: 388-92.
       [http://dx.doi.org/10.2355/tetsutohagane1955.89.4_388]

[79]   Matsunaga H, Takagi M, Kogiku F. Fundamental properties of blocks which set steel slag by hydration reaction and biofouling build-up properties on exposure to marine environment. Tetsu To Hagane 2003; 89: 454-60.
       [http://dx.doi.org/10.2355/tetsutohagane1955.89.4_454]

[80]   Yamamoto T, Suzuki M, Oh SJ, Matsuda O. Release of phosphorus and silicon from steelmaking slag and their effects on growth of natural phytoplankton assemblages. Tetsu To Hagane 2003; 89: 482-8.
       [http://dx.doi.org/10.2355/tetsutohagane1955.89.4_482]

[81]   Yamamoto M, Fukushima M, Liu D. Effect of humic substances on iron elusion in the method of restoration of seaweed beds with steelmaking slag. Tetsu To Hagane 2011; 97: 159-64.
       [http://dx.doi.org/10.2355/tetsutohagane.97.159]

[82]   Matsunaga K. When the forest disappears, the sea dies –Ecology links land to sea. 2nd ed. Tokyo: Kodansha 2010; p. 104.

[83]   Zhang X, Atsumi H, Matsuura H, Tsukihashi F. Influence of gluconic acid on dissolution of si, p and fe from steelmaking slag with different composition into seawater. ISIJ Int 2014; 54: 1443-9.
[http://dx.doi.org/10.2355/isijinternational.54.1443]

[84]   Zhang X, Matsuura H, Tsukihashi F. Enhancement of the dissolution of nutrient elements from steelmaking slag into seawater by gluconic acid. J Sustain Metal 2015; 1: 134-43.
[http://dx.doi.org/10.1007/s40831-015-0013-9]

[85]   Hayashi A, Tozawa H, Shimada K, *et al.* Effects of the seaweed bed construction using the mixture of steelmaking slag and dredged soil on the growth of seaweeds. ISIJ Int 2011; 51: 1919-28.
[http://dx.doi.org/10.2355/isijinternational.51.1919]

[86]   Ishiwatari R, Yonebayashi K, Miyajima T. Humic Substances in the environment –characteristics and research methods–, japanese humic substances society. Tokyo: Sankyo Publishing 2008.

[87]   Honda H, Tsuchida T. Application of improved dredged soil with steelmaking slag to artificial tidal flat. JFE Technical Report 2017; 40: 31-7.

[88]   Zhang X, Matsuura H, Tsukihashi F. Dissolution mechanisms of steelmaking slag–dredged soil mixture into seawater. J Sust Metal 2016; 2: 123-32.
[http://dx.doi.org/10.1007/s40831-015-0040-6]

[89]   Yazawa Y, Hamada A, Yoshida T, Sasaki K, Fujiyama R, Takeda H. The link between material transport and ecosystem by fulvic-fe complex in the Boso Peninsula 1: Interrelationship between land-use and nutrients in the Obitsu River Basin. Bull Soc Sea Water Sci Jpn 2011; 65: 223-38.

[90]   Sakai Y, Toyoda K, Horiie M. Journal of the Japanese National Committee on Large Dams 2015; 232: 110-6.

[91]   Yamamoto T, Hoshika S, Noguchi H, Oyamada K, Moriwaki T. Proceedings of the 4th Symposium on Marine Environment, Biology and Coastal Environment Restoration Technology. 47.

[92]   Zhang X, Matsuura H, Tsukihashi F. Dissolution behavior of steel making slag components into seawater and the role of organic acid. Symposium on "New function search for steelmaking slag and application technology development in the marine environment. Osaka, Japan Iron and Steel Institute, 7-13

[93]   Anpo M, Yosimura W, Nagasaka S, Yoshimura E. Search for marker genes in iron deficiency state of Susabinori (*P. Yezoensis*). Symposium on "New function search for steelmaking slag and application technology development in the marine environment. Osaka, Japan Iron and Steel Institute, 2011, 14-15

[94]   Daigo I, Kurimoto S, Matsuno T, Adachi Y. Possibility of Utilization of Steelmaking Slag in Ocean and evaluation of carbon absorption effect of algae. Symposium on "New function search for steelmaking slag and application technology development in the marine environment. Osaka, Japan Iron and Steel Institute, 2011, 16-21

[95]   Guide to using converter steelmaking slag in sea area, Japan Iron and Steel Federation. 2008.

[96]   Miki O, Ueki C, Kato T. 45th Annual Meeting of Japan Society on Water Environment. Hokkaido. 2011; pp. A-09-1.

[97]   Miyata Y, Takahashi T, Yabuta K, Numata T, Sasaki Y. Proceedings of the 4th Symposium on Marine Environment, Biology and Coastal Environment Restoration Technology. July 2005, 121

[98]   Taniguchi A, Suzuki Y, Nishino Y, Sato T, Sikato Y. Proceedings of the 10th Symposium on Marine Environment, Biological and Coastal Environment Restoration Technologies. September 2011, 33

[99]   Yamamoto T, Asaoka S. Symp the Environment Committee of the Japanese Society of Fisheries Science. 2011.

[100]   Asaoka S, Yamamoto T, Hayakawa S. Removal of Hydrogen Sulfide Using Granulated Coal Ash. J J Soc Water Environ 2009; 32: 363-8.
[http://dx.doi.org/10.2965/jswe.32.363]

[101] Asaoka S, Hayakawa S, Kim KH, Takeda K, Katayama M, Yamamoto T. Combined adsorption and oxidation mechanisms of hydrogen sulfide on granulated coal ash. J Coll Interface Sci 2012; 377(1): 284-90.
[http://dx.doi.org/10.1016/j.jcis.2012.03.023] [PMID: 22487226]

[102] Schoonen MAA, Barnes HL. Reactions forming pyrite and marcasite from solution: I. Nucleation of FeS_2 below 100°C. Geochim Cosmochim Acta 1991; 55: 1495-504.
[http://dx.doi.org/10.1016/0016-7037(91)90122-L]

[103] Canfield DE, Raiswell R, Bottrell S. The reactivity of sedimentary iron minerals toward sulfide. Am J Sci 1992; 292: 659-83.
[http://dx.doi.org/10.2475/ajs.292.9.659]

[104] Yao W, Millero FJ. Oxidation of hydrogen sulfide by hydrous Fe(III) oxides in seawater. Mar Chem 1996; 52: 1-16.
[http://dx.doi.org/10.1016/0304-4203(95)00072-0]

[105] Yao W, Millero FJ. The rate of sulfide oxidation by δMnO_2 in seawater. Geochim Cosmochim Acta 1993; 57: 3359-65.
[http://dx.doi.org/10.1016/0016-7037(93)90544-7]

[106] Rozan TF, Taillefert M, Trouwborst RE, *et al.* Iron-sulfur-phosphorus cycling in the sediments of a shallow coastal bay: Implications for sediment nutrient release and benthic macroalgal blooms. Limnol Oceanogr 2002; 47: 1346.
[http://dx.doi.org/10.4319/lo.2002.47.5.1346]

[107] Canfield DE. Reactive iron in marine sediments. Geochim Cosmochim Acta 1989; 53: 619-32.
[http://dx.doi.org/10.1016/0016-7037(89)90005-7] [PMID: 11539783]

[108] Thamdrup B, Fossing H, Jorgensen BB. Manganese, iron and sulfur cycling in a coastal marine sediment, Aarhus bay, Denmark. Geochim Cosmochim Acta 1994; 58: 5115-29.
[http://dx.doi.org/10.1016/0016-7037(94)90298-4]

[109] Giordani G, Bartoli M, Cattadori M, Viaroli P. Sulphide release from anoxic sediments in relation to iron availability and organic matter recalcitrance and its effects on inorganic phosphorus recycling. Hydrobiologia 1996; 329: 211-22.
[http://dx.doi.org/10.1007/BF00034559]

[110] Heijs SK, Jonkers HM, van Gemerden H, Schaub BEM, Stal LJ. The buffering capacity towards free sulphide in sediments of a coastal lagoon (bassin d'arcachon, France)—the relative importance of chemical and biological processes. Estuar Coast Shelf Sci 1999; 49: 21-35.
[http://dx.doi.org/10.1006/ecss.1999.0482]

[111] Kanaya G, Kikuchi E. Relationships between sediment chemical buffering capacity and H_2S accumulation: comparative study in two temperate estuarine brackish lagoons. Hydrobiologia 2004; 528: 187-99.
[http://dx.doi.org/10.1007/s10750-004-2342-8]

[112] Kanaya G, Kikuchi E. Influences of anthropogenic eutrophication on estuarine benthic ecosystems. Chikyu Kankyo 2011; 16: 33-44.

[113] Hayashi A, Watanabe T, Kaneko R, *et al.* Decrease of sulfide in enclosed coastal sea by using steelmaking slag. ISIJ Int 2013; 53: 1894-901.
[http://dx.doi.org/10.2355/isijinternational.53.1894]

[114] Hayashi A, Asaoka S, Watanabe T, *et al.* Mechanism of suppression of sulfide ion in seawater using steelmaking slag. ISIJ Int 2014; 54: 1741-8.
[http://dx.doi.org/10.2355/isijinternational.54.1741]

[115] Miyata Y, Hayashi A, Kuwayama M, Yamamoto T, Urabe N. Reduction test of hydrogen sulfide in silty sediment of fukuyama inner harbor using steelmaking slag. ISIJ Int 2015; 55: 2686-93.
[http://dx.doi.org/10.2355/isijinternational.ISIJINT-2015-114]

[116]   Miyata Y, Hayashi A, Kuwayama M, Yamamoto T, Tanishiki K, Urabe N. A field experiment of sulfide reduction in silty sediment using steel-making slag. ISIJ Int 2016; 56: 2100-6.
[http://dx.doi.org/10.2355/isijinternational.ISIJINT-2015-622]

CHAPTER 20

Elution of Rehabilitation Material Made from Steelmaking Slag for Coastal Environments

Hiroyuki Matsuura[*]

Department of Materials Engineering, The University of Tokyo, Tokyo, Japan

Abstract: Sea desertification has become a serious problem in sea coasts worldwide. Owing to Japan's long sea coast, fishery is one of its key industries; thus, sea desertification is a crucial problem for the Japanese society. The positive effect of steelmaking slag containing humic substances or soil in sea coast on the rehabilitation of damaged coastal environments has been phenomenologically confirmed. This study has focused on elucidating this phenomenon, and it has been clarified that the coexistence of organic substances and steelmaking slag supported the stable dissolution of nutrients from steelmaking slag by chelating the inorganic ions.

Keywords: Chelation reaction, Nutrient dissolution, Organic acid, Sea desertification.

INTRODUCTION

The widely spread modern-integrated steel production process consists of ironmaking using a blast furnace and steelmaking using a basic oxygen furnace. Using such pyrometallurgical processes, various by-products are obtained, including ironmaking or steelmaking slag. For each tonne of hot metal and crude steel, approximately 300 kg ironmaking and 150 kg steelmaking slag are produced, respectively [1]. Steelmaking slag is generated *via* various oxidation processes to remove elements that are harmful for steel, particularly P, by using highly basic components. Steelmaking slag is rich in FeO, CaO, SiO_2, and P_2O_5, and therefore, it has been considered a good candidate material to continuously supply iron to seawater for the rehabilitation of the severely damaged coastal environment, the so-called sea desertification area.

Dissolved Fe(II) ions are immediately oxidized into Fe(III) ions in seawater, as illustrated in Fig. (**1**) [2, 3]. Since the solubility of ferric hydroxide in seawater is

[*] **Corresponding author Hiroyuki Matsuura:** Department of Materials Engineering, The University of Tokyo, Tokyo, Japan; Tel; +81-3-5841-7146; Fax; +81-3-5841-7156; E-mail: matsuura@material.t.u-tokyo.ac.jp

Toshiyuki Takahashi (Ed.)

quite low, the effect of adding slag to seawater is limited. To overcome this drawback, various organic materials have been mixed with steelmaking slag.

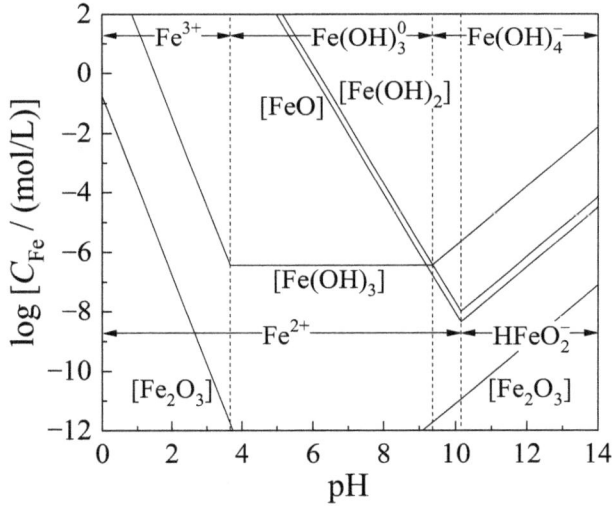

Fig. (1). Solubility diagram of Fe.

Fisherpersons have been familiar with the phenomenological effect of organic materials on the promotion of seaweed growth; however, the detailed mechanism of this process has not been clarified until recently. Various experiments at various seacoast sites have revealed that the addition of different organic substances to steelmaking slag was effective to efficiently supply iron as soluble ions *via* the chelation reactions between the Fe(II) and/or Fe(III) ions and organic acids [4 - 6].

The author recently conducted laboratory-scale dissolution experiments to clarify the effect of the dredged sea bottom soil, which is used to reconstruct the seaweed bed by mixing it with steelmaking slag and modifying its physical strength and chemical stability [7].

To determine the influence of slag composition on their dissolution behavior, several artificial slags were prepared, and their compositions are summarized in Table **1**. Artificial slags were produced by melting mixtures of prescribed percentages of reagent grade FeO, CaO, SiO_2, $CaH(PO_4) \cdot 2H_2O$, and Al_2O_3 in steel crucibles at 1723 K under Ar atmosphere for 60 min followed by quenching. Subsequently, the obtained slag lumps were pulverized to obtain particles smaller than 150 μm.

Table 1. Artificial slags compositions.

Slag	FeO (Mass%)	CaO (Mass%)	SiO$_2$ (Mass%)	Al$_2$O$_3$ (Mass%)	P$_2$O$_5$ (Mass%)	%CaO/%SiO$_2$ (-)
A	10	55.0	27.5	5	2.5	2.0
B	20	48.3	24.2	5	2.5	2.0
C	30	41.7	20.8	5	2.5	2.0
D	30	45.0	22.5	-	2.5	2.0

Sea bottom soil dredged from the inner harbor of the Seto Inland Sea, Japan, was used as an additive, and its composition is listed in Table **2**. The soil consisted of very fine particles and mud, and its water content was approximately 300%. The X-ray diffraction analysis of the dried soil indicated the presence of SiO$_2$ and NaCl phases.

Table 2. Composition of dredged sea bottom soil.

Constituent	Content (Mass%)
Water	78.5
Organic substances	1.1
Ca	0.2
Mg	0.3
Fe	0.6
Al	0.3
P	0.1
SiO$_2$	7.4

The slag-soil mixture was prepared by mixing slag powder and as-received soil (mass ratio of 1:8) and subsequently aging the mixture in a 250 cm^3 polyethylene bottle. The mixture was hermetically sealed and cured for 24 h.

Table **3** summarizes the compositions of the artificial seawater samples. While two types of artificial sweater were used, their main components and compositions were similar. The artificial seawater was equilibrated using air *via* aeration for 24 h.

Dissolution experiments were conducted by adding 100 cm^3 artificial seawater into a polyethylene bottle containing the slag-soil mixture and shaking the bottle using a 20 ± 5 mm wide shaking machine at the speed of 160 cycles per min at 298 K. The pH and oxidation-reduction potential of the solution were measured immediately after shaking, followed by filtering the solution using a 0.45 μm pore size membrane filter.

Table 3. Composition of artificial seawater samples (mg/L).

Constituent	Tomita Pharmaceutical MARINE ART SF-1	Yashima Pure Chemicals AQUAMARINE
NaCl	22100	24534.0
$MgCl_2 \cdot 6H_2O$	9900	11111.5
Na_2SO_4	3900	4094.0
$CaCl_2 \cdot 2H_2O$	1500	1535.0
KCl	610	694.5
$NaHCO_3$	190	201.0
KBr	96	100.5
$Na_2B_4O_7 \cdot 10H_2O$	78	-
$SrCl_2 \cdot 6H_2O$	-	42.5
H_3BO_3	-	27.0
$SrCl_2$	13	-
NaF	3	3.0
LiCl	1	-
Others	0.12	-

The concentrations of all filtrates; Ca, Mg, Si, and Al of were measured using inductively coupled plasma-optical emission spectrometry (ICP-OES), while the concentration of Fe was analyzed using inductively coupled plasma mass spectrometry (ICP-MS) or 1,10-phenanthroline spectrometry. For very low concentrations of Fe, the extraction of Fe and desalination of the sample solutions were performed using solid-phase extraction chelating cartridges before analyzing the sample using ICP-MS. The concentration of P was analyzed using ICP-OES or molybdenum blue spectrophotometry.

Fig. (**2**) illustrates the variations in seawater pH with shaking time. The addition of slag to seawater resulted in the significant increase in pH immediately after the beginning of shaking. This phenomenon was attributed to the excessive dissolution of CaO from slag as expressed by reaction (1),

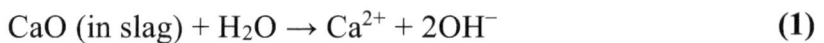

$$\text{CaO (in slag)} + H_2O \rightarrow Ca^{2+} + 2OH^- \tag{1}$$

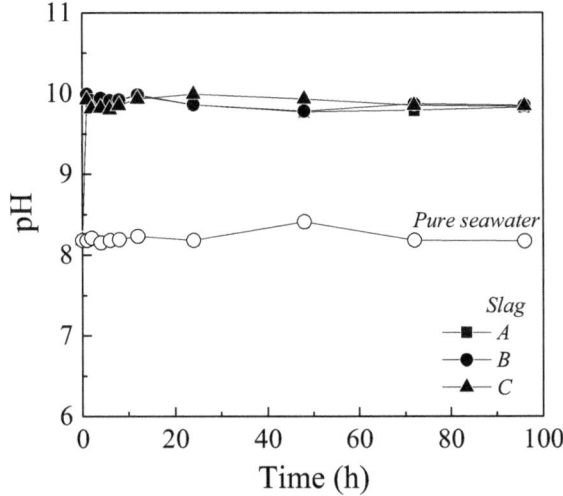

Fig. (2). Changes in pH of shaking solution with time during slag dissolution.

Based on the solubility diagrams of Ca and Mg compounds in seawater presented in Fig. (**3**), the precipitation reaction (2) occurred, which provided the buffer effect required to maintain the constant pH of seawater [3]. This $Mg(OH)_2$ precipitation reaction should be avoided because seawater becomes cloudy, which might be misinterpreted as the dissolution of toxic substances from rehabilitation materials.

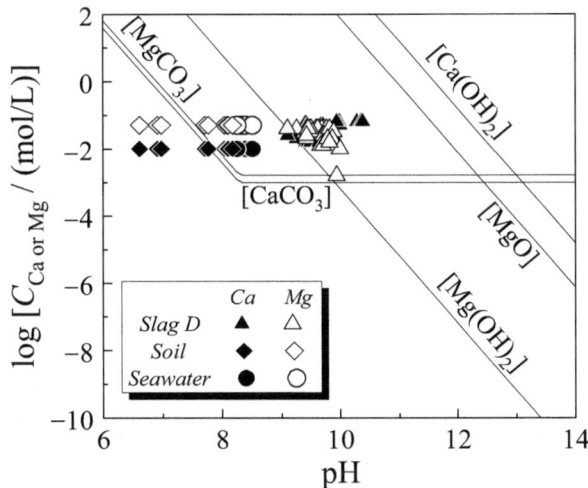

Fig. (3). Solubility diagram of Ca and Mg in seawater with experimental results.

$$CaO \text{ (in slag)} + Mg^{2+} + H_2O \rightarrow Mg(OH)_2 \text{ (s)} + Ca^{2+} \qquad (2)$$

To prevent the intensive dissolution of CaO, steelmaking slag is normally stabilized to low solubility $CaCO_3$ by the hydration or carbonation of CaO. The simultaneous stabilization of muddy sea bottom soil and steelmaking slag by kneading and curing has also been reported. The dissolution of steelmaking slag after kneading and curing was measured to clarify the influence of the mixed muddy sea bottom soil on the dissolution of major elements in steelmaking slag.

Fig. (4) presents the pH variations during the shaking experiments of the slag-soil mixture in seawater. The addition of dredged soil caused the pH of the solution to decrease; however, the inhibitory effect for the increase in pH was not significant. The changes in pH could be affected by the slag and soil chemistry, slag/soil ratio, and curing time. As depicted in Fig. (4), the influence of slag chemistry was insignificant. Fig. (5) illustrates the pH changes measured using slag-soil mixtures exhibiting various curing times. The addition of dredged soil slightly lowered the pH, while the curing time did not affect the pH.

The dissolution of the major elements in the slag or slag-soil mixtures presented considerable different trends. Fig. (6) illustrates the comparison of Ca dissolution from slag and slag-soil mixture. While mixing the slag with soil only slightly lowered the pH, it significantly increased the dissolution of Ca. CaO in steelmaking slag is commonly tightly bound to acidic components, mainly SiO_2 and Al_2O_3. However, mixing and curing the slag with soil would break those bonds owing to the presence of water and hydrated SiO_2 in soil. As illustrated in Fig. (7), the dissolution of Si and Al was also increased by the addition of soil.

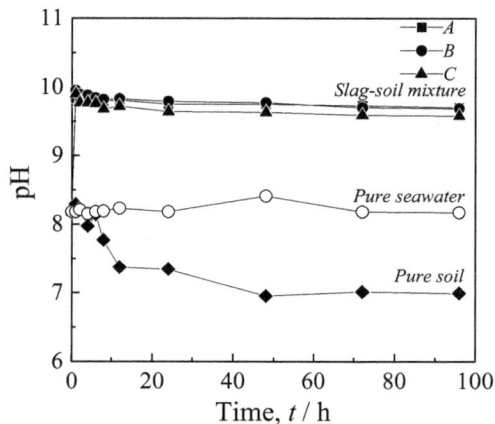

Fig. (4). Change in pH of shaking solution with time during slag-soil mixture dissolution.

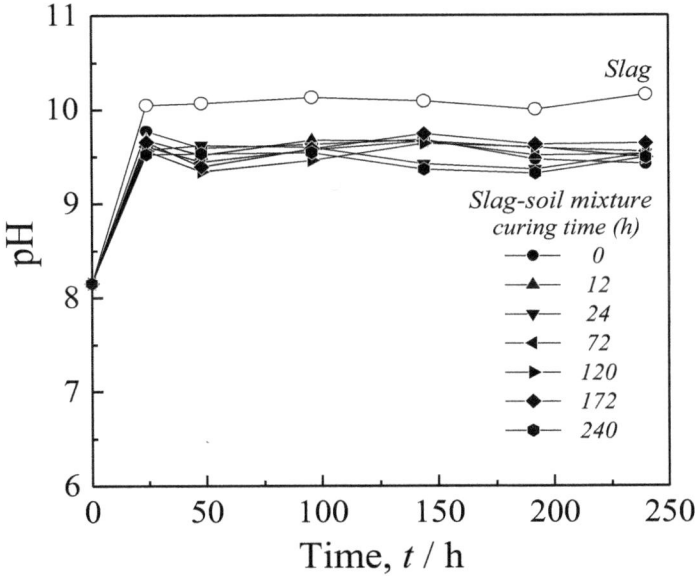

Fig. (5). Effect of curing time of slag (C)-soil mixture on pH changes of shaking solution with time.

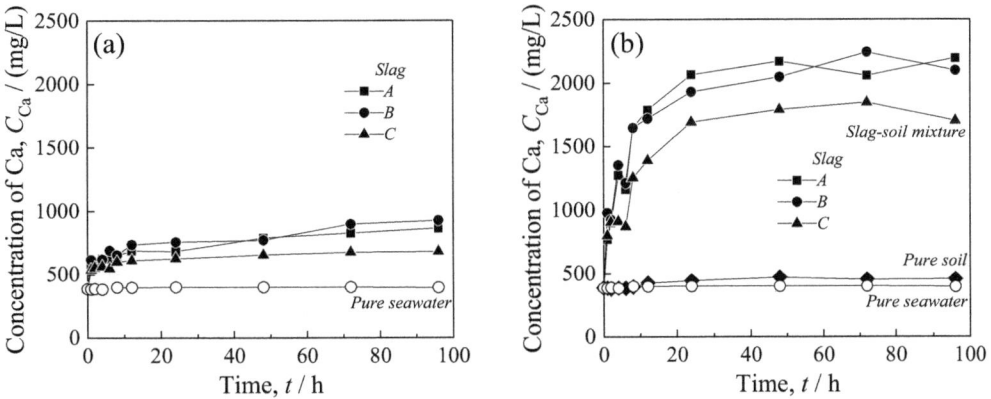

Fig. (6). Comparison of changes in concentration of Ca of shaking solution with time using (**a**) slag or (**b**) slag-soil mixture.

The concentration of Fe in solution after the shaking experiment was still minimal, as illustrated in Fig. (**8**). While the concentration of Fe in seawater is several μg/L on average, the concentration of Fe after shaking ranged between 5 and 25 μg/L. Although the slightly increased dissolution was confirmed, its effect was insignificant.

From above results, further investigations were conducted to quantitatively elucidate the effect of organic acids on the dissolution of Fe from steelmaking slag [8, 9]. Gluconic acid, which is a metabolized product of glucose and is present in various natural products, was selected in the present study. Since gluconic acid is well known to act as chelator in basic aqueous solutions, and many of its thermodynamic properties for chelation reactions have been reported in the literature, it was used in the present study as a model organic acid. Shaking experiments were conducted in the manner described above, using seawater containing 0.12 to 0.50 g/L gluconic acid. The initial pH of the seawater was adjusted to be approximately 8.2 by neutralizing it using NaOH solution.

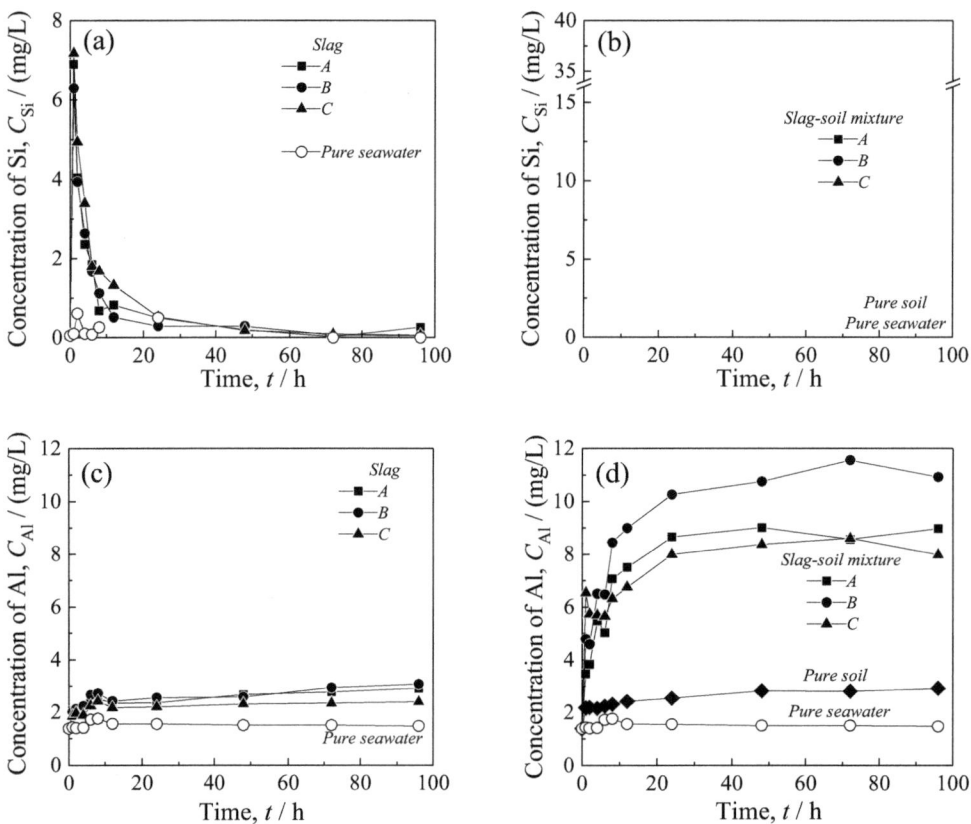

Fig. (7). Effect of soil mixing on dissolution of (a) and (b) Si and (c) and (d) Al in seawater.

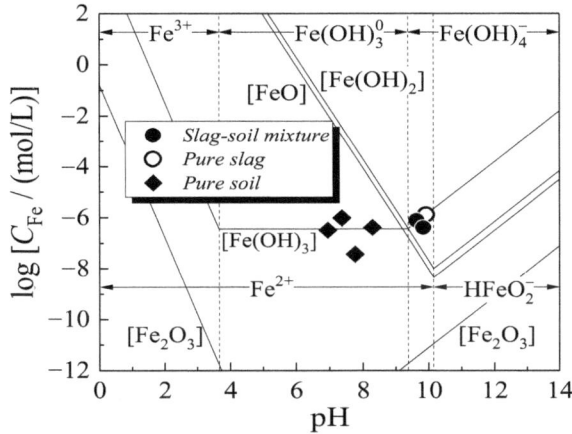

Fig. (8). Effect of soil mixing on dissolution of Fe from slags (**A**)–(**C**) in seawater.

The changes in the pH and concentration of Ca of the shaking solutions with time are illustrated in Fig. (**9**). The concentration of Ca increased as the gluconic acid content of seawater increased, while the pH remained constant. As presented in Fig. (**10**), the concentrations of Ca and Mg slightly increased when gluconic acid was added, compared to the experiment where no gluconic acid was used, and this was attributed to chelation reactions (3) and (4) for Ca and Mg, respectively. It was considered that reaction (2) reached equilibrium in solution regardless of the presence of gluconic acid. Since chelated Ca and Mg ions did not affect reaction (2), the total Ca and Mg concentrations increased as gluconic acid was added, as depicted in Fig. (**10**), where GH_4^- is the gluconate anion.

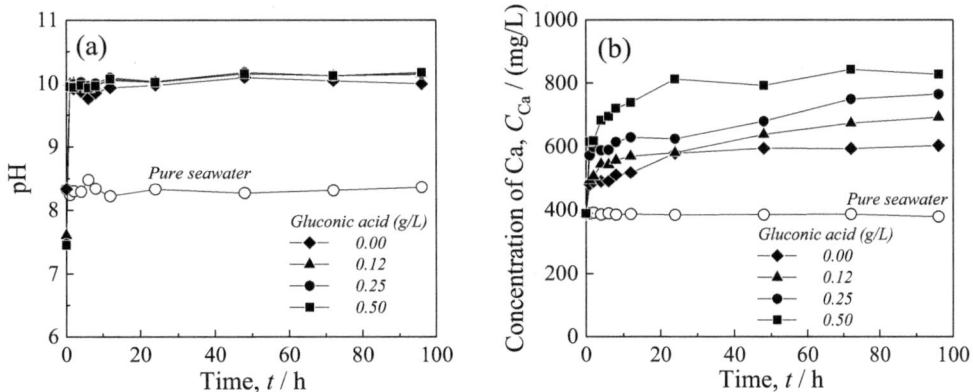

Fig. (9). Effect of gluconic acid addition on changes in (**a**) pH and (**b**) concentration of Ca in seawater with time (slag D).

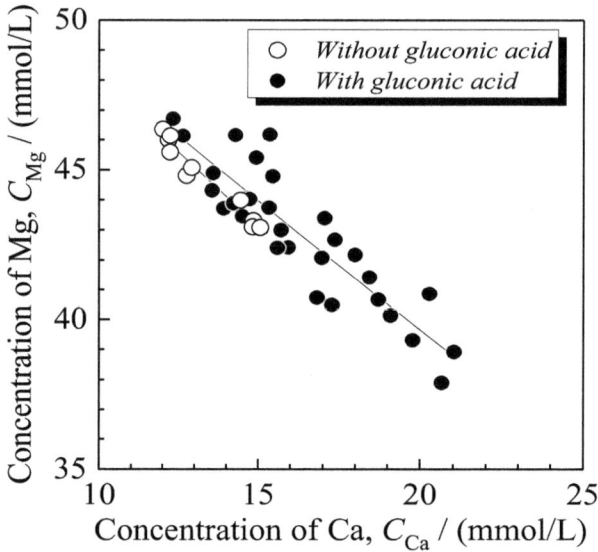

Fig. (10). Relationship between concentrations of Ca and Mg in seawater with or without gluconic acid (Slag D).

$$Ca^{2+} + GH_4^- \rightarrow Ca(GH_4)^+ \qquad \log K_{(3)} = 1.22 \ (298 \ K) \ [10] \qquad \textbf{(3)}$$

$$Mg^{2+} + GH_4^- \rightarrow Mg(GH_4)^+ \qquad \log K_{(4)} = 0.70 \ (298 \ K) \ [10] \qquad \textbf{(4)}$$

Dissolution of Si was also increased by the addition of gluconic acid, as illustrated in Fig. (11). Since no chelation reactions between Si-bearing ions and gluconic acid have been reported, the effect of gluconic acid would be explained by the chelation of eluted Ca and prevention of the precipitation of the Ca-Si-O gel [11]. However, chelation reactions between Si-bearing ions and gluconic acid could take place, and therefore further studies would be necessary to elucidate thermochemical properties.

Fig. (12) illustrates the changes in the concentration of Fe with time. The concentration of Fe in seawater is several µg/L. The addition of gluconic acid considerably increased the dissolution of Fe into seawater. Several chelation reactions for the Fe(II) and Fe(III) ions have been reported, as expressed by reactions (5)-(9). The formation of chelated Fe ions stabilized the dissolved Fe and the concentration of Fe exceeded its solubility limit illustrated in Fig. (1). The efficient supply of nutrient elements from steelmaking slag would be maximized by the effective and continuous supply of organic acids to steelmaking slag.

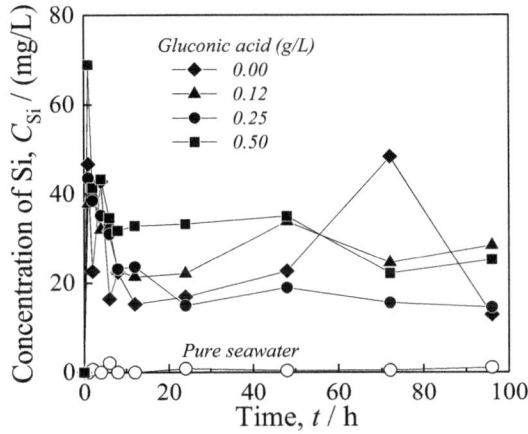

Fig. (11). Effect of gluconic acid addition on changes in concentration of Si in seawater with time (Slag D).

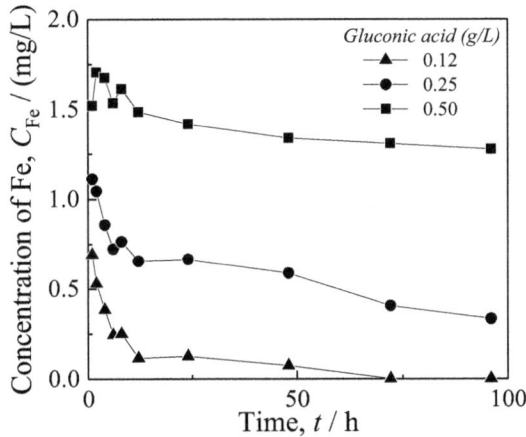

Fig. (12). Effect of gluconic acid addition on changes in concentration of Fe in seawater with time (Slag D).

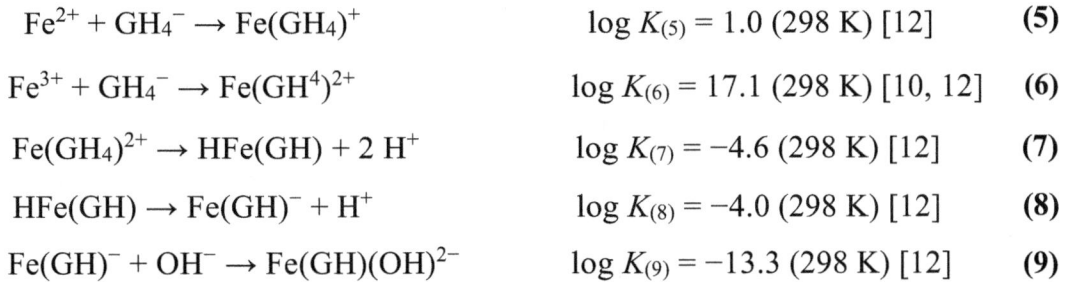

$$Fe^{2+} + GH_4^- \rightarrow Fe(GH_4)^+ \qquad \log K_{(5)} = 1.0 \ (298 \ K) \ [12] \qquad (5)$$

$$Fe^{3+} + GH_4^- \rightarrow Fe(GH^4)^{2+} \qquad \log K_{(6)} = 17.1 \ (298 \ K) \ [10, 12] \qquad (6)$$

$$Fe(GH_4)^{2+} \rightarrow HFe(GH) + 2 \ H^+ \qquad \log K_{(7)} = -4.6 \ (298 \ K) \ [12] \qquad (7)$$

$$HFe(GH) \rightarrow Fe(GH)^- + H^+ \qquad \log K_{(8)} = -4.0 \ (298 \ K) \ [12] \qquad (8)$$

$$Fe(GH)^- + OH^- \rightarrow Fe(GH)(OH)^{2-} \qquad \log K_{(9)} = -13.3 \ (298 \ K) \ [12] \qquad (9)$$

CONSENT FOR PUBLICATION

Not applicable.

CONFLICT OF INTEREST

The author confirms that this chapter contents have no conflict of interest.

ACKNOWLEDGEMENTS

Declare none.

REFERENCES

[1]   Nippon Slag AssoNippon Slag Association. Annual Statistical Report of Iron and Steelmaking Slag FY. 2018.

[2]   Futatsuka T, Shitogiden K, Miki T, Nagasaka T, Hino M. Dissolution behavior of nutrition elements from steelmaking slag into seawater. ISIJ Int 2004; 44: 753-61.
        [http://dx.doi.org/10.2355/isijinternational.44.753]

[3]   Zhang X, Matsuura H, Tsukihashi F. Dissolution mechanism of various elements into seawater for recycling of steelmaking slag. ISIJ Int 2012; 52: 928-33.
        [http://dx.doi.org/10.2355/isijinternational.52.928]

[4]   Yamamoto M, Hamasuna N, Fukushima M, *et al.* Recovery from barren ground by supplying slug and humic substances. J Jpn Inst Energy 2006; 85: 971-8.
        [http://dx.doi.org/10.3775/jie.85.971]

[5]   Yamamoto M, Fukushima M, Liu D. Effect of humic substances on iron elusion in the method of restoration of seaweed beds with steelmaking slag. Tetsu To Hagane 2011; 97: 159-64.
        [http://dx.doi.org/10.2355/tetsutohagane.97.159]

[6]   Hayashi A, Tozawa H, Shimada K, *et al.* Effects of the seaweed bed construction using the mixture of steelmaking slag and dredged soil on the growth of seaweeds. ISIJ Int 2011; 51: 1919-28.
        [http://dx.doi.org/10.2355/isijinternational.51.1919]

[7]   Zhang X. Matsuura, H., Tsukihashi, F. Dissolution mechanisms of steelmaking slag-dredged soil mixture into seawater. J Sustain Metall 2016; 2: 123-32.
        [http://dx.doi.org/10.1007/s40831-015-0040-6]

[8]   Zhang X, Atsumi H, Matsuura H, Tsukihashi F. Influence of gluconic acid on dissolution of si, p and fe from steelmaking slag with different composition into seawater. ISIJ Int 2014; 54: 1443-9.
        [http://dx.doi.org/10.2355/isijinternational.54.1443]

[9]   Zhang X. Matsuura, H., Tsukihashi, F. Enhancement of the dissolution of nutrient elements from steelmaking slag into seawater by gluconic acid. J Sustain Metall 2015; 1: 134-43.
        [http://dx.doi.org/10.1007/s40831-015-0013-9]

[10]   Sawyer DT. Metal-gluconate complexes. Chem Rev 1964; 64: 633-43.
        [http://dx.doi.org/10.1021/cr60232a003]

[11]   Suito H, Inoue R. Dissolution behavior and stabilization of fluorine in secondary refining slags. ISIJ Int 2002; 42: 921-9.
        [http://dx.doi.org/10.2355/isijinternational.42.921]

[12]   Pecsok RL, Sandera J. The gluconate complexes. II. The ferric-gluconate system. J Am Chem Soc 1955; 77: 1489-94.
        [http://dx.doi.org/10.1021/ja01611a025]

CHAPTER 21

Effects of Eluate Components from Steelmaking Slag on Microalgae

Toshiyuki Takahashi[*]

Department of Chemical Science and Engineering, National Institute of Technology (KOSEN), Miyakonojo College, Miyakonojo, Japan

Abstract: This chapter describes effects of an eluate from steelmaking slag on phytoplankton, particularly *Chlorella* species as model phytoplankton. Eluates used to assess toxicity in this study were derived from slag using a leaching test condition based on JIS K 0058-1. Electric arc furnace slag in particular was used as a slag. In this study, we also developed methods to estimate not only the number of algal cells by hemocytometry but also their physiological activities *via* flow cytometry. After a leaching test, the concentrations of eluates from slag were almost lower than those of environmental quality standards. In addition to an elemental analysis, a *Chlorella*-based bioassay was performed on the eluates. After the treatment of algae with eluates, algae were analyzed by microscopy and flow cytometry. As a result, treatment with eluates induced neither death nor growth inhibition. This treatment was not extremely toxic but rather induced algal growth and a vigorous physiological state.

Keywords: *Chlorella*, Chlorophyll, Electric arc furnace slag, Environmental quality standards, Flow cytometry, Fluorescence, Phytoplankton.

INTRODUCTION

Just like the coral reefs, a seaweed bed supports ocean biodiversity. However, episodic depletion of seaweeds in coastal areas is lately a severe problem in Japan and other parts of the world. This phenomenon is expressed in several phrases like barren ground, reef burning, *isoyake* (used for withered algae in Japanese) [1], *getnogum* (used as a coined word for coast plus melting in Korean) [2], algal deterioration, and a sea desert. The cause for a barren ground is difficult to assess because several factors such as marine terrain, oceanographic properties, biocenosis, and coastal zone management of each area have an influence on the risk of barren ground. Whatever the reason for barren ground, it results in a decre-

---
[*] **Corresponding author Toshiyuki Takahashi:** Department of Chemical Science and Engineering, National Institute of Technology, Miyakonojo College, Miyakonojo, Japan; Tel: +81-986-47-1219; Fax: +81-986-47-1231; E-mail: mttaka@cc.miyakonojo-nct.ac.jp

ase of marine-ecosystem services. To conserve and control marine ecosystems, accurate evaluation of seaweeds is necessary.

Considering the decrease in marine-ecosystem services and aiming to improve them, researchers have implemented some pilot approaches; an example is the use of steelmaking byproducts as rehabilitation materials [3, 4]. Steelmaking byproducts, iron and steel slag contain metallic components from steel industries. When slag is introduced into aquatic environments, its metal components leak into the surrounding environment. Therefore, it is necessary to elucidate the effects of the metallic eluate from slag on aquatic organisms.

Using the unicellular alga *Chlorella* as a model microorganism, this study presents evaluation of effects of an eluate from slag on algae. This study evaluated not only an impact of the eluate from slag on algal cell numbers using microscopy-based hemocytometry but also the influence on the algal properties such as viability according to flow cytometry (FCM) as a high-throughput assay [5 - 8].

STEEL SLAG SAMPLES, THE ELUATE, AND THE ASSAYS

Iron and steel slag from the blast furnace slag and steelmaking slag have been produced constantly as steel industrial byproducts. Here, steelmaking slag included converter slag and electric arc furnace slag (EAF slag hereafter in this chapter). All blast furnace slag (23,970,000 tons/year in Japan from April 2017 to March 2018) is recycled for uses, such as steelmaking slag base and cement (as a soil aggregate). However, 210,000 tons/year (April 2017 to March 2018) of steelmaking slag in Japan (87,000 tons of converter slag and 123,000 tons of EAF slag) ultimately ends up at landfill sites [9]. With consideration for resource recycling, new applications for slag not only on land but also in aquatic environments [10, 11] have been recently anticipated.

Panel A in Fig. (**1**) presents composition of EAF slag as an example [5]. Uses of slag in the environment have been generally restricted by several laws for environmental pollution according to the environmental quality standards (EQS). Slag utilization in an aqueous environment has been subjected to rigid control relative to that on land. It is in large part due to leakage of some soluble metals from slag into aquatic environments. Only for information, panel B in Fig. (**1**) shows eluate composition after immersion of each slag for 6 h in an acidic solution adjusted to pH 6: a leaching test based on JIS K 0058-1 (Test method for chemicals in slags Part 1: Leaching test method) [5 - 8, 12 - 14]. To prevent the influence of the eluate from slag on ecological systems, it is necessary to examine harmful effects of the slag eluate on organisms before using slag in the environment.

A)

mass%	FeO	SiO$_2$	CaO	Al$_2$O$_3$	MgO	MnO	Cr$_2$O$_3$	ZnO	NiO	CuO
Slag A	0.74	44.1	33	5.39	7.68	4.09	3.29	0.01	0.06	0.024
Slag B	35.1	19.2	20.8	15.2	4.1	5.1	0.43	0.071	0.028	0.025

B)

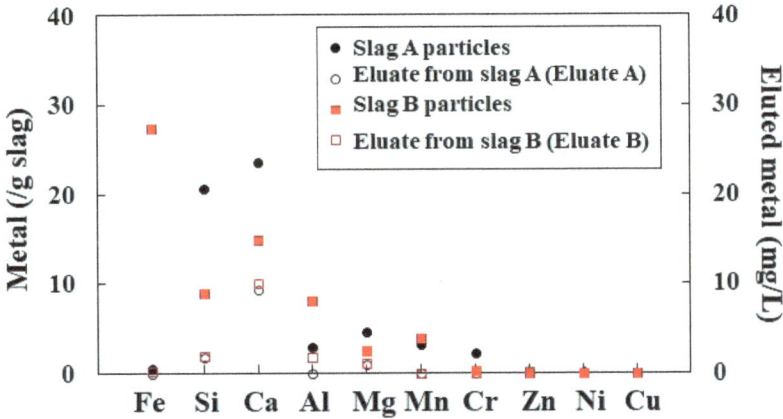

Fig. (1). Chemical compositions of electric arc furnace slag (EAF) slag particles and their eluates. (**A**) Chemical composition of two types of EAF slag, stainless-steel slag (designated as slag **A**) and common steel one (slag **B**), are presented [5]. Here, all Fe or Cr compounds are described as FeO or Cr$_2$O$_3$, respectively because it is difficult to distinguish FeO and Cr$_2$O$_3$ from plain elemental Fe and Cr in particles of stainless-steel slag. Briefly, slag A includes greater amounts of SiO$_2$, CaO, and Cr$_2$O$_3$ than slag B but contains less FeO than slag B does. (**B**) Main metal contents of each slag's particles (black circles and red squares), and results (white circles and squares) of a leaching test based on JIS K 0058-1 using each slag.

Shifting our perspective on slag, slags also contain several essential elements for organisms such as Ca, Mg, and trace elements. In fact, Ca [15] and Zn [16] respectively are involved in signal transduction in cells and gene expression as part of DNA-binding proteins.

Using the above leaching test and EQS for soil and water pollution, potential environmental risks of both slag particles and of the corresponding eluate were assessed based on the results of elemental analysis rather than a bioassay. The chemical concentrations of metal effluents from slags (Fig. **1**), incidentally, were almost lower than the EQS for pollution and the EQS for effluent and drinking water [5 - 8]. Each value of EQS has been derived from the sensitivity of model organisms to each element. Therefore, exposure of organisms to complicated compositions containing multiple elements such as an eluate from slag has not always been considered in EQS. An effect of an eluate from slag on organisms is not obvious and has been confirmed only by concentration verification using chemical analysis without any bioassay. To prevent field trials of slag from directly polluting any aqueous environment, *in vitro* assessments are useful for evaluating the influence of an eluate from slag on organisms.

FEATURES OF A UNICELLULAR ALGA FOR *IN VITRO* ASSESSMENTS OF ELUATE EFFECTS ON AQUATIC ORGANISMS

In addition to the important roles of phytoplankton in aquatic ecosystems, several microalgae have been recently used for the development of broadly diverse biotechnologies such as industrial processes including chemical production, health, food, agricultural, environmental applications, and biofuels [17]. Whatever the reason for microalgal use for industrial applications, the guarantee of algal status is particularly important because microalgae are highly sensitive to environmental changes such as temperature, pH, and some contaminants (*e.g.*, chemicals) [17, 18].

Some organisms like *Chlorella* as a model microbe in phytoplankton are well supported by the methods for examining cell behaviors including proliferation, the cell cycle, and metabolic pathways [8, 17, 19]. All algae including *Chlorella* sp., for instance, have one or several chloroplasts for photosynthesis in their cytoplasm. Although chlorophyll exposed to excitation light emits red fluorescence (Fig. **2**), chlorophyll is generally sensitive to the physiological factors such as heat, acidity, and the treatment of algae with some hazardous chemicals.

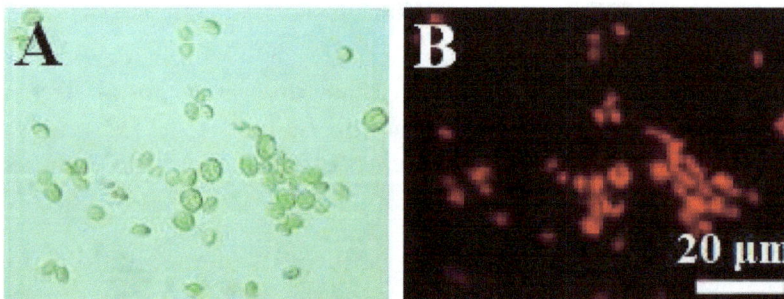

Fig. (2). A bright-field image of *Chlorella*-like algae isolated from ciliate *Paramecium bursaria* (**A**) and the corresponding fluorescence image of algae irradiated by an excitation light for chlorophyll (**B**).

These factors can inactivate chlorophyll and eventually result in chlorophyll degradation. In fact, red fluorescence derived from chlorophyll decreases after the heating of algae (Fig. **3A-D**). It is noteworthy that yellow fluorescence was greater simultaneously with a decrease of red fluorescence (Fig. **3E**). Thus, a change of the yellow fluorescence can be regarded as an index of various statuses of algae such as dying algae rather than vigorous algae, which are associated directly with strong red fluorescence [5, 14]. To evaluate toxicity and effects of eluates from slag samples on aquatic organisms, *Chlorella kessleri* (having the properties described above) served as a model organism for a cell-based assay in this study [5, 14].

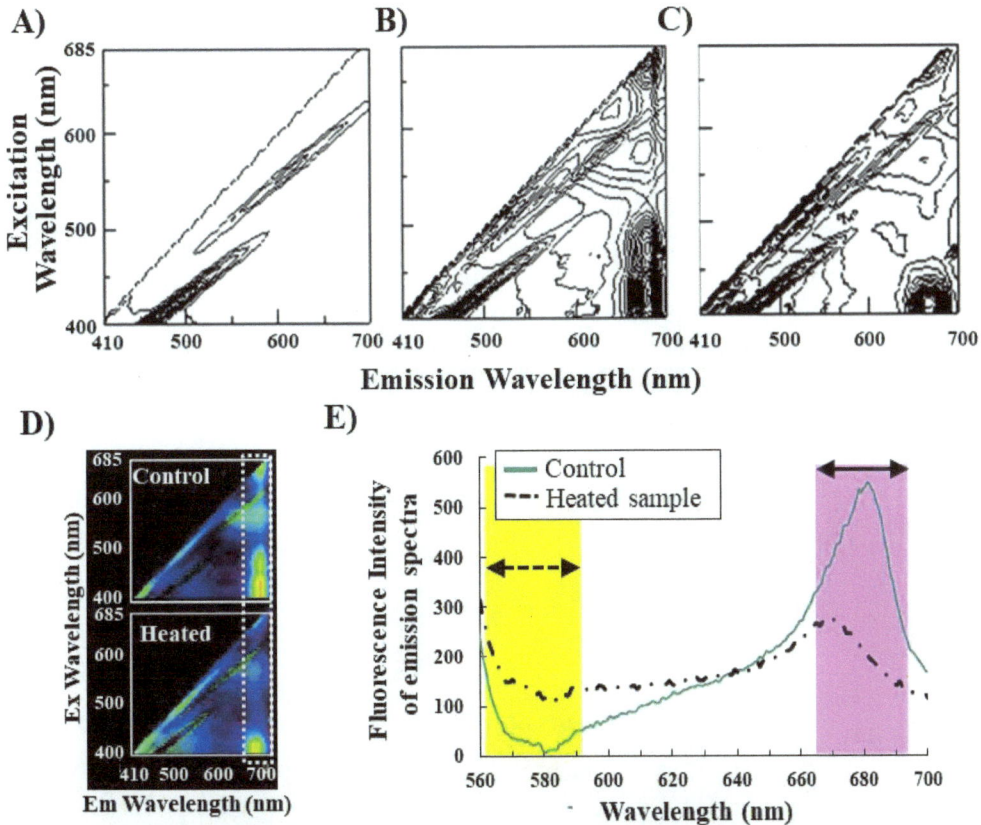

Fig. (3). Changes in optical properties of *Chlorella*-like algae.
(**A–C**) Contour plots show a 3D fluorescence excitation-emission matrix (3D matrix), spectrographs of the medium (**A**) and *Chlorella*-like algae without (**B**) or with heat treatment (**C**) [19]. (**D**) Difference spectra between 3D matrix images of control algae (**B**) and those of the medium alone (**A**), and between those of heated algae (**C**) and those of the medium alone (**A**) [19]. An area enclosed within a white dotted line shows fluorescence corresponding approximately to chlorophyll fluorescence. Briefly, chlorophyll fluorescence of algae decreased after heat treatment. (**E**) Emission spectra of algae are shown without (green curve, designated as a control) or with heat treatment (black dotted curve, heated sample) [20]. Herein, the heated sample serves as an indicator of dead algae. Yellow and pink areas denote each range of detection for the yellow or the red fluorescence channel employed for the subsequent flow cytometry (FCM) (See Fig. **5**).

IN VITRO EVALUATION OF ELUATES FROM SLAG ON ALGAL CELL NUMBERS USING A MICROSCOPE-BASED METHOD

To evaluate the effects of each EAF slag on algae, a leaching test was first carried out to elute metal components from slag [5 - 8]. After elution for 6 h, the solution eluted from slag was passed through a 0.45 μm pore filter to eliminate slag particles. The filtered eluate from the slag (designated as eluate A or eluate B; Fig. (**1B**) was subjected to the following experiments in a cell-based assay. *C. kessleri* (initial concentration of 10^4 algal cells/ml) was cultured in a medium (CA medium in this study; pH 7.2) supplemented with an eluate under natural white fluorescent light at 23°C ± 2°C and on a light-dark cycle (12 h light/12 h dark). After the treatment of algae for 1 week, algal cell numbers in each condition were quantified on a hemocytometer. As a result, algal cell numbers increased depending on the concentration of an eluate, up to 30 vol%, with each eluate (Fig. **4**). By contrast with conditions at lower concentrations of the eluate, the number of algal cells was approximately constant at higher concentrations of the eluate than 30 vol%. In any event, these data seem to imply that the eluates caused no obvious cytotoxicity, such as death of algae.

Fig. (4). Effects of respective eluates on algal growth in an algal bioassay.
The graph shows the number of algal cells (± standard error) after treatment with each eluate (black and white symbols respectively denote stainless-steel slag and common steel one) (See Fig. **1**) [5]. Here, the graph shows a relation between the proliferation ratio (% of control [without an eluate]) and eluate contents (vol%).

EVALUATION OF THE INFLUENCE OF ELUATES FROM SLAG ON ALGAL STATUS USING FLOW CYTOMETRY

To evaluate the effects of the eluate from each EAF slag on algal status, *C. kessleri* treated with each eluate was analyzed on a flow cytometer [5].

Before mentioning the effects of eluates on algae, this section introduces the features of FCM briefly. FCM, a powerful tool especially in cell biology, can provide single-cell optical information about microbes or about several animal and plant cells. The information obtained from FCM includes cell size and several metabolic parameters. In addition to an arbitrary unidimensional histogram from FCM data, clear patterning graphs with multidimensional parameters can help us understand correlations among several parameters [8]. Although several parametric signals can be detected in a cell suspension, this study focused on three parameters (Fig. **5**). The first is a forward scattering signal (FSS) for cell size to discriminate target cells from measurement noise signals; the second and the third are red fluorescence of chlorophyll of algae and yellow fluorescence for the detection of dead algae, respectively (Fig. **3E**).

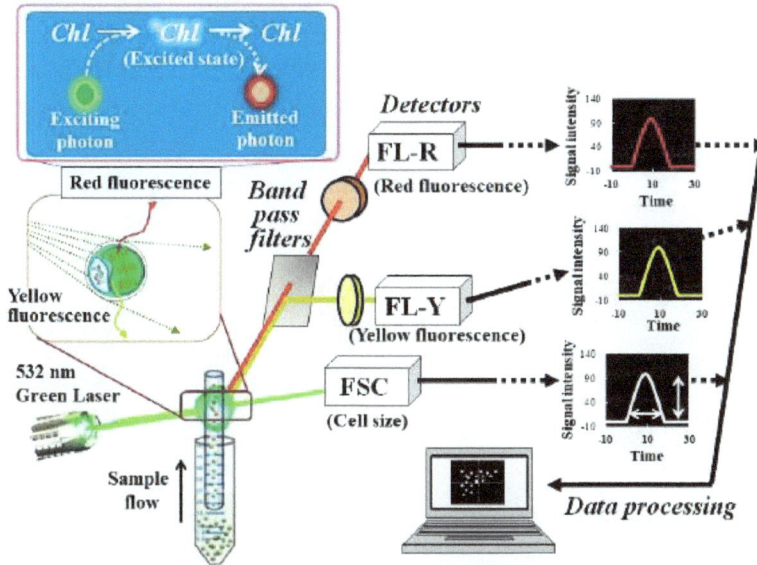

Fig. (5). Schematic view of FCM analysis of microalgae.
This study involved a flow cytometer equipped with a green laser operating at 532 nm for evaluating the effects of eluates from slag on algae. Algae were analyzed one by one through a rectangular capillary with a 100 μm round bore for several optical properties. Although FSSs were collected to ascertain cell size, the autofluorescence of chlorophyll and the corresponding yellow fluorescence were detected in the red fluorescence channel through a 680/30 nm band pass filter and in the yellow fluorescence channel through a 576/28 nm band pass filter, respectively. The explanatory diagram for the red fluorescence shows the molecular behavior of chlorophyll in algae irradiated by an excitation laser as an example. Herein, the words '*Chl*' and '**Chl*' respectively indicate chlorophyll in the ground state and the one in the excited state.

In analogy with the estimation by hemocytometry (Fig. **4**), algae treated with one of the eluates for 1 week were subjected to FCM analysis. Here, medium containing 50 vol% of one of the eluates was used for FCM analysis. To distinguish target algal signals from signals of mere measurement noise, signals above the culture medium alone in terms of FSS signals were first selected as algal signals. Next, these signals gated by FSS were respectively reanalyzed for red fluorescence as chlorophyll, and for the corresponding yellow fluorescence as a barometer for an unhealthy alga different from a vigorous one. For comparison of vigorous algae with other statuses of algae and for determination of an indicator of dead algae, algae treated with heat were prepared as described in Fig. (**3E**). Scatter plots based on the signals derived from algae are shown in the graph of red fluorescence intensity *vs.* the corresponding yellow one (Fig. **6**). For descriptive purposes of distribution patterns, each graph in Fig. (**6**) is divided into four subareas (designated as subareas I–IV) [5]. Here, this segmentation is mainly set up according to algal viability: subarea I represents vigorous algae; subarea II denotes dead and other types of algae; subarea III is designated for algae with low red fluorescence intensity; and subarea IV for the remaining algae other than those in subareas I–III. As a result, a few signals from control algae were distributed in subareas III and IV. Summing up the graph patterns in Fig. (**6**), signals in subareas I and II, respectively, constitute an indicator of living algae and that of dead algae.

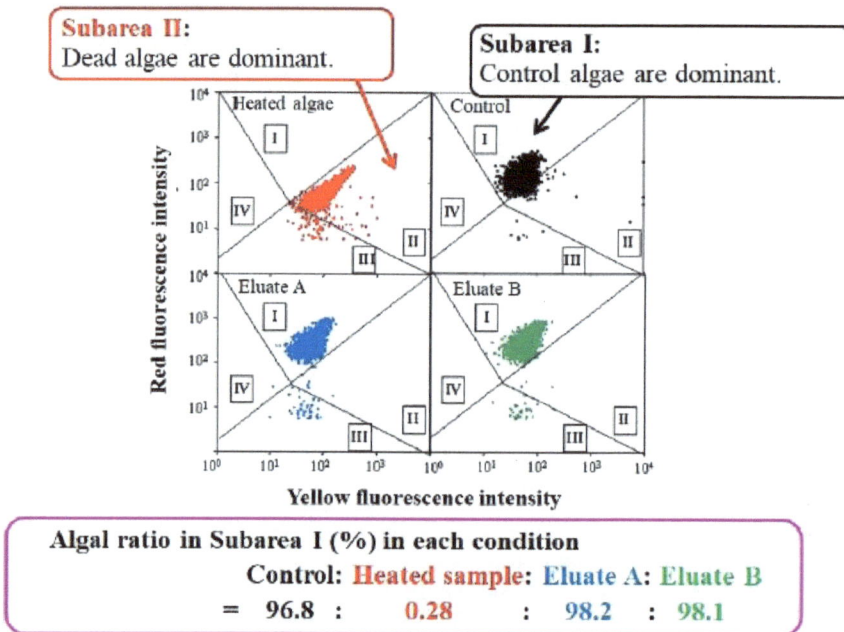

Fig. (**6**). Differences in dot patterns among the test conditions and their quantification.

To decipher the effects of eluates from slags on algae, optical characteristics of algae were analyzed by FCM in detail (Fig. **6**). The dot pattern of control algae correlated with the red fluorescence intensity but not with the yellow fluorescence intensity. Their values from control algae converged from 10^2–10^3 in the red channel and 10^1–10^2 in the yellow channel in the plot of red fluorescence intensity *vs.* yellow one. In contrast to control algae, the dot pattern of heated algae was significantly associated with both the red and the yellow fluorescence intensity. These values from heated algae huddled within a range of 10^1–10^2 in the red channel and 10^1–10^3 in the yellow channel in the 2-dimensional graph. It is noteworthy that a dot distribution pattern of algae treated with one of eluates from slag samples was not significantly different from that of the control group. It differed from control algae in that the dot patterns of algae treated with eluates shifted slightly upward relative to the control.

The description at the bottom of Fig. (**6**) shows the results of quantitative analysis of algae in subarea I under each condition. It is interesting to note that the percentage of algae in subarea I was slightly greater under test conditions with each eluate from slag samples than that in the control. Consequently, this result suggests that components of the eluate from slag samples did not give rise to algae stress directly, similarly to the result of the algal growth test (Fig. **4**). In general, excessive time would be required for studies involving microscopic observation. As compared to a microscopy-based study, approaches based on FCM are more time-efficient. Thus, both harmfulness of the eluate from slags and their useful effect on algae can be determined simultaneously and rapidly by FCM in the case of this metallic toxicity to algae. This assessment system, which estimates chlorophyll fluorescence by FCM, is expected to be applicable to other phytoplankton including other algae and aquatic plants because of their possession of chlorophyll [6].

EFFECTS OF ELUATES FROM SLAG ON ALGAE

As mentioned above (Fig. **1B**), components of eluates from slag samples contain several metals such as copper, zinc, and aluminum. They were present in small amounts in each eluate tested in this study but have been subject to rigid control by several laws and regulations such as EQS. As expected, their metals are not present in the CA medium for algae in this study. Aluminum, for instance, has been reported to serve as a growth inhibitor for plants [21, 22]. These elements including aluminum and other metals, however, did not affect algal growth directly or show lethal toxicity toward algae (Figs. **3** and **6**) [5]. These results revealed that eluates from the steel slag samples tested in this study induced neither cell death nor growth inhibition. The addition of eluates from slag was associated with an increase in algal cell numbers and algal health status rather than a decrease.

However, it should be noted that this study does not necessarily prove safety and efficacy of all slag types and their eluates. Conforming to the leaching conditions based on JIS K 0058-1, this study used only eluates obtained by elution from slag for 6 h, which did not contain any slag particles. If slag is employed for a specific purpose in an aqueous environment, slag must continue to react with water for more than 6 h. Then, the eluate as a product might contain higher concentrations of metals than that used in this study. Moreover, elution behavior from slag is dependent on the origin of each slag and its chemical composition. Therefore, further accumulation of data without or with a condition including slag particles is needed to evaluate both toxicity of slag to organisms in aquatic environments and advantages of its physicochemical properties for organisms' metabolic pathways.

CONCLUSION

Considering the decrease in marine-ecosystem services as a result of depletion of seaweeds, some pilot approaches are necessary for conserving and improving a seaweed-based marine ecosystem. These approaches, however, sometimes cause environmental problems. To conserve marine ecosystems and prevent byproducts such as slag from contaminating an aquatic environment excessively, an easy laboratory scale experiment like a cell-based assay might help us understand the properties and effects of these approaches before any field trial.

Influences of pollution on phytoplankton are generally related to higher-order biota in many ways because phytoplanktons constitute the bedrock of aquatic ecosystems. To evaluate toxicity of an eluate from slag, a bioassay using *Chlorella* as a model of both alga and phytoplankton was carried out in this study. This study also presents a method for estimating not only the number of algal cells *via* microscopy-based hemocytometry but also their algal activities by FCM. Consequently, a slag eluate within concentrations used in this study caused algal cells to increase in number in comparison with the control condition without any eluate. Moreover, this method based on FCM can be a powerful tool for the rapid and simple evaluation of the effects of an eluate from slag on physiological activities of algae other than algal cell numbers. This method is expected to support high-throughput screening and help to develop solutions minimizing the risk of barren ground in coastal areas.

CONSENT FOR PUBLICATION

Not applicable.

CONFLICT OF INTEREST

The author confirms that this chapter contents have no conflict of interest.

ACKNOWLEDGEMENTS

Declare none.

REFERENCES

[1]   Fujita D. Current Status of 'Isoyake' in the World. Fisheries Eng 2002; 39: 41-6.

[2]   Kim Y-D, Hong J-P, Song HL, Park MS, Moon TS, Yoo HI. Studies on technology for seaweed forest construction and transplanted Ecklonia cava growth for an artificial seaweed reef. J Environ Biol 2012; 33(5): 969-75.
        [PMID: 23734467]

[3]   Hayashi A, Watanabe T, Kaneko R, *et al.* Decrease of sulfide in enclosed coastal sea by using steelmaking slag. Tetsu-to-Hagané 2012; 98: 57-64.
        [http://dx.doi.org/10.2355/tetsutohagane.98.207]

[4]   Fujimoto K, Kato T, Ueki C, Tsutsumi N. Sea forest creation with utilizing by-product slag of steelmaking process (development of regeneration technology for seaweed bed). Nippon Steel Sumitomo Metal Techn Rep 2011; 391: 206-9.

[5]   Takahashi T, Yokoyama S. Bioassay of components eluted from electric arc furnace steel slag using microalgae *Chlorella*. ISIJ Int 2016; 56: 1495-503.
        [http://dx.doi.org/10.2355/isijinternational.ISIJINT-2015-539]

[6]   Takahashi T. Application of phytoplankton. In: Kanematsu H, Barry DM, Eds. Corrosion Control and Surface Finishing - Environmentally Friendly Approaches. Springer Japan 2016; pp. 213-24.
        [http://dx.doi.org/10.1007/978-4-431-55957-3_19]

[7]   Takahashi T. Quality assessment of microalgae exposed to trace metals using flow cytometry. In: Shiomi N, Waisundara VY, Eds. Superfood and Functional Food - Development of Superfood and its Role in Medicine. InTech Open 2017; pp. 29-45.
        [http://dx.doi.org/10.5772/65516]

[8]   Takahashi T. Efficient interpretation of multiparametric data using principal component analysis as an example of quality assessment of microalgae. In: Gemiei M, Ed. Multidimensional Flow Cytometry Techniques for Novel Highly Informative Assays. InTechOpen 2018; pp. 81-97.
        [http://dx.doi.org/10.5772/intechopen.71460]

[9]   Nippon Slag Association. Iron and Steel Slag. Tekko Sulagu Toukeinenpou 2018; FS-163: 2-10.

[10]   Miyata Y, Sato Y, Shimizu S, Oyamada K. Environmental Improvement in the Sea Bottom by Steelmaking Slag. JFE Techn Rep 2009. No. 13, 41–45

[11]   Kato T, Kosugi C, Kiso E, Torii K. Application of Steelmaking Slag to Marine Forest Restoration. Nippon Steel & Sumitomo Metal Technical Rep 2014. No. 399, 79-84

[12]   Takahashi T, Ogura Y, Ogawa A, Kanematsu H, Yokoyama S. An effective and economic strategy to restore acidified freshwater ecosystems with steel industrial byproducts. J Water Environ Technol 2012; 10: 347-62.
        [http://dx.doi.org/10.2965/jwet.2012.347]

[13]   Yokoyama S, Suzuki A, Izaki M, Umemoto M. Elution behavior of electronic arc furnace oxidizing slag into fresh water. Tetsu-to-Hagané 2009; 95: 434-43.
        [http://dx.doi.org/10.2355/tetsutohagane.95.434]

[14]   Takahashi T, Yokoyama S. Bioassay of components eluted from electric arc furnace steel slag using microalgae *Chlorella*. Tetsu-to-Hagané 2015; 101: 506-14.
        [http://dx.doi.org/10.2355/tetsutohagane.TETSU-2014-130]

[15]   Solomon EP, Gerg LR, Martin DW. Endocrine Regulation In: BIOLOGY fifth edition, Saunders College Publishing. 1999. Chapter 47, 1018-1044.

[16]   Solomon EP, Gerg LR, Martin DW. Gene Regulation: The Control of Gene Expression In: BIOLOGY fifth edition, Saunders College Publishing 1999. Chapter 13, 287-303.

[17]   Takahashi T. Applicability of automated cell counter with a chlorophyll detector in routine management of microalgae. Sci Rep 2018; 8(1): 4967.
[http://dx.doi.org/10.1038/s41598-018-23311-8] [PMID: 29563559]

[18]   Kaplan D. Absorption and Adsorption of Heavy Metals by Microalgae. In: Richmond A, Hu Q, Eds. Handbook of Microalgal Culture: Applied Phycology and Biotechnology Second Edition. John Wiley & Sons, Ltd 2013; pp. 602-11.
[http://dx.doi.org/10.1002/9781118567166.ch32]

[19]   Takahashi T. Simultaneous evaluation of life cycle dynamics between a host *paramecium* and the endosymbionts of *paramecium bursaria* using capillary flow cytometry. Sci Rep 2016; 6: 31638.
[http://dx.doi.org/10.1038/srep31638] [PMID: 27531180]

[20]   Takahashi T. Direct evaluation of endosymbiotic status in Paramecium bursaria using a capillary flow cytometer. Cytometry A 2014; 85(11): 911-4.
[http://dx.doi.org/10.1002/cyto.a.22562] [PMID: 25160605]

[21]   Bose J, Babourina O, Rengel Z. Role of magnesium in alleviation of aluminium toxicity in plants. J Exp Bot 2011; 62(7): 2251-64.
[http://dx.doi.org/10.1093/jxb/erq456] [PMID: 21273333]

[22]   Zheng L, Lan P, Shen RF, Li WF. Proteomics of aluminum tolerance in plants. Proteomics 2014; 14(4-5): 566-78.
[http://dx.doi.org/10.1002/pmic.201300252] [PMID: 24339160]

CHAPTER 22

Development of Regeneration Technology for a Seaweed Bed

Chika Kosugi[1,*] and **Toshiaki Kato**[1,2]

[1] *Advanced Technology Research Laboratories, Nippon Steel Corporation, Futtsu, Japan*

[2] *Nippon Steel Eco-Tech Corporation, Tokyo, Japan*

Abstract: Reduction of seaweed beds in coastal areas, called *isoyake* in Japanese, is a serious problem in Japan, which results in the disappearance of habitats for many fishes and the decline of coastal fisheries. Many factors may contribute to the reduction of seaweed beds, including seawater temperature rising, grazing by herbivores, deficiencies of nutrients (nitrogen, phosphorus) and iron, among others. We examined fertilization by selected some nutrients and minerals, with a particular focus on iron, and developed a fertilizer, Vivary™ Unit. Vivary™ Unit is a 1:1 (*v:v*) mixture of the steelmaking slag and humus soil. Steelmaking slag, which is a by-product of the steelmaking process, is a source of iron, and artificial humus soil contains chelators such as humic acid. To establish a regeneration technology for seaweed beds, we verified the role of iron in seaweed growth and the bioavailability of Vivary™ Unit with several experiments in various scales. *In vitro* culture was used to demonstrate the synthesis of chlorophyll *a* in *Pyropia yezoensis* thalli and the requirement of iron in gametophyte maturation in *Saccharina japonica* var. *religiosa*. The effects of Vivary™ Unit were demonstrated through two experiments. (1) a cultivation test of *P. yezoensis* in a mesocosm facility, (2) a field test in Hokkaido, Japan. In the cultivation of *P. yezoensis* the elution of nitrogen, phosphorus, silica and iron from Vivary™ Unit were revealed, and *P. yezoensis* grew only in the mesocosm with Vivary™ Unit. In the field test in Hokkaido, seaweed beds, especially *S. japonica* var. *religiosa*, were restored following fertilization *via* a buried Vivary™ Unit. We indicated the effect of dissolved iron supplied by the buried Vivary™ Unit by examining the correlation between EC (electrical conductivity) and dissolved iron compared with the correlation between EC and dissolved silica.

Keywords: Chlorosis, Converter slag, Fertilizer, Humus soil, iron, *isoyake*, Mesocosm, *Pyropia yezoensis*, Reduction of seaweed beds, Saccharinales, *Saccharina japonica* var. *religiosa*, Seaweed bed restorations, Steelmaking slag, Vivary™ Unit.

* **Corresponding author Chika Kosugi:** Advanced Technology Research Laboratories, Nippon Steel Corporation, Futtsu, Japan; Tel: +81-70-3914-4684; Fax: +81-439-80-2745; E-mail: kosugi.4qp.chika@jp.nipponsteel.com

Toshiyuki Takahashi (Ed.)

INTRODUCTION

Reduction of seaweed beds in coastal area, called *isoyake* in Japanese, has been recognized in Japan for more than a century [1]. Seaweed beds buffer wave action and protect against sun, and thus play an important role as a habitat for many fishes and shellfishes. Therefore, the reduction of seaweed beds results not only in deterioration of the coastal environment, but also in decline on the fishery. Many factors may affect the reduction of seaweed beds, including increasing seawater temperature, grazing by herbivores such as sea urchins, and deficiencies in iron and nutrients such as nitrogen and phosphorus [2]. Many measures have been taken to restore marine primary producers, namely phytoplankton and seaweeds, and to create seaweed beds. These measures include seaweed transplantation [3, 4], the placement of adhesive substrates (*e.g.* concrete blocks) [2, 3], the removal of herbivores (sea urchins and seaweed-eating fishes) [5], fertilization with nitrogen and phosphorus [6, 7], and fertilization with iron [8 - 10].

Iron is a significant trace metal for seaweed growth. Seaweeds require iron for various processes, including photosynthesis, chlorophyll synthesis, respiration, and mitochondria electron transport [11]. At the pH of seawater (8.0-8.3), ferric ions combine with hydroxyl ions to form ferric hydroxide, which is relatively insoluble. Therefore, almost all iron exists as part of complexes with natural chelators or ligands such as humic acid, which appear to enter the system through sediment resuspension, rainfall, river inputs, or runoff [11]. In Hokkaido, Japan, the iron supply along the rivers has been decreasing due to environmental changes, such as deforestation and dam construction [12, 13].

In this study, we focused on the iron deficiency in seawater and developed a fertilizer, which is a 1:1 (*v:v*) mixture of steelmaking slag and humus soil, named Vivary™ Unit [14]. A by-product of the steelmaking process, converter slag contains a high concentration of bivalent iron (FeO). The steelmaking slag (converter slag) is used as the iron source in Vivary™ Unit, and artificial humus soil, made from the fermented chips of waste wood, provides chelators, such as humic acid. We tried to reproduce the iron supply process from the mountain through the river by fertilizing the coastal area with Vivary™ Unit. Previously, we revealed the bioavailability of Vivary™ Unit to seaweed with the experiments at various scales, including flask scale culture experiments [14], mesocosm tests [15 - 17], and field tests [8 - 10]. In this chapter, we introduce: (1) experiments examining the fundamental biological iron effects on *Pyropia yezoensis* and *Saccharina japonica* var. *religiosa*, (2) a *P. yezoensis* culture test using mesocosms, and (3) a field test at Hokkaido, Japan.

IRON EFFECTS ON SEAWEEDS

Chlorosis in the thalli of *Pyropia yezoensis* is primarily caused by nitrogen and phosphorus deficiencies, but the iron deficiency is also a known factor [18]. Thalli of *P. yezoensis* were cultured in media in three groups: without nitrogen (N-free), phosphorus (P-free) or iron (Fe-free). All thalli were gradually discoloured, except for the control group, which was kept in enriched seawater PES [19]. Photosynthetic pigments (chlorophyll *a*, phycoerythrin, phycocyanin) notably decreased, as indicated by absorption spectra and light microscopic observation (Fig. **1**). The degree of the pigments reduction varied in accordance with the deprived nutrients, which were revealed by comparing the height of peaks on the absorption spectra. The N-free group showed the greatest reduction in photosynthetic pigments, and the Fe-free group showed the least reduction.

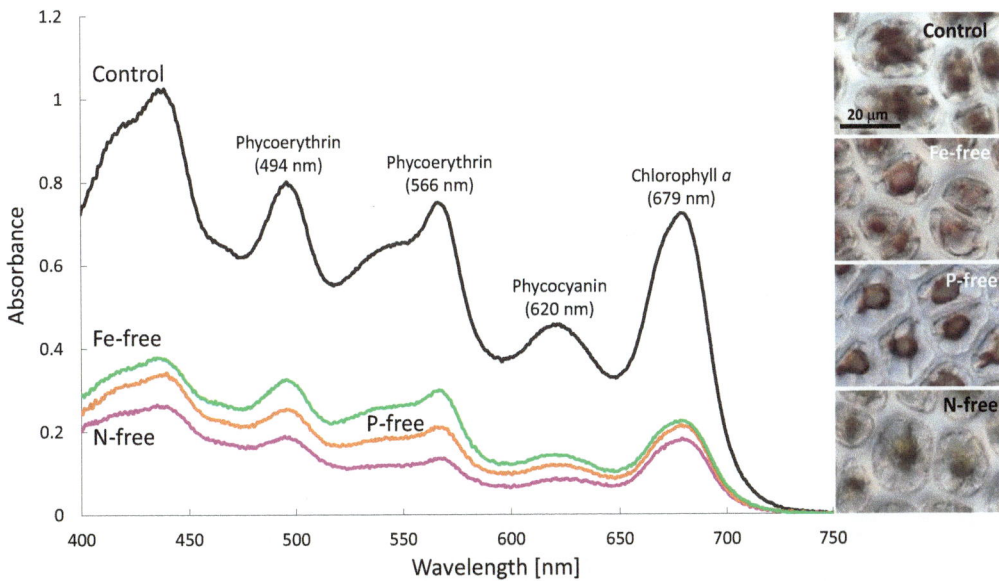

Fig. (1). Absorption spectra of *Pyropia yezoensis* thalli in the nutrient deficient media and light microscopic images of each thallus.

Also, the pattern of chlorosis differed among the nutrients deficiency groups. The PE/Chl. *a* ratio was determined *via* each absorbance peak of phycoerythrin (A_{566}) and chlorophyll *a* (A_{679}) (Table **1**). The PE/Chl. *a* ratio of the control and P-free groups were similar, implying that the photosynthetic pigments proportionally decreased in the P-free group, maintaining the balance of pigments. The PE/Chl. *a* ratio of the N-free group was lower than the control, meaning that phycoerythrin decreased more than chlorophyll *a*. On the other hand, the PE/Chl. *a* ratio of the Fe-free group was larger than the control, which showed a greater reduction of chlorophyll *a* through iron deficiency. This result revealed that iron plays a role in chlorophyll *a* synthesis in *P. yeozoensis* thalli.

Table 1. PE/Chl. *a* ratio (A_{566}/A_{679}) of the absorption spectra of *P. yezoensis* thalli in the nutrient deficient media.

Control (PES)	N-free	P-free	Fe-free
1.04	0.75	1.00	1.33

In addition, iron plays a very important role in Saccharinales's life cycle (Fig. **2**). Microscopic female/male gametophytes form oogonia or spermatogonia as they mature, and each germ cell releases an egg or sperms. After fertilization has occurred between the egg and sperm, sporophytes are produced. In the maturation of gametophytes, iron is crucial. When female and male gametophytes were cultured in iron-free media (0 mg/L), oogonia and spermatogonia did not form (Fig. **3**). In iron-added media (0.5, 1, 2 mg/L), gametophytes formed and sporophytes germinated as juvenile thalli (Fig. **3**). These results showed the iron requirement in the maturation of Saccharinales's gametophytes [14, 19 - 21].

MESOCOSM TEST OF VIVARY™ UNIT

Mesocosm experiments were carried our at the mesocosm facility in the company's research centre, Futtsu, Japan. The facility is equipped with water tanks in a glasshouse (Fig. **4**). The bioavailability of Vivary™ Unit to seaweed was demonstrated by the cultivation of *Pyropia yezoensis* in this facility [15]. Initially, both water tanks used for the mesocosm were filled with seawater, which was circulated without adding fresh seawater during the experiment. Laver seed net on which numerous small thalli had germinated, were set up in each tank, and 60 kg of fertilizer was placed into one side of the experimental tank, while the other tank contained no fertilizer as a control. Water samples were taken three times per week, and tested for dissolved nutrient concentrations (inorganic nitrogen, phosphorus), dissolved silica, and dissolved iron. All analysed samples in this chapter were filtered through a 0.45 μm nitrocellulose membrane.

In the experimental mesocosm tank, nutrients (inorganic nitrogen, phosphorus), silica, and iron were added. In brief, these elements eluted from Vivary™ Unit (Fig. **5**). In the control tank, inorganic nitrogen, phosphorus, silica, and iron gradually decreased.

P. yezoensis thalli grew only in the tank with Vivary™ Unit (Fig. **6**). At the beginning of the cultivation, thalli had grown on the both seed-meshes, but *P. yezoensis* thalli in the control tank (no fertilizer) did not continue to grow. On the other hand, *P. yezoensis* thalli in the mesocosm tank with Vivary™ Unit grew over 10 cm in length by the end of the experiment. Although slight chlorosis occurred in the experimental tank after 51 days, their colour was recovered after the additional supplement with 30 kg of fertilizer. Through this experiment, we confirmed the effects of Vivary™ Unit on the growth and colour recovery in *P. yezoensis* thalli.

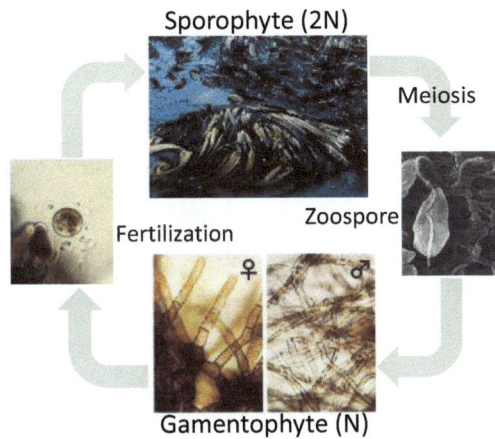

Fig. (2). Life cycle of Saccharinales.

FIELD TEST OF VIVARY® UNIT

The field test was carried out on the Shaguma coast of Mashike-town in Hokkaido, Japan, in October 2004, as a demonstration experiment of Vivary™ Unit (Fig. **7**) [9, 10]. The coastal ground in the experimental area consisted of boulder 200-500 mm in diameter, and the water depth was less than 1.5 m to at least 50 m off the shoreline. At the experimental site, 39 bags of Vivary™ Unit were buried in three ditch rows parallel to the shoreline. The length and depth of each ditch was 26 and 0.8 m, respectively. A reference site was designed at the location of 108 m away from the experimental site.

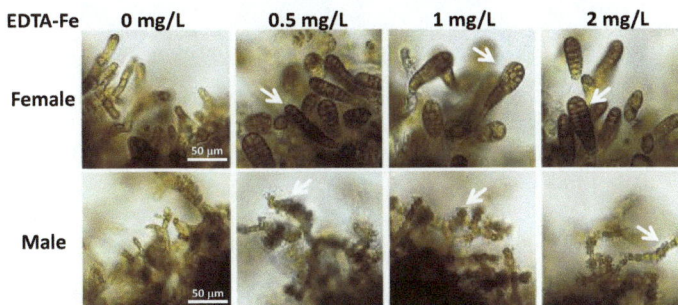

Fig. (3). Effect of EDTA-Fe on female and male gametophytes of *Saccharina japonica* var. *religiosa*. Arrows show juvenile thalli in female and spermatogonia in male.

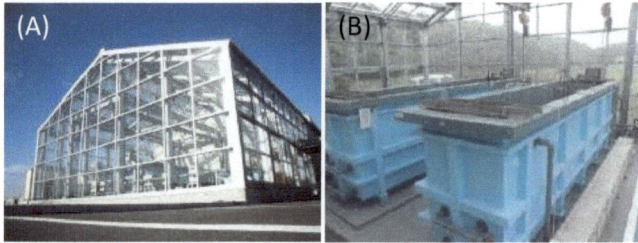

Fig. (4). Photographs of mesocosm facility.
(**A**) The mesocosm facility. (**B**) The mesocosm tanks.

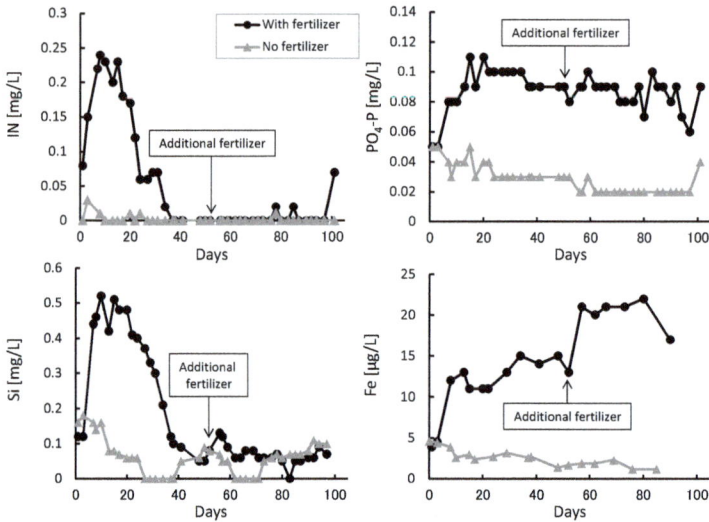

Fig. (5). The time course changes of inorganic nitrogen (IN), phosphorus (PO_4-P), silica (Si) and iron (Fe) in mesocosm tanks (With fertilizer and No fertilizer).

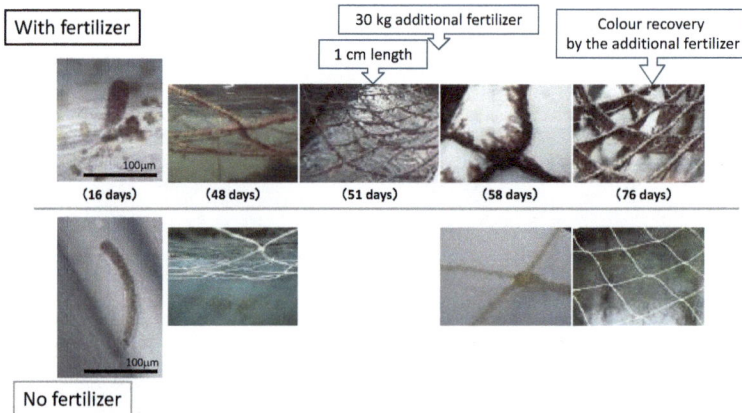

Fig. (6). The time course changes of the growth of *P. yezoensis* thalli. Upper photos show experimental tank with fertilizer, bottom photos show control tank with no fertilizer.

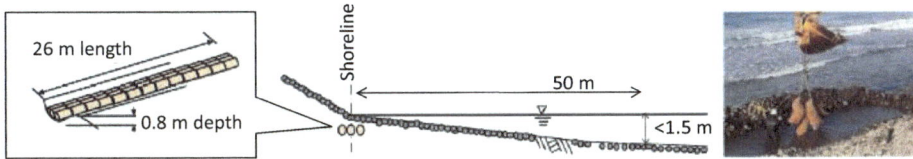

Fig. (7). Outline of the experimental coastal area and the photograph of construction scenery.

Before the experiment, only a small amount of seaweed was observed in this coastal area (Fig. **8A**). However, nine months after burying the fertilizer unit, seaweed, mainly *S. japonica* var. *religiosa*, had grown more thickly at the experimental site than at the reference site (Fig. **8B**).

Fig. (8). Photographs of experimental field. **(A)** Coastal ground before the experiment in July 2004. **(B)** Seaweed growth nine months after burying the fertilizer unit, in July 2005.

Fig. (**9**) shows the differences in wet weights between the experimental and reference sites from 2005 to 2007 using a box-and-whisker plot. The wet weight of *S. japonica* var. *religiosa* at the experimental site was greater than that at the reference site with a distance of 25 m from the shoreline. This result indicated that installation of the slag-humus soil fertilizer was effective for restoring seaweed beds.

To elucidate the relationship between the dissolved iron supplied by the fertilizer unit and the recovery of seaweed, analyses of seawater (for dissolved iron and silica concentrations) were carried out in June 2007 in the experimental area. Electrical conductivity (EC) values ranged from 39.0 to 48.5 ms/cm, indicating that fresh water from a river near the experimental area strongly influenced the quality of coastal seawater. A distinct negative correlation was found between EC and dissolved silica, which was assumed to be related to the dissolved silica in the nearby river (Fig. **10A**). On the other hand, dissolved iron and EC were not correlated (Fig. **10B**). The correlation between EC and dissolved iron was not clear in comparison with the correlation between EC and dissolved silica. The distribution of iron concentration cannot be explained by the influence of the river. This suggests that other factors besides the influx of fresh water greatly affect the iron distribution in the experimental area. The source of iron was thought to be the fertilizer.

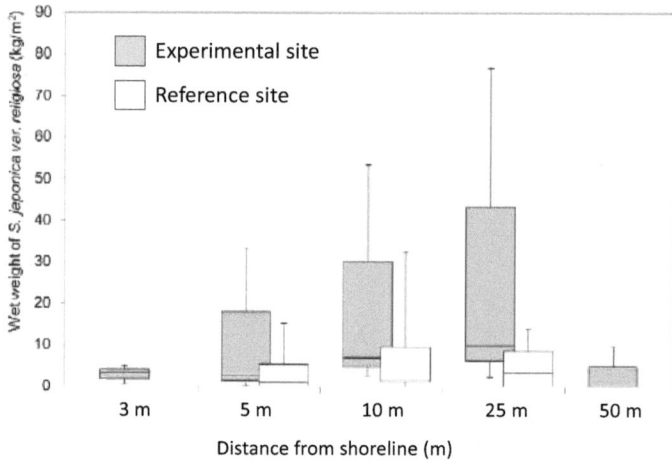

Fig. (9). Box-and-whisker graph of differences in the wet weight of *S. japonica* var. *religiosa* between the experiment site and reference site [10].

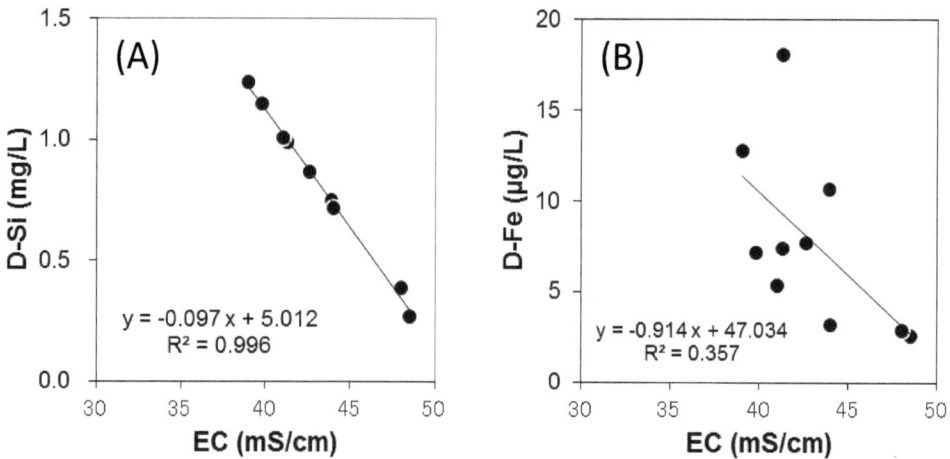

Fig. (10). Relationship between electrical conductivity (EC) and other water quality parameters. (**A**) Dissolved silica concentration. (**B**) Dissolved iron concentration.

As shown in Fig. (**11**), the maximum dissolved iron concentration was 18.1 µg/L at a location 3 m offshore of the experimental site and dissolved iron decreased as distance from the shoreline increased.

These results strongly suggested that the fertilizer unit of slag and humus soil had the greatest effect on the iron concentration. Therefore, iron supplied from slag and humus soil could be the main factor responsible the restoration of seaweed beds. The seaweed beds in the test field have continued flourishing even now (Fig. **12**), suggesting that the vegetation that has grown thus far is functioning as a stock enhancement of seaweeds.

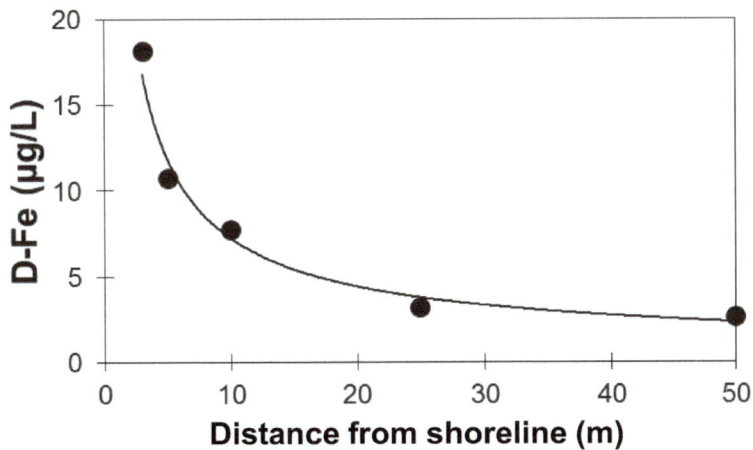

Fig. (11). Distribution of dissolved iron concentration at the experimental site.

Fig. (12). Photograph of experimental field in June 2013 [22].

CONSENT FOR PUBLICATION

Not applicable.

CONFLICT OF INTEREST

The authors confirm that this chapter contents have no conflict of interest.

ACKNOWLEDGEMENTS

Declare none.

REFFERENCES

[1]   Fujita D. Current Status of 'Isoyake' in the World. Fisheries Engineering 2002; 39: 41-6. [in Japanese].

[2]   Fisheries Agency. Guideline for Countermeasures against Barren Grounds. Tokyo, Japan: Kaitei Isoyake Taisaku Guideline 2015. [in Japanese]

[3]   Terawaki T, Hasegawa H, Arai S, Ohno M. Management-free techniques for restoration of *Eisenia* and *Ecklonia* beds along the central Pacific coast of Japan. J Appl Phycol 2001; 13: 13-7. [http://dx.doi.org/10.1023/A:1008135515037]

[4]   Carney LT, Waaland JR, Klinger T, Ewing K. Restoration of the bull kelp *Nereocystis luetkeana* in nearshore rocky habitats. Mar Ecol Prog Ser 2005; 302: 49-61. [http://dx.doi.org/10.3354/meps302049]

[5]   Watanuki A, Aota T, Otsuka E, *et al.* Restoration of kelp beds on an urchin barren: removal of sea uerchins by citizen divers in southwestern Hokkaido. Suisan Sougou Kenkyuu Senta Kenkyuu Houkoku 2010; 32: 83-7.

[6]   Ogawa H, Fujita M. The effect of fertilizer application on farming of the seaweed *Undaria pinnatifida* (Laminariales, Phaeophyta). Phycol Res 1997; 45: 113-6. [http://dx.doi.org/10.1111/j.1440-1835.1997.tb00070.x]

[7]   Agatsuma Y, Endo H, Yoshida S, *et al.* Enhancement of *Saccharina* kelp production by nutrient supply in the Sea of Japan off southwestern Hokkaido, Japan. J Appl Phycol 2014; 26: 1845-52. [http://dx.doi.org/10.1007/s10811-013-0196-z]

[8]   Yamamoto M, Hamasuna N, Fukushima M, *et al.* Recovery from barren ground by supplying slag and humic substances. J Jpn Inst Energy 2006; 85: 971-8. [in Japanese]. [http://dx.doi.org/10.3775/jie.85.971]

[9]   Yamamoto M, Fukushima M, Kiso E, *et al.* Application of iron humates to barren ground in a coastal area for restoring seaweed beds. J Chem Eng of Jpn 2010; 43: 627-34. [http://dx.doi.org/10.1252/jcej.43.627]

[10] Yamamoto M, Kato T, Kanayama S, Nakase K, Tsutsumi N. Effectiveness of iron fertilization for seaweed bed restoration in coastal areas. J Water Environ Technol 2017; 15: 186-97. [http://dx.doi.org/10.2965/jwet.16-080]

[11] Hurd CL, Harrison PJ, Bischof K, Lobban CS. Seaweed Ecology and Physiology. 2nd ed., Cambridge: Cambridge University Press 2014. [http://dx.doi.org/10.1017/CBO9781139192637]

[12] Matsunaga K, Nishioka J, Kuma K, Toya K, Suzuki Y. Riverine input of bioavailable iron supporting phytoplankton growth in Kesennuma Bay (Japan). Water Res 1998; 11: 3436-42. [http://dx.doi.org/10.1016/S0043-1354(98)00113-4]

[13] Matsunaga K, Suzuki Y, Kuma K, Kudo I. Diffusion of Fe(II) from an iron propagation cage and its effect on tissue iron and pigments of macroalgae on the cage. J Appl Phycol 1994; 6: 397-403.
[http://dx.doi.org/10.1007/BF02182156]

[14] Fujimoto K, Kato T, Ueki C, Tsutsumi N. Sea forest creation utilizing by-product slag of steelmaking process (development of technology for regeneration of seaweed bed). Nippon Steel Techn Rep 2012; 101: 208-11.

[15] Ueki C, Kato T, Miki O. Mesocosm experiment for fertilizer made from steel-making slag and humus soil with growth of *Porphyra yezoensis, nori.* J Adv Marine Sci Technol Society 2011; 17: 49-55. [in Japanese].

[16] Ueki C, Kumagai T, Fujita D. Effects of an iron-enriched fertilizer on marine algae in Toyama Bay deep seawater. Deep Ocean Water Res 2012; 13: 7-16. [in Japanese].

[17] Kosugi C, Kumagai T, Kobayashi M, Fujita D. Effects of an iron-enriched fertilizer on *Saccharina japonica* in thank with running Toyama Bay deep seawater. Deep Ocean Water Research 2016; 16: 1-13. [in Japanese].

[18] Ueki C, Murakami A, Kato T, Saga N, Motomura T. Effects of nutrient deprivation on photosynthetic pigments and uletrastructure of chloroplasts in *Porphyra yezoensis.* Nippon Suisan Gakkaishi 2010; 76: 375-82.
[http://dx.doi.org/10.2331/suisan.76.375]

[19] Provasoli L. Media and prospects for the cultivation of marine algae. Watanabe, A, Hattori, A (Eds) Cultures and Collections of Algae Proceedings of the US-Japan Conference, Hakone, Japan, September 1966. , Hakone, Japan: Japanese Society of Plant Physiology 1968; p. 63-75.

[20] Motomura T, Sakai Y. Effect of chelated iron in culture media on oogenesis in *Laminaria angustata.* Nippon Suisan Gakkaishi 1981; 47: 1535-40.
[http://dx.doi.org/10.2331/suisan.47.1535]

[21] Motomura T, Sakai Y. Regulation gametogenesis of *Laminaria* and *Desmarestia* (Phaeophyta) by iron and boron. Jpn J Phycol 1984; 32: 209-15.

[22] Kato T, Kosugi C, Kiso E, Torii K. Application of steelmaking slag to marine forest restration. Nippon Steel Sumitomo Metal Techn Rep 2015; 109: 79-84.

SUBJECT INDEX

www.ingramcontent.com/pod-product-compliance
Lightning Source LLC
Chambersburg PA
CBHW050813220326
41598CB00006B/195